HISTOCHEMICAL and CYTOCHEMICAL METHODS of VISUALIZATION

Methods in Visualization

Series Editor: Gérard Morel

HISTOCHEMICAL and CYTOCHEMICAL METHODS of VISUALIZATION

Edited by
Jean-Marie Exbrayat

CRC Press
Taylor & Francis Group
Boca Raton London New York

CRC Press is an imprint of the
Taylor & Francis Group, an **informa** business

CRC Press
Taylor & Francis Group
6000 Broken Sound Parkway NW, Suite 300
Boca Raton, FL 33487-2742

First issued in paperback 2016

© 2013 by Taylor & Francis Group, LLC
CRC Press is an imprint of Taylor & Francis Group, an Informa business

No claim to original U.S. Government works

ISBN 13: 978-1-138-19917-0 (pbk)
ISBN 13: 978-1-4398-2222-7 (hbk)

Library of Congress Cataloging-in-Publication Data

Histochemical and cytochemical methods of visualization / editor, Jean-Marie Exbrayat.
       p. ; cm. -- (Methods in visualization)
    Includes bibliographical references and index.
    ISBN 978-1-4398-2222-7 (hardcover : alk. paper)
    I. Exbrayat, J. M., editor of compilation. II. Series: Methods in visualization.
    [DNLM: 1. Histocytochemistry--methods--Laboratory Manuals. 2. Cells--ultrastructure--Laboratory Manuals. 3. Microscopy--methods--Laboratory Manuals. 4. Tissues--anatomy & histology--Laboratory Manuals. QS 525]

    QR63
    579.078--dc23   2013017592

Visit the Taylor & Francis Web site at
http://www.taylorandfrancis.com

and the CRC Press Web site at
http://www.crcpress.com

Contents

SECTION 1 *Histological and Histochemical Methods with Light Microscopy*

SECTION 2 *Cytological Methods in Electron Microscopy*

SECTION 3   Image Quantification

Series Preface

Visualizing molecules inside organisms, tissues, or cells continues to be an exciting challenge for cell biologists. With new discoveries in physics, chemistry, immunology, molecular biology, analytical methods, and so forth, limits and possibilities are expanded, not only for older visualizing methods (photonic and electronic microscopy) but also for more recent methods (confocal and scanning tunneling microscopy). These visualization techniques have gained so much in specificity and sensitivity that many researchers are considering an expansion from in-tube to *in situ* experiments. The application potentials are intensifying not only in pathology applications but also in more restricted applications such as tridimensional structural analysis or functional genomics.

This series addresses the need for information in this field by presenting theoretical and technical information on a wide variety of related subjects, including *in situ* techniques, visualization of structures, localization and interaction of molecules, and functional dynamism *in vitro* or *in vivo*.

The tasks involved in developing these methods often deter researchers and students from using them. To overcome this, the techniques are presented with supporting materials such as governing principles, sample preparation, data analysis, and carefully selected protocols. Additionally, at every step we insert guidelines, comments, and pointers on ways to increase sensitivity and specificity, as well as to reduce background noise. Consistent throughout this series is an original two-column presentation with conceptual schematics, synthesizing tables, and useful comments that help the user to quickly locate protocols and identify limits of specific protocols within the parameter being investigated.

The titles in this series are written by experts who provide to both newcomers and seasoned researchers a theoretical and practical approach to cellular biology and empower them with tools to develop and optimize protocols and to visualize their results. The series is useful to the experienced histologist as well as to the student confronting identification or analytical expression problems. It provides technical clues that could only be available through long-time research experience.

Gérard Morel, Ph.D., D.Sci.
Series Editor

Editor

Jean-Marie Exbrayat, Ph.D., D.Sc., is a professor at Catholic University of Lyon (France) where he manages the Laboratory of General Biology. He is also directeur d'etudes (professor) at the Ecole Pratique des Hautes Etudes, where he manages the Laboratory of Comparative Reproduction and Development.

Dr. Exbrayat earned his M.S. and Ph.D. in 1974 and 1977, respectively, from Montpellier University (France). He was appointed assistant then master-assistant of biology at Catholic University in Lyon in 1978, and became doctor of sciences in 1986. He became a professor in 1987 and also directeur d'etudes (professor) at the Ecole Pratique des Hautes Etudes in 1991.

He teaches animal biology, histology and embryology, and history and philosophy of sciences. He was the head of the School of Histology for 25 years, and dean of the Faculty of Sciences (1997 until 2005).

Dr. Exbrayat is a member of several scientific societies. He has published more than 300 papers and 12 books. His current major interests include the variations of genital tracts and endocrine organs during reproductive cycles related to external and internal factors, and embryonic development in both lower and higher vertebrates, as well as the history and philosophy of sciences, especially evolutionary biology.

Contributors

Claire Brun, Ph.D.
Université de Lyon
Lyon, France
cbrun@univ-catholyon.fr

Chantal de Chastellier, Ph.D., H.D.R.
Centre d'Immunologie de Marseille-Luminy
Marseille, France
dechastellier@ciml.univ-mrs.fr

Léon Espinosa, Ph.D.
Centre Commun de Quantimétrie (CCQ)
Lyon, France
leon.espinosa@ifr88.cnrs-mrs.fr

Jeanne Estabel, Ph.D.
Wellcome Trust Sanger Institute
Cambridge, United Kingdom
je2@sanger.ac.uk

Françoise Giroud, Ph.D.
Université Joseph Fourier
La Tronche, France
francoise.giroud@imag.fr

Marie-Paule Montmasson
Université Joseph Fourier
La Tronche, France
marie-paule.montmasson@imag.fr

Elara Moudilou, Ph.D.
Université de Lyon
Lyon, France
emoudilou@univ-catholyon.fr

Nicolas Thelen, Ph.D.
Université de Liège
Liège, Belgium
nthelen@ulg.ac.be

Marc Thiry, Ph.D.
Université de Liège
Liège, Belgium
mthiry@ulg.ac.be

Yves Tourneur, Ph.D., D.Sci.
Université de Lyon
Lyon, France
yves.tourneur@univ-lyon1.fr

Acknowledgments

The editor thanks Dr. Gérard Morel, the series editor of *Methods in Visualization*, who gave numerous suggestions for improving and expanding the manuscript. The editor also thanks the contributors who provided chapters: Dr. Claire Brun and Dr. Elara Moudilou, whose chapter discusses apoptosis visualization; Dr. Chantal de Chastellier, for the chapter on ultrastructural cytoenzymology; Dr. Jeanne Estabel, who wrote the chapter about enzyme histochemistry; Dr. Françoise Giroud and Marie-Paule Montmasson, who provided both chapters devoted to visualization of cell proliferation and *in situ* high-speed tissue proteomics; and Dr. Marc Thiry and Dr. Nicolas Thelen, whose chapter is devoted to acetylation methods. Dr. Yves Tourneur and Dr. Léon Espinosa are thanked for the part of the book that covers image quantification in histology and cytology.

The editor also thanks those with whom he works daily, especially Marie-Thérèse Laurent, a skilled technician whose ability and effectiveness contributed significantly to the improvement of methods used in the laboratory.

Thank you also to Barbara Norwitz who accepted publication of this book and Jill Jurgensen for her help and counsel in the preparation of a definitive form of manuscript.

Introduction

During the first half of the 20th century, histochemical methods were leading to the development of biochemistry. After the conception of the electron microscope at the end of the 1940s, histochemical methods were adapted to this new means of investigation and the cytochemical methods originated. Histochemistry is essentially based upon a peculiar *in situ* staining of chemical compounds, and cytochemistry is only based upon contrast and darkness, as only gray levels are available with electron microscopy. After the 1980s, microscopic observations were not often used, and were even sometimes considered to be obsolete methods of investigation. However, the usefulness of these methods is being considered again at their exact value, but manuals with useful techniques remain rare. They are strictly and separately devoted to light microscopy (LM) or electron microscopy (EM) techniques; and some researchers, technicians, and students are deprived of these means when they want to use such a method. In addition, the development of computers now allows for more precise quantification of images, giving a new interest to the study of histological and cytological material in biological as well as in medical sciences.

The purpose of this book is to give the essential techniques that can be used for histochemical investigations in both light and transmission electron microscopy. The text gathers several classical techniques used in light microscopy since the 1920s, classical techniques used for transmission electron microscopy, and several useful specialized methods. There is also a part of the book which covers image analysis of sections. This book will be useful for experienced and novice researchers, technicians, and students.

The book is organized into three sections. The first section is devoted to techniques in light microscopy with classical methods of visualization (Chapter 1), histochemical (Chapter 2) and histoenzymatic (Chapter 3) methods, and methods used to visualize cell proliferation (Chapter 4), apoptosis (Chapter 5), and high-speed tissue proteomics (Chapter 6). This section ends with Chapter 7 in which preparation of products for light microscopy is described. The second section concerns the cytochemical methods used in electron microscopy with general (Chapter 8) and cytochemical techniques (Chapter 9), and more specialized methods such as cytoenzymology (Chapter 10) and acetyl methods (Chapter 11). Chapter 12 provides methods of preparation of products for electron microscopy. The third section provides an overview of image analysis (Chapter 13). Color illustrations are provided in a color insert.

Section 1

Histological and Histochemical Methods with Light Microscopy

1 Classical Methods of Visualization

Jean-Marie Exbrayat

CONTENTS

1.1   INTRODUCTION

Methods for direct observation of tissues and cells are important in biological studies. They have been applied to botany, zoology, and human biology for healthy as well as pathological organisms. These various methods are based upon physical and chemical principles. They are used to obtain the finest possible knowledge of tissue and cell components. These techniques are used from light to electron microscopy, from the direct observation of sections to their automatic analysis with a computer. Histochemical and cytological methods are the result of a long history of experimentation.

1.1.1   THE PRECURSORS

For a long time, man has known both the complexity and anatomy of living organisms: a large number of animal and vegetal species, and several specialized tissues and organs of the organisms. But the intimate structure of living organisms began to be precisely studied during the 17th century. At this time, the first microscopes were built. Traditionally, Dutchman Antoon Van Leeuwenhoeck (1632–1723) is given credit as the discoverer of the first microscope, but Hooke and Fontana also built light microscopes. At this time, the first observations of living cells were published (Van Leeuwenhoeck [1776], Malpighi [1628–1694], Swammerdam [1637–1680]). Quickly, Corti observed moving organelles into the cells: it was the first observation of chloroplasts. Cell theory was first given by Schleiden and Schwann (1839) and became definitive with Virchow (1858). According to cell theory, all living organisms are constituted with cells, each one corresponding to the smallest living unit and possessing all living attributes (Schleiden [1804–1881] and Schwann [1810–1882]), and each cell is coming from another cell (Virchow [1821–1904] wrote: "Omnis cellula ex cellula").

1.1.2   HISTOLOGY AND CYTOLOGY

Microscopes improved over time. During the 19th century, a new science appeared: the histology or science of tissues. At the end of the 19th century, it was possible to observe several characteristics of the cells: cytology or science of the cell was born.

These new sciences of observation were the result of the work of both physicists specializing in optics and biologists. The aberrations of optic lens were corrected; magnification was first limited to 500, then 1200 was attempted, and is now about 2000 in the most performing microscopes. The light source, indispensable to observe the tissues by transparency, was very much improved. Daylight or light given by petrol lamps oriented by a mirror was replaced by an incandescent lamp incorporated to the microscope. Today, halogen lamps with a low voltage advantageously replaces the most powerful incandescent lamps. These improvements allowed the observations of tissue and cell details with more precision. Contrast of tissue components was improved by means of staining methods. First, vital dyes were incorporated into the cells, but this method was limited by the quick death and consequent degradation of tissues. The best results were given by the observation of dead tissues and cells preserved by means of physical or chemical processes, sectioned to become thin and transparent, and stained with selective dyes. Today the tissue is always immobilized in a state as close as possible to the living one with chemicals (formalin, alcohol, etc.); then it is embedded in a wax (paraffin, gelatin, plastic wax, etc.). Then, it is sectioned into 4 to 10 μm thick sections. For that, microtomes have been perfected. These engines are equipped with a blade on which the wax block goes on with a known distance. Sections are then pasted onto a glass slide, then cleared of wax and stained. The simplest staining uses a single dye; bichromic staining uses both a basic dye for the nuclei, and an acidic one for cytoplasm and connective fibers. For a long time, trichromic staining (Mallory's trichroma, 1900, or Masson's trichroma, 1912) were always used; they were used to stain selectively nuclei, cytoplasms, and collagen fibers.

More precise techniques based upon the detection of chemicals belonging to cells and tissues were perfected: PAS for carbohydrates, use of lysochromes for lipids, detection of specific molecular groups of proteins, nucleic acids, minerals, and so forth. Histochemistry was born. The enzymatic activities have also been detected, giving the science called histoenzymology. Using photographic emulsion, histoautoradiography permits localized radioactive precursors to be incorporated into the organism, giving a dynamic aspect to histology and cytology.

Then microscopes using fluorescence were perfected. The light source of such a microscope is a mercury vapor lamp giving UV light. Under this microscope, some elements of cells emit a characteristic primary fluorescent allowing their localizations; the use of fluorescent dyes gives a secondary fluorescence.

Microscope with phase contrast was born in about the 1940s. This new microscope permitted to increase the content of fresh cells and tissues, avoiding both fixation and staining. This technique and its improvements permitted to increase the knowledge of cells by studying its behavior, avoiding the artifacts related to the observation of preserved and stained tissue.

1.1.3 Immunocytology

The first works devoted to the perfection of qualitative methods based upon the antigen–antibody reaction were done in the 1930s. Coons and his collaborators (1941, 1942, 1950) published the first true immunocytological methods. Several techniques based on the use of fluorescent antibodies were perfected at this time. The advantage of such methods was their large precision. Histochemical techniques allow characterization of the main chemical family at which the detected component belongs; histoenzymology allows one to visualize the enzymatic activities by means of a stained product from the combination of the enzyme on the section and a substrate given by the user. Immunocytology allows precise detection of the molecules on the surface of tissue. Thus, several methods have been perfected: direct immunocytology using both a primary specific antibody and a secondary species-specific antibody for labeling. Several methods for amplifications have been perfected; the fluorescent dye is often replaced by an enzyme or several molecules of the same enzyme. Today, these methods can be considered classical.

1.1.4 *IN SITU* HYBRIDIZATION

The large development of molecular biology from the 1980s has consequently given new tools to the biologist. The application of molecular biology methods to histology and cytology was born about 30 years ago. The general principle of *in situ* hybridization is based upon the application of a labeled complementary probe on the gene sequence researched. These techniques now allow to visualize both DNA and RNA and to detect mutations, or gene expression in cells.

1.1.5 ELECTRON MICROSCOPY

Both transmission and scanning electron microscopy have been used since the 1930s. Several descriptions about the evolution of these techniques will be given in the section devoted to electron microscopy.

1.1.6 QUANTITATIVE METHODS

Since the 1980s, new methods for observation and analysis appeared, using a computer to quantify the images. Flow cytometry is used to study cells one by one, using the light absorbed or emitted by the cells, the natural fluorescence, or that of a specific fluorescent dye. Flow cytometry also permits one to pick out the cells according to the chosen parameters.

In automatic quantitative analysis, the observations are done by a camera, then analyzed by means of a computer. This method allows colorimetric studies, spatial reconstruction, and measuring without subjectivity.

1.2   GENERAL PRINCIPLES OF HISTOLOGY AND HISTOCHEMISTRY

To perform a histological study, tissues and cells must be submitted to a series of operations allowing preservation and visualization of organic components with a light microscope. For electron microscopy study, tissues also must be submitted to a series of similar operations (see Chapter 8).

Histology gives a general view of tissue structure, often using empirical methods. With histochemistry methods, the nature of chemical components of tissues and cells can be precise, according to a specific staining. In this case, staining reactions are known with sufficient precision and one may modify parameters according to the component researched.

The general protocol used for histological and histochemical study can be divided into the following main parts:

- Sampling
- Fixation
- Dehydration, clarification (dehydration and clarification are not performed when organs are embedded with hydrophilic waxes)
- Impregnation, embedding
- Sectioning (sections can be done with embedded or frozen organs)
- Adhesion of section on a slide
- Staining (a lot of staining methods can be used for histological as well as histochemical study)
- Dehydration; it is useless (and even wrong) in the case of mounting with a hydrophilic medium
- Mounting (can be done with hydrophobic or hydrophilic medium)
- Observation

Each step can differ from one sampling to another, according to the nature of tissue studied, the results expected, and the technique used.

This section is devoted to the main protocols used for a histological study. Some of these are also available for histochemistry.

1.3   TISSUE PREPARATION

1.3.1   SAMPLING TYPES

All material with a biological origin can be isolated to visualize its molecular components: histological sections, smears, karyotypes, monolayer, or suspension cell cultures.

1.3.1.1   Tissue Dissection

Organ or piece dissection before fixation must be done with care to preserve the integrity of tissue and cells for study.

- Avoid alteration from autolysis and putrefaction by fixating as quickly as possible after the death of the organism. It is possible to cool the pieces if they cannot be fixed immediately. However, fixation done on cooled material never gives very good results.
- Avoid alteration caused by dissecting tools by cautious use of these tools. To obtain clear fractures, dissection by means of razor blade is preferred to dissection with scissors.
- Avoid osmotic lysis by washing tissues into an isoosmotic physiological liquid and not water.
- Avoid tissue drying during dissection by working in a wet medium.

1.3.1.2   Cell Cultures

1.3.1.2.1   Monolayer Cell Cultures

Monolayer cell cultures can be cultivated in several types of flasks. Cell treatment varies depending on the purpose: to observe entire cells directly upon the culture support or to obtain semithin sections. The flask types that are used in monolayer cell culture vary: Rous's boxes; microplates; Leighton's tubes, containing a detachable slide on which cell culture occurs; and Lab-Tek, which is a culture box from whose bottom can be detached to observe the cells directly with a microscope.

- Fixation of monolayer cell cultures
  - Empty the cell culture medium.
  - Place the fixative. All fixatives can be used whatever the cell type.
  - Let the cell remain in contact with the fixative for several hours. Fixation duration depends on the cell type and the fixative. Because of the thickness of monolayer cell cultures, this duration can be shortened (15 min for instance).
- Preparation of cell culture to obtain semithin sections
  - Empty the cell culture medium.
  - Cover the monolayer with 0.25% trypsin at 37°C. Trypsin must be used with care; if use is prolonged, plasmic membranes and finally cells will be destroyed.
  - Quickly observe the evolution of cells by means of a reversed microscope.
  - Incubate at 37°C.
  - When the cell culture begins to slip, stop the trypsin action with cell culture medium.
  - Dissociate cell masses by aspirating and pressing back the cell suspension (20 to 30 times).
  - Centrifuge the cell suspension at 600 g for 10 min. Centrifugation must be done with care to avoid mechanical destruction of cells.
  - Decant and rinse the bottom by adding suspension to the fixative.

1.3.1.2.2   Suspension Cell Cultures

- Remove cells from the vessel.
- Centrifuge the suspension at 600 g for 10 min. Centrifugation must be done with care to avoid mechanical destruction of cells.
- Decant and rinse the bottom by adding suspension into the fixative.

1.3.2   FIXATION

1.3.2.1   Definition

Fixation (or preservation) consists of preparing an organism, an organ, a tissue, or even a cell to be held, in death, in a state as close as possible to the living state.

1.3.2.2   General Principles

1.3.2.2.1   Importance of Fixation

In every histological operation, the fixation is the first stage that is indispensable for good preservation of cell and tissue components. These components must remain in a state as close as possible to the life state. Fixation must be performed in a way that allows visualization of these components by histological or histochemical methods. The process must not eliminate the components, and it must not react with the atomic groups that will be used to visualize the structure. Fixation must preserve the tissue morphology in a recognizable form, which is essential to good localization of researched molecules.

1.3.2.2.2   Effects of Fixation

The effects of fixation on the tissues are numerous:

- Immobilization of cell components after extraction from the natural environmental medium
- Inhibition of cell autolysis that is due to enzyme liberation after degradation of lysosomic membranes
- Inhibition of putrefaction due to microorganisms
- Tissue hardening—If hardening is reasonable and permits one to obtain histological sections, the fixative is called "tolerant." If hardening is too great and does not permit obtaining sections, the fixative is called "intolerant."
- Modification of tissue refraction index that permits observing them even before staining
- Rendering insoluble cell substances
- Modification of tissue volume as a function of the fixative—Modification of cell volume after a fixative action must be considered, especially for morphometric studies. Using the same fixative to obtain comparative results is recommended.
- Effects on dye affinity—Acetic acid, for example, separates proteins from nucleic acids, which can lead to the extraction of the latter. Use of formalin often decreases the intensity of certain staining.

1.3.2.3   Chemical Action of Fixative

A good histological fixative is a good fixative of proteins because the cell or tissue structure is linked to these macromolecules. Chemical fixatives react on reactive protein groups by establishing bridges between these molecules.

Formaldehyde ($H_2C = O$) reacts with amines, imines, guanidyls, hydroxyls, sulfhydryls, carboxyls, peptic binding, and aromatic nuclei. It establishes methylene bridges between protein molecules. At room temperature, formaldehyde is a gas. Formalin, which is usually used, is a solution of formaldehyde in water. Trade formalin sold at 30% or 40% in water is used in a diluted solution:

- Formalin, 10 mL
- Water, 100 mL

1.3.2.4 Different Fixative Types

1.3.2.4.1 Definitions

Fixatives can be simple, often constituted of only one molecule. They also can be constituted by a mixture of simple fixatives. Among simple fixatives, some of them are denaturant. Their action induces a protein denaturation that lends a reticulated aspect to cytoplasm. Other fixatives are not denaturant and the aspect of cytoplasm is homogeneous. Denaturant fixatives are still called coagulant.

1.3.2.4.2 Coagulant Fixatives

- Ethanol—Ethanol (70% to 100%) is mainly used to preserve mineral elements. It is also used as a component of fixative mixtures (Carnoy's fluid, for instance) used for protein and nucleic acid histochemistry. Ethanol is a good fixative used for method based on detection of carbonyls obtained after acidic hydrolysis. Tissues are hardened with its use.
- Picric acid—Picric acid is also a coagulant fixative used as an adjuvant into fixative mixtures (Bouin's fluid, Halmi's fluid). Its yellow color does not interfere with staining sections with dyes. If the yellow color of section obtained from tissues fixed by a fluid containing picric acid is cumbersome, the color can be eliminated before section staining by using alkaline baths, such as lithium carbonate. Be careful to separate the section. The color can also be eliminated by use of 70% ethanol or running water only.
- Mercury chloride—Mercury chloride (or sublimate; $HgCl_2$) is also used for several mixtures. It penetrates quickly but not as deeply. Tissues are hardened by its use. It also can give a precipitate onto the sections. Mercury chloride is very corrosive. Avoid contact with metal dissecting tools. It is possible to eliminate mercury chloride by rinsing sections with lugol, then with sodium hyposulfite.

1.3.2.4.3 Noncoagulant Fixatives

- Formaldehyde—Formaldehyde can be used alone or in a mixed fixative. If used alone, it is often helpful to neutralize or to buffer it to avoid formation of formic acid by oxidization. Formaldehyde, which is used as formalin, penetrates quickly into tissues that are not excessively hardened. This fixative permits a relatively long preservation of tissues but it can modify staining qualities. At room temperature, formaldehyde is a gas. Formalin is a solution of 37% or 40% formaldehyde in water. This solution is diluted to obtain 10% formaldehyde before use. In some countries, formalin is considered as a carcinogenic substance. Several substitutive products begin to be available to replace it, but their use is not still well developed.
- Osmium tetroxide—Osmium tetroxide is a very good cytological fixative that reacts with the lipids, more particularly phospholipids that belong to the cell membrane structure. It permits cell proofing. It is also a powerful oxidant that must be avoided in histochemical reactions. Osmium tetroxide penetrates very little in tissues that are hardened, making it impossible to obtain sections. It is also used to fix smears. Osmium tetroxide is also called osmic acid by histologists, but this name is chemically incorrect because this molecule has no acidic characteristic. Osmium tetroxide is also used to fix tissues for electron microscopy (see Chapter 8).
- Acetic acid—Acetic acid is also used into fixative mixtures. It quickly penetrates into the tissues and stabilizes nucleoproteins. If it is used at too high a concentration, it can separate nucleic acids and nucleoproteins. It is a good fixative for the nucleus if it is used at a concentration between 0.3 and 5 mL in 100 mL of fixative.

1.3.2.4.4   Fixative Mixtures

Fixative mixtures are most often used. The effects of several simple fixatives are additive. Below is a list of the fixative mixtures most often used for histochemistry and histology (see the following sections in Chapter 7). Other fixatives also exist.

- Alcohol–formalin: See Section 7.1.1.1.
- Baker's fluid: See Section 7.1.1.2.
- Bouin's fluid: See Section 7.1.1.3.
- Bouin–Hollande's fluid: See Section 7.1.1.4.
- Carnoy's fluid: See Section 7.1.1.5. Must be prepared extemporary.
- Flemming's fluid: See Section 7.1.1.6. Must be prepared extemporary.
- Formalin: See Section 7.1.1.7.
- Formalin–calcium: See Section 7.1.1.8. See also Baker's fluid.
- Neutral formalin: See Section 7.1.1.9.
- Salt formalin: See Section 7.1.1.10. 0.1 M, pH 7.
- Buffered formalin: See Section 7.1.1.11.
- Halmi's fluid: See Section 7.1.1.12.
- Helly's fluid: Also called Zenker formalin. See Section 7.1.1.13.
- Heidenhain's Susa: See Section 7.1.1.14.
- Zenker's fluid: See Section 7.1.1.15.

1.3.2.4.5   Fixation Duration

The duration must be adapted to the thickness of the piece (5 to 10 mm for photonic microscopy). To improve the fixation of large pieces, prefixing can be done for 24 to 48 h into formalin, before postfixation with Bouin's fluid for 1 to 24 hours.

- Bouin's fluid: 24 to 48 h
- Carnoy's fluid: 4 h
- Formalin: Indefinitely

1.3.2.5   Methods for Chemical Fixation

1.3.2.5.1   Precautions

Whatever the fixative and the cell or tissue element that is to be visualized, the fixation must answer to several imperatives and several precautions must be taken. The presence of blood on the surface of tissues also causes blood cells (leukocytes, red blood cells) to be carried to areas where they are not usually present.

- Avoid getting blood or mucus on the tissue surface. These biological elements may become hard after coagulation or they may polymerize and constrict fixative penetration.
- Use a sufficient volume of fixative. One method is to enclose the piece in a bag of gauze and hang it to the mouth of the bottle.
- If fixation is of a long duration, exchange the fixative periodically with a new solution. An equilibrium is established between intra and extra tissue fluids and the fixative. In the end, the fixative will not penetrate, and tissue and cell components will not be correctly immobilized.
- Use relatively wide bottles. Pieces must not be stuck on the bottle wall above the fixative level.

1.3.2.5.2   Fixation by Immersion

To fix by immersion, immerse the tissue fragment in the fixative as quickly as possible. To avoid autolysis and putrefaction, the organ or tissue fragment must be extracted very quickly after the animal's (or vegetation's) death. Do not forget to label each fragment.

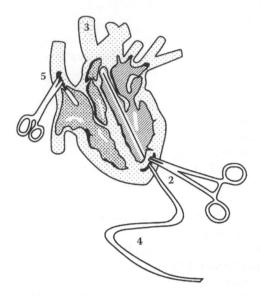

FIGURE 1.1 Perfusion. (1) Open the thoracic cage (not shown in the figure). (2) Incise the left ventral ventricle. (3) Introduce the nozzle. (4) Clamp the nozzle by means of a Mohr's grip. (5) Open the right auricle. (From Morel, G., 1998, *Hybridation in situ*, Polytechnica, Paris. With permission.)

1.3.2.5.3   Fixation by Perfusion

Perfusion is used for the fixation of tissues that are particularly fragile (Figure 1.1). The animal can be entirely subjected to the perfusion. Fixation by perfusion can be used for the brain, for instance.
    To perform a perfusion:

- Anaesthetize the animal
- Open the thoracic cage
- Remove the pericardium
- Incise the left ventricle near the top of the heart
- Insert the nozzle into the aorta
- Hold the nozzle by means of a Mohr grip
- Open the right auricle
- Rinse with physiological buffer
- Perfuse with fixative
- After perfusion, dissect the organs and plunge them into fixative

1.3.2.5.4   Fixation with Osmium Tetroxide Vapor

The fixation with osmium tetroxide vapor method is essentially used to fix a blood smear.

- Place the fixative in the bottom of a bottle. 2% osmium tetroxide in distillated water is usually used.
- Close the bottle.
- Quickly place the smear above the fixative for several seconds. It is obvious that the face of the slide with the smear is placed above the fixative. Fixation can be 10 sec to 3 min.

1.3.2.5.5   Fixation for Semithin and Ultrathin Sections

1.3.2.5.5.1   Principle   Semithin sections are often used to localize the useful part of the tissue before to obtain ultrathin sections for the electron microscope. For this, fixation must be very precise to avoid unsatisfactory images. Fixative must be isoosmotic to tissue fluids; it must be at the same pH and it must possess the same ionic composition. Other fixative preparation modes for electron microscopy exist, especially those using another buffer (i.e., cacodylate buffer). They can also be used for tissue preparation before visualizing cell components on semithin sections. It is also possible to use only paraformaldehyde or glutaraldehyde alone.

To accomplish this, the tissue is generally first preserved in a mixture of formaldehyde and glutaraldehyde to polymerize proteins. Then it must be postfixed by osmium tetroxide to impermeabilize cell membranes by reaction with phospholipidic bilayer. (See also Chapter 8.)

One method for fixation is given here.

1.3.2.5.5.2   Glutaraldehyde/Paraformaldehyde Fixative
- Sörensen's buffer (PBS) 0.2 M, pH 7.4
  - Monosodium phosphate (sol. n°1)
    - Monosodium phosphate, 3.12 g
    - Distilled water, 100 mL
  - Disodium phosphate (solution n°2)
    - Disodium phosphate, 7.16 g
    - Distilled water, 100 mL
  - Buffer
    - Solution n°1, 19 mL
    - Solution n°2, 81 mL
- Glutaraldehyde
  - Glutaraldehyde, 4 g; Glutaraldehyde is prepared from a solution already diluted to 25% or 50%.
  - Distilled water, 100 mL
- Paraformaldehyde. To prepare: Dissolve paraformaldehyde in tepid water, 3 min; heat for 20 min at 80°C; sodium hydroxide, 2 drops
  - Stock solution
    - Paraformaldehyde, 4 g
    - Distilled water, 10 mL
  - Working solution
    - Mother solution, 2 g
    - Distilled water, 100 mL
- Preparation of fixative
  - Glutaraldehyde 4%, 100 mL
  - Paraformaldehyde 2%, 100 mL
  - Buffer 0.2 M, pH 7.4, 200 mL

Fixative can be stored for several days at 4°C.

1.3.2.5.5.3   Osmium Tetroxide
- Mother solution
  - Osmium tetroxide, 0.5 g; Osmium tetroxide is available in phial containing 0.5 g
  - Distilled water, 25 mL
- Working solution
  - Mother solution, 10 mL
  - Buffer 0.2 M, pH 7.4, 10 mL

1.3.2.6   Physical Fixation
Beyond chemical fixation by means of molecules that react with tissue components, some methods are based exclusively on physical principles.

1.3.2.6.1 Cryodessication

In cryodessication a fresh tissue is quickly frozen, then dried under vacuum at a very low temperature. Tissue water goes directly from the solid state to the gaseous state. The tissue fragment is then directly embedded in melted wax. Cryodessication has a lyophilization phase that avoids tissue components diffusion. Tissue molecules do not undergo any chemical modification.

1.3.2.6.2 Freezing–Dissolution

In freezing–dissolution a fresh tissue is quickly frozen, then the ice is dissolved with absolute ethanol. Embedding is then done in melted wax. In freezing microtome, the temperature of both blade and tissue is very low. In cryotome, the microtome is entirely contained in a cold enceinte, the "cryostat."

1.3.2.6.3 Classic Fixation by Cold

In certain cases, fresh tissue is directly fixed by cold, then cut with a freezing microtome or a "cryotome." Freezing temperature is generally between –20°C and –40°C.

1.3.2.6.4 Chemical and Cold Fixation

Tissues that were fixed by chemical fixative may be cut without embedding by means of a freezing microtome or a cryotome.

1.3.2.7 Holding Fluids

Preserved tissues can be stored in holding fluids:

- Bouin's fluid preservation: Ethanol 70%. Ethanol must be avoided if lipids must be visualized.
- Formalin preservation: Formalin
- Carnoy's fluid preservation: Butanol. Butanol must be avoided if lipids must be visualized.

1.3.3 EMBEDDING

1.3.3.1 Paraffin Embedding

1.3.3.1.1 Principle

Paraffins, from Latin *parum affinis* ("little reactive") are alkanes. They are organic molecules exclusively constituted of carbon and hydrogen with no functional group that can react with cell or tissue molecules. Therefore, hard blocks can be obtained, allowing them to be cut. Paraffins are insoluble in water, so it is not possible to immerse a tissue containing water directly in paraffin. This tissue must first be dehydrated by ethanol (or acetone). Paraffin is not soluble into ethanol or acetone, so the tissue must be immersed in an intermediate substance that is soluble in the solvent and in paraffin. Among these substances are toluene, xylene, chloroform, and warm butanol. This operation is called clarification because tissue fragments become transparent.

Tissue is then immersed in liquid paraffin for several hours. This stage is the impregnation. After changing the paraffin bath, the organ is immersed into liquid paraffin that is contained in a recipient at room temperature. When paraffin is solid, the embedding is finished and the organ can be cut with a microtome. Impregnation is done in a steam room with a temperature sufficiently warm to maintain the paraffin in a liquid phase. Fusion temperature of paraffin used for histology is generally comprised between 50°C and 65°C. Paraffin with a low fusion point is used for smooth tissue, and paraffin with a high fusion point is used for hard tissues.

Receptacles include Leuckart's bars (Figure 1.2) or cassettes. Leuckart's bars are arranged to form a receptacle into which liquid paraffin is poured. The tissue piece is immersed in the paraffin. After several minutes, a solid block forms that is ready to cut. Do not forget to label each block.

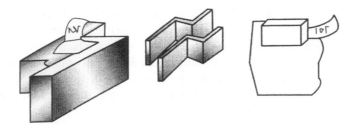

FIGURE 1.2 Leuckart's bars for embedding. (From Morel, G., 1998, *Hybridation in situ*, Polytechnica, Paris. With permission.)

1.3.3.1.2 Protocols

- Dehydration—Pieces preserved in Bouin's fluid or formalin
  - Ethanol 70%, 4 h. The duration of baths in the different ethanol solutions can be increased: 24 h for each bath in ethanol 95% and 4 h in each bath in absolute (100%) ethanol. Conversely, in certain cases, the duration of each bath can be decreased (only 1 h in each one but use 40°C).
  - Ethanol 96%, 2 × 12 h
  - Ethanol 100%, 2 × 4 h. Organs that are preserved in Carnoy's fluid are directly immersed in butanol. The duration of butanol baths can also be modified. A lengthened stay that can reach 24 h and more is useful for embedding. Butanol can also allow preservation of tissue fragments before embedding.
- Clarification
  - Butanol, 2 × 12 h
- Paraffin impregnation
  - Leave the tissue in melted paraffin for 4 to 12 h depending on the tissues: 4 h for liver, kidney, spleen, and lung, and 12 h for the other tissues. The temperature used is the paraffin melting point. Duration of fixation is indicative; it is recommended to test it for new tissue studied.
- Embedding
  - The impregnated tissue is embedded in a paraffin block that is made by means of a mold (Leuckart's bars, embedding cases). In the case of a manipulation mistake, it is sometimes useful to embed the pieces again. For that, immerse the cut block in melting paraffin. When the piece is melted out of its solid paraffin encasement, make the block again. The block is ready for cutting.

1.3.3.1.3 Paraplast Embedding

Paraplast is useful to obtain thinner sections than with paraffin. It also allows very hard tissues or organs possessing parts of different hardness to be cut.

1.3.3.1.3.1 Principle Paraplast is a mixture of natural paraffins and synthetic polymers. It has the good qualities of resistance and elasticity. The embedding technique is similar to that employed with paraffin.

1.3.3.1.3.2 Protocol

- Dehydration—Pieces preserved in Bouin's fluid or formalin
  - Ethanol 70%, 4 h
  - Ethanol 96%, 2 × 12 h
  - Ethanol 100%, 2 × 4 h

The duration of baths in the different ethanol solutions can be increased: 24 h for each bath in ethanol 95% and 4 h in each absolute (100%) ethanol. Conversely, in certain cases, these baths can be decreased (only 1 h in each bath but use 40°C).

Organs that are preserved in Carnoy's fluid are directly immersed in butanol. The duration of the butanol bath can also be modified. A lengthened stay that can reach 24 h and more is useful for embedding. Butanol can also allow preservation of tissue fragments before embedding.

- Clarification
  - Butanol, 2 × 12 h
- Paraplast impregnation
  - Leave the tissue in melted paraplast for 4 to 12 h depending on the tissue type: 4 h for liver, kidney, spleen, lung and 12 h for other tissues. The temperature used is the paraffin melting point. Duration of fixation is indicative; it is recommended to test it for new tissue studied.
- Embedding
  - The impregnated tissue is embedded in paraplast that is formed with a mold (Leuckart's bars, embedding cases, etc.). In case of a manipulation error, it is sometimes useful to embed the pieces again. For that, immerse the cut block in melted paraplast. When the piece is melted out of its solid paraplast encasement, make the block again. The block is ready for cutting.

1.3.3.2   Celloidin Embedding

Celloidin embedding has been widely used to study eyes, which are always difficult to cut because of the presence of both a hard structure (the crystalline lens) and smooth structures (vitreous and aqueous humor). This medium has allowed the embedding of large organs, such as human brain.

1.3.3.2.1   Principle

Celloidin (nitrocellulose) embedding is a simple technique that does not require any warming. Like paraplast, this embedding medium possesses a rigid and elastic consistency that allows hard tissues or tissues with varying consistency to be cut in thin sections. It also allows embedding and sectioning of very large pieces. Celloidin embedding is a slow method that needs several weeks. It does not allow sections thinner than 10 μm to be made. It is also difficult to obtain serial sections by this technique. Another inconvenience is the necessity to stock blocks in ethanol at 70%.

1.3.3.2.2 Protocol

- Dehydration—Pieces preserved by Bouin's fluid or formalin. Dehydration must be particularly carefully done.
  - Ethanol 70%, 4 h
  - Ethanol 96%, 2 × 12 h
  - Ethanol 100%, 2 × 4 h
  - Organs that are preserved in Carnoy's fluid are directly immersed in butanol.
- Embedding—Organ embedding is done in a paper or card mold.
  - Ethanol/ether, 24 h
  - 2% celloidin in ethanol/ether 1/1, 7 days
  - 4% celloidin in ethanol/ether 1/1, 7 days
  - 7% celloidin in ethanol/ether 1/1, 7 days
  - Permit the ethanol/ether mixture to evaporate under a vacuum bell containing a cup with sulfuric acid. Sulfuric acid is a powerful dehydrating agent. The embedding mixture contains 100% ethanol and ethyl ether.

- Ethanol 70%, or formalin vapors. Ethanol or formalin vapors are obtained by putting the preparation under a vacuum bell in which formalin or ethanol is present. This operation hardens the block.
- Bolles-Lee's fluid, 5 min. Bolles-Lee's fluid is used for clearing the block.
  - Chloroform, 100 mL
  - Cedar oil, 200 mL
- Blocks are stored in 70% ethanol.

1.3.3.3 Double Embedding: Celloidin and Paraffin

1.3.3.3.1 Principle

Celloidin and paraffin double embedding is used for histological treatment of organs or entire small animals that contain numerous small holes. This method is especially used to study insects, small crustaceans, and lungs. Preserved and dehydrated pieces are impregnated with celloidin, then by paraffin. Celloidin enters the holes that paraffin cannot reach. Subsequent encasement with paraffin then forms a block with all the classic qualities of paraffin.

1.3.3.3.2 Protocol

- Dehydration
  Pieces preserved by Bouin's fluid or formalin
  - Ethanol 70%, 4 h
  - Ethanol 96%, 2 × 12 h
  - Ethanol 100%, 2 × 4 h
  Organs that are preserved in Carnoy's fluid are directly immersed in butanol.
  - Acetone (propanone), 2 × 30 min
- Celloidin impregnation
  - Add methyl salicylate until pieces fall to the bottom of the receptacle.
  - Add celloidin 1% in methyl salicylate until pieces fall to the bottom of the receptacle.
- Paraffin impregnation—Pieces are directly immersed in a liquid paraffin bath for 4 to 12 h, like a normal paraffin embedding.
- Embedding—Pieces are embedded in liquid paraffin that is contained in a mold. After cooling, a block is formed, as after a normal paraffin embedding.

1.3.3.4 Gelatin Embedding

Gelatin embedding also allows one to obtain frozen sections. This type of embedding is used for fragile or small objects. The preserved fragments are immersed in lukewarm gelatin solutions at increasing concentrations (10%, then 20% gelatin in water). The solution and fragments are then placed into a mold and hardened with formalin.

1.3.3.5 Double Embedding: Agar-Agar and Paraffin

Agar-agar and paraffin double embedding is particularly useful for oocytes, or small eggs or also pituitary organs, which must be sectioned according to an orientated draft.

1.3.3.5.1 General Principle

Agar-agar and paraffin double embedding is used to cut small tissue fragments at ambient temperature that are difficult to embed in a classic paraffin block. This embedding has all the advantages of normal paraffin embedding. In the technique, preserved and dehydrated tissue is first embedded with gelose, and then with paraffin.

1.3.3.5.2 Protocol

- Reagents

- • Agar-agar gel
  - – Add agar-agar, 1.3 g
  - – Rinse with boiling distilled water, 100 mL
  - – Let agar-agar dissolve
  - – Add formalin 2.5%
- • Paraffin
- • Agar-agar embedding
  - • Spread a 1 to 2 mm thick agar-agar layer
  - • Place the pieces to be embedded on the layer
  - • Cover the pieces with melted agar-agar
  - • Let cool
  - • Immerse in ethanol 70% for 30 min
  - • Cut the block
- • Dehydration
  Pieces preserved by Bouin's fluid or formalin:
  - • Ethanol 70%, 4 h
  - • Ethanol 96%, 2 × 12 h
  - • Ethanol 100%, 2 × 4 h

  The duration of baths in the different ethanol solutions can be increased: 24 h for each bath in ethanol 95% and 4 h in each absolute (100%) ethanol. Conversely, in certain cases, these baths can be decreased (only 1 h in each bath but use 40°C).

  Organs that are preserved in Carnoy's fluid are directly immersed in butanol. The duration of butanol bath can also be modified. A lengthened stay that can reach 24 h and more is useful for embedding. Butanol can also allow preservation of tissue fragments before embedding.

- • Clarification
  - • Immerse in butanol, 2 × 12 h
- • Paraffin impregnation. Duration of impregnation depends on the tissue type: 4 h for liver, kidney, spleen, and lung, and 12 h for other tissues. Duration of fixation is indicative; it is recommended to test it for new tissues studied. The temperature used is the melting point of paraffin. In case of a manipulation error, it is sometimes useful to embed the pieces again. For that, immerse the cut block in melted paraffin. When the piece is melted out of its solid paraffin encasement, make the block again.
  - • Leave the tissue in melted paraffin for 4 to 12 h
- • Embedding
  - • Embed the impregnated tissue in a paraffin block that is formed with a mold (Leuckart's bars, embedding cases, etc.). The block is ready for cutting.

1.3.3.6   Resin Embedding

1.3.3.6.1   General Principle

Resin embedding is used primarily for electron microscopy. This type of embedding can also be used for photonic microscope. Blocks that are formed are particularly hard, allowing one to obtain semithin sections (0.5 to 1 μm) and to increase the precision of images.

1.3.3.6.2   Epon Embedding

1.3.3.6.2.1   Principle   In certain cases (electron microscopy, semithin sections), it is necessary to make blocks in a hard resin that allows very thin sections (0.5 to 1 μm). For that, one can use epoxy waxes that polymerize at about 60°C. Epon is a synthetic polymer that comprises several monomer molecules, which are associated by a cross-linker. Polymerization is performed with a catalyst. By using several monomer proportions, one can obtain blocks of different hardness. It is necessary to use a microtome equipped with a glass knife. These waxes cannot be sectioned with a steel razor without damage to both the tissue and razor.

1.3.3.6.2.2 Protocol
- Reagents
  - Epikote 812
  - DDSA (dodecyl succinic anhydride)
  - MNA (methyl nadic acid)
  - DMP-30 (2-4-6-tridimethylaminoethyl phenol)
  - Propylene oxide, also called 1-2-epoxypropane, is a solvent for embedding wax.
- Dehydration—The dehydration stage is indispensable because waxes are not hydrosoluble. The process must be performed particularly quickly to preserve cell and tissue structures.
  - Ethanol 30%, 10 min
  - Ethanol 50%, 10 min
  - Ethanol 70%, 2 × 10 min
  - Ethanol 95%, 10 min
  - Propylene oxide (4°C), 10 min
- Quick dehydration
  - Ethanol 70%, 2 × 10 min
  - Ethanol 95%, 2 × 10 min
  - Propylene oxide, 10 min at 4°C

Work under a hood. Avoid contact with skin.
- Impregnation and embedding medium
  - Epon A
    - Epikote 812, 31 mL
    - DDSA, 50 mL
  - Epon B
    - Epikote 812, 50 mL
    - MNA, 44 mL
  - Embedding medium—Epon A and B proportions can be modified according to the hardness that is desired for the block. For a soft block, increase the Epon B proportion. To obtain a hard block, increase the Epon A proportion.
    - Epon A, 40 mL
    - Epon B, 60 mL
    - DMP 30, 1.7 mL
- Substitution medium
  - Embedding medium, 50 mL
  - Propylene oxide, 50 mL
- Substitution
  - Substitution medium, 1 h at room temperature (RT)
- Impregnation
  - Impregnation medium, 12 h at RT
- Embedding—Embedding is done in molds with different forms. Fill the mold with embedding medium and dispose the object to be studied. Do not forget to label.
  - Embedding medium, 2 h at 37°C
  - Embedding medium, 3 days at 60°C

1.3.3.6.3 Durcupant Embedding
- Reagents
  - Araldite CY 212
  - Araldite HY 964

- • Dibutyl phthalate
- • Accelerator DY 064
- Dehydration—The dehydration stage is indispensable because waxes are not hydrosoluble. It must be particularly quick to preserve cell and tissue structures.
  - • Ethanol 70%, 2 × 15 min
  - • Ethanol 90%, 2 × 15 min
  - • Ethanol 95%, 2 × 15 min
  - • Ethanol 100%, 2 × 20 min
- Impregnation and embedding medium
  - • Araldite I
    - – Araldite CY 212, 15 mL
    - – Hardener HY 964, 5 mL
    - – Dibutyl phthalate, 1 mL
  - • Araldite II
    - – Araldite CY 212, 10 mL
    - – Hardener HY 964, 10 mL
    - – Dibutyl phthalate, 1 mL
  - • Embedding medium
    - – Araldite II, 21 mL
    - – Accelerator DY 064, 0.35 mL
  - • Impregnation medium
    - – Araldite I, 50 mL
    - – Ethanol 100%, 50 mL
- Impregnation
  - • Impregnation medium, 30 min
  - • Araldite I, 12 h at 60°C
  - • Araldite II, 24 h at 60°C
- Embedding
  - • Embedding is done in mold with different forms. Fill the mold with embedding medium, dispose the object to be studied. Do not forget to introduce a label.
  - • Embedding medium, 72 h at 20°C

1.3.3.7 Treatment and Embedding of Hard Tissues

1.3.3.7.1 Different Hard Tissues

There are two categories of hard tissues. The hardness of one is linked to scleroproteins (an example is the chitin of insects or crustaceans). The hardness of the other is linked to calcium (bones are examples). Tissue treatment will vary according to the cell or tissue component that is responsible for the hardness.

1.3.3.7.2 Noncalcified Hard Tissue

The embedding is similar to that for standard soft tissue. To improve the inclusion consistency and to facilitate sectioning, one can use clearing agents that are not hardeners, such as amyl acetate or cedar oil. Paraffin baths can be lengthened.

1.3.3.7.3 Calcified Hard Tissues

Histological study of calcified tissues can be done from decalcified preparations, but it is not a general rule.

1.3.3.7.3.1 Decalcification   Never decalcify fresh tissues. Decalcifying agents can be used on preserved tissue only. These agents can be included in the fixative. They also can be used after preservation. See Chapter 7, Section 7.1.1.1.

- Decalcifying agents
  - Nitric acid 5% to 7.5% in water, alone or with an ethanol–formalin fixative
  - Hydrochloric acid 4% to 8% in water
  - Citric acid 7% in water
  - Sulfosalicylic acid 6% to 8% in water
  - Trichloracetic acid
- It is also possible to use chelating agents as EDTA or solutions as RDO. The easiest way to appreciate decalcification is to push a pin into the tissue to test its hardness. One can also detect calcium in the fixative by use of a chelator that provides a colored reaction. Radiography is a more elaborated means, but it is also the least used.
- It is also possible to use ion exchange resin or electrophoresis.
- Decalcifying effect is stopped when decalcification is judged sufficient.
- Preparation of calcified tissues—This method is used to study bones, even though the sections are thick.
  - Thin slices can be cut with a band saw, then they are placed on a glass plate with an abrasive paste.
  - Tissues can be cut with a freezing microtome.
  - Inclusion can be done with a special wax.

1.3.3.7.3.2 Bone Embedding
- Fixation—All fixatives can be used unless they contain a decalcifying substance, that is, an acidic element.
- Dehydration and substitution
  - Ethanol 100%, 48 h
  - Toluene, 48 h
- Impregnation and embedding—MMA (methyl methacrylate) is used. It is polymerized by use of benzoyl peroxide.
  - MMA 1
    - MMA, 100 mL
    - Dibutyl phthalate, 25 mL
  - MMA 2
    - MMA 1, 100 mL
    - Benzoyl peroxide, 1 mL
  - MMA 3
    - MMA 1, 100 mL
    - Benzoyl peroxide, 2.5 mL
- Protocol
  - MMA 1, 48 h
  - MMA 2, 48 h
  - MMA 3, 48 h
- Embedding
  - MMA 3, 72 h at 37°C
Blocks are transparent.

1.3.4 SECTIONS

1.3.4.1 Paraffin, Paraffin–Celloidin, or Gelatin–Paraffin Blocks

1.3.4.1.1 Microtome Sections

Blocks are put on the stage of a vertical microtome (Minot's microtome). The sections are usually retailed to obtain 4 to 7 μm thickness.

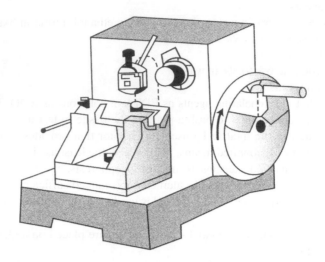

FIGURE 1.3   Vertical microtome.

The block can be positioned on the stage in several manners. In modern microtomes, encased material is placed on the stage with a pair of clamps. In the oldest microtomes, the paraffin block is stuck to the stage by warming the stage with an alcohol lamp and applying it against the paraffin block. In other microtome types, the clamps can be used to grip the block of wax. Sections are usually cut 4 to 7 μm thick. Section thickness is ranged on the microtome. On modern types, the thickness is directly given in micrometers. In the oldest models, the thickness is given in round fractions of the endless screw that constitutes the forward movement system of the microtome. The block is submitted to a vertical up-and-down motion. When it goes down, a section is cut by a steel razor. The block then goes up behind the blade. The blade can be straight (for dense and hard tissues) or with a concave face in front of the outer side (for soft and less dense tissues). Sections can be directly taken from the blade with a brush. In certain microtome types, they can be removed onto a belt conveyor. Avoid steel objects, such as scalpels and grips. Generally, the microtome movement is manipulated by hand. This allows the histologist to vary the cutting speed. Some microtomes are automatic. Vertical microtomes are also called "Minot's microtomes" (Figure 1.3). Sometimes horizontal microtomes can also be used.

1.3.4.1.2   Cutting the Block

Before the paraffin block is placed on the stage, it must be prepared. The paraffin around the object to be sectioned must be removed, leaving the object enclosed in a trapezoidal paraffin block. The lower and upper sides of the block must be parallel. During cutting, the lower side of the second section sticks to the upper side of the first section. If the two sides are not parallel, the resulting ribbon of sections will be curved, which can make it difficult to mount a series of sections on a slide (see Figure 1.4).

1.3.4.1.3   Difficulties

Several difficulties occasionally occur during cutting operations. Some problems, their causes, and their solutions are given below.

- Ribbon is not formed.
  - Blade is blunt. Displace the blade laterally, or replace the blade.
  - Sections are thick. Reduce the thickness of the section by manipulating the advance of the microtome.

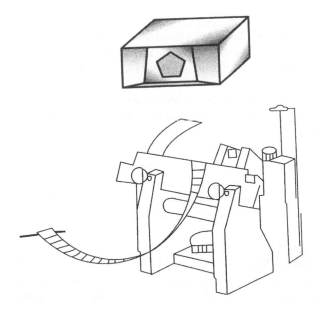

FIGURE 1.4 Cutting the block. (From Martoja, R., and M. Martoja-Pierson, 1967, *Initiation aux techniques de l'histologie animale*, Masson, Paris. With permission.)

- • Paraffin is too hard. Reembed by melting the paraffin and embedding the object in a new wax with a lower melting point.
  - • Room temperature is too low. Warm the blade and the specimen with a lamp, or drive the specimen into tepid water or blow on the specimen (in this case, verify the object is locked in the higher position).
  - • Angle between object and blade is too open. Range the blade pitch.
  - • Other reasons. Hold sections with a thin brush as they are formed.
- • Ribbon is curved.
  - • Upper and lower sides are not parallel. Cut the block again to obtain parallel sides.
  - • Upper and lower sides are not parallel to the razor edge. Range the block position.
  - • Razor edge is irregular or blunt. Displace the razor laterally or take another razor.
  - • Block sides are not at the same temperature because of light, a warming source, and so on that give different hardness to the wax. Put the microtome where it is not submitted to temperature variations.
- • The thickness of sections are different.
  - • Angle between specimen and blade is too small, section cannot be obtained. Range the blade pitch.
  - • Angle between razor and object is too great. Change the blade pitch.
  - • Microtome vibrates. To avoid vibrations, lock the screws of the microtome.
  - • Section is too hard. Immerse the block in water; this operation will make it soft.
  - • Wax is too soft. Embed the specimen again with a paraffin having a higher melting point.
- • Sections are compressed.
  - • Razor edge is blunt. Displace the razor laterally or take another razor.
  - • Temperature is too high. Immerse the block in very cool water before cutting.
  - • Angle between specimen and blade is too small. Range the blade pitch.
  - • Section speed is too important. Decrease the section speed.
  - • Tissue fragment is compressed but not the wax. Embed again.
  - • Wax pieces are stuck on the razor edge. Carefully clean the two sides of the razor with a solvent of paraffin.

- Sections are torn.
  - Tissue fragment has not been well dehydrated. Embed the piece again.
  - Tissue fragment is soft. Embed the tissue again in paraffin with a higher melting point.
  - Block is opalescent, which is linked to the fact that water is included in the paraffin. Embed the piece again.
  - Tissue is too hard. Double embed with celloidin and paraffin.
  - Wax has cooled very slowly. Embed again with a slow cooling of paraffin.
- Ribbon is striated.
  - Razor is scratched. Displace the razor or change the razor.
  - Razor edge is dirty. Carefully clean the two sides of the blade with a wax solvent.
  - Angle between the object and the blade is too open. Range the blade pitch.
  - Sections are damaged by particles included in the wax. Embed again with filtered paraffin.
  - Tissue fragment is too voluminous. Embed again with celloidin or celloidin and paraffin.
- Sections stick on the blade.
  - Temperature is too high. Immerse the block in very cool water before cutting.
  - Razor is scratched. Wait for a decrease in room temperature or displace the razor.
  - Angle between specimen and blade is too small. Change the razor or change the blade pitch.
- Section fly away.
  - There is static electricity in the air. Increase humidity with a flame or with a receptacle full of water.
- Sections are vibrated.
  - The block and/or the blade are not very locked. Lock the microtome screws.
  - The block is too hard. Reembed by melting the paraffin and embedding in a new wax with a lower melting point.
  - Tissue is particularly calcified. Double embed with celloidin and paraffin.
  - Angle between the object and the blade is too open. Reembed after decalcification or change the blade pitch.

1.3.4.2   Celloidin Sections

Blocks are cut with a special horizontal microtome. Sectioning is done in ethanol 70%.

1.3.4.3   Sections for Plastic Waxes

Blocks of plastic wax are placed on the stage of a microtome with a mandrel or between two clamps. The knife is made of glass with a special form. A concave side is directed to the exterior, which allows one to remove the sections easily before placing them on a slide. Plastic blocks can easily be cut automatically.

1.3.4.4   Bone Sections

Sections of wax-embedded bones are cut with a heavy horizontal microtome. Sections are obtained by use of a tungsten carbide blade.

1.3.4.5   Frozen Sections

Frozen sections are done by use of a cryotome. It is a vertical microtome that is installed in a cooled room, the cryostat. The blade used is stored into this room and the temperature of the blade is that used for section. The piece of tissue is cooled. Then it is directly embedded in a wax that is liquid at laboratory temperature and solid at the low temperatures that are used for sections. This operation is done on the stage of the cryotome. Sections are obtained with a classic microtome for paraffin sections. Sections can be thin (5 μm). They are directly collected on the blade with a slide where they are spontaneously stuck.

1.3.5 ADHESION OF SECTIONS

1.3.5.1 Paraffin and Double-Embedded Sections
Several methods can be used to adhere paraffin and double-embedded sections.

1.3.5.1.1 Water Adhesion
Water adhesion is the easiest method of sticking. Sections are disposed on the surface of a 37°C water bath. Then they are picked up on a slide that is slipped under the section. The slides are then warmed at 37°C. To obtain very efficacious adhesion, it is necessary to remove oils from the slides by washing them with ethanol and hydrochloric acid (1 vol/1 vol). Then they must be rinsed with distilled water. It is also possible to purchase washed and oil-free slides.

1.3.5.1.2 Glycerin–Albumin Adhesion
- Preparation with egg white
  - Mix an equal weight of egg white and glycerin. Use a kitchen mixer to emulsify albumin molecules.
  - Add 0.5% sodium salicylate or thymol
  - Filter slowly. This sticking agent must be preserved in a refrigerator.
- Meyer's preparation
  - Ovalbumin powder, 1 g
  - Distilled water, 100 mL
  - Let dissolve
  - Glycerin, 100 mL
  - Filter
  - Sodium salicylate (or thymol), 1 g
- Albuminous water. Prepare just before use.
  - Distilled water, 20 mL
  - Albuminous solution, 1 mL
  - In certain cases, it is useful to increase the sticking power using more than 1 mL of albumin. In other cases, the sticking power can be decreased by using less than 1 mL of albumin.
- First protocol
  - Warm albumin water in a Petri dish, 40°C to 50°C
  - Place the sections on the albumin water
  - Allow sections to be perfectly plated
  - Slip the sections on clean wet slices
  - Eliminate excess liquid
  - Dry the slides horizontally for 4 h at 40°C. As in the case of water sticking, it is necessary to use perfectly clean slides. For that, slides are washed with ethanol and hydrochloric acid (1 vol/1 vol). Then they must be rinsed with distilled water, or purchase washed and oil-free slides.
  - Or dry the slides vertically for 2 h at 60°C.
  - Dispose slides on the stage at 40°C or 50°C. As in the case of water sticking, it is necessary to use perfectly clean slides. For that, slides are washed with ethanol and hydrochloric acid (1 vol/1 vol). Then they must be rinsed with distilled water, or purchase washed and oil-free slides.
- Second protocol
  - Place the clean dry slides on the stage at 40°C to 50°C
  - Put some drops of albumin water on the slide
  - Place the sections on the sticking reagent
  - Allow sections to be perfectly plated

- Eliminate excess liquid
- Dry the slides horizontally for minimum of 4 h at 40°C

1.3.5.1.3   Adhesion with Gelatinous Water

Adhesion with gelatinous water is the same as the previous one, but albumin water is replaced with a solution of 1% gelatin (warm distilled water is added until gelatin dissolution). As in water adhesion, it is necessary to use perfectly clean slides. For that, clean with ethanol and hydrochloric acid (1 vol/1 vol). Then they must be rinsed with distilled water, or purchase washed and oil-free slides.

1.3.5.1.4   Adhesion on Gelatinized Slides

- Gelatinous water
  - Distilled water, 500 mL at 60°C
  - Gelatin, 2.5 g
  - Leave overnight at 58°C
  - Add a pinch of chrome alum (fungicide)
- Protocol—Dip slides in gelatinous water, two times, then let them dry vertically for 2 to 3 h.
  - As for water adhesion, it is necessary to use perfectly cleaned slides. For that, clean with ethanol and hydrochloric acid (1 vol/1 vol). Then they must be rinsed with distilled water, or purchase washed and oil-free slides.
  - Repeat the operation
  - Store slides at +4°C
  - Stick the sections as above, replacing albumin water with distilled water

1.3.5.2   Adhesion of Collodion Sections

1.3.5.2.1   General Principles

Sticking collodion sections before staining is particularly difficult because of embedding mass elimination. In the majority of cases, sections are first stained by floating on dye baths. They are then adhered to the slides. As for water adhesion, it is necessary to use perfectly cleaned slides. For that, slides are washed with ethanol and hydrochloric acid (1 vol/1 vol). Then they must be rinsed with distilled water, or purchase washed and oil-free slides.

1.3.5.2.2   Before Staining

- Maximow's method
  - Wet section with ethanol 70%
  - Unpleat the section on the razor
  - Put the section on a glycerin–albumin coat slide. As for water adhesion, it is necessary to use perfectly cleaned slides. For that, slides are washed with ethanol and hydrochloric acid (1 vol/1 vol). Then they must be rinsed with distilled water, or purchase washed and oil-free slides.
  - Blot the section with paper filter
  - Cover with oil of clove
  - Let rest for 10 min
  - Ethanol 95%, 10 min
  - Ethanol 100%, 2 × 10 min
  - Ethanol ether (1/1) until the dissolution of celloidin
  - Stock in ethanol 70%
- Celloidin adhesion—Sections can be adhered by a solution of celloidin in an ethanol–ether mixing (1/1). As for water adhesion, it is necessary to use perfectly cleaned slides. For

that, slides are washed with ethanol and hydrochloric acid (1 vol/1 vol). Then they must be rinsed with distilled water, or purchase washed and oil-free slides.

- Gelatin adhesion—Sections can also be stuck with gelatin.
- Stick the sections with gelatinous water at 1%.
- Eliminate celloidin oil of cloves, ethanol–ether solution, and 100% ethanol. As for water adhesion, it is necessary to use perfectly cleaned slides. For that, slides are washed with ethanol and hydrochloric acid (1 vol/1 vol). Then they must be rinsed with distilled water, or purchase washed and oil-free slides.
- Sections are ready to be stained.

1.3.5.2.3   After Staining
- Convey the sections in the staining reagents
- Stick sections with 2% celloidin solution or stick sections with 1% gelatinous water
- Eliminate celloidin with oil of cloves, ethanol–ether solution and 100% ethanol. As for water adhesion, it is necessary to use perfectly cleaned slides. For that, slides are washed with ethanol and hydrochloric acid (1 vol/1 vol). Then they must be rinsed with distilled water, or purchase washed and oil-free slides.

1.3.5.3   Adhesion of Plastic Wax Sections
Plastic wax sections are adhered with water by proceeding as indicated earlier. It is also possible to put the sections on distilled water; then let this water evaporate on a warming stage. As for water adhesion, it is necessary to use perfectly cleaned slides. For that, slides are washed with ethanol and hydrochloric acid (1 vol/1 vol). Then they must be rinsed with distilled water, or purchase washed and oil-free slides.

1.3.5.4   Adhesion of Frozen Sections
Frozen sections are recovered directly from the blade by apposition of the slide, which has been refrigerated. As for water adhesion, it is necessary to use perfectly cleaned slides. For that, slides are washed with ethanol and hydrochloric acid (1 vol/1 vol). Then they must be rinsed with distilled water, or purchase washed and oil-free slides.

1.3.5.5   Adhesion of Bone Sections
Sections of plastic wax-embedded bones are stained by conveying them directly into staining baths. It is only after this operation that sections can be mounted. Sections are adhered on slides by certain waxes.

1.3.6   DEPARAFFINING AND DEHYDRATION

1.3.6.1   Principle
Before they can be stained by dye or by a histochemical reactive in aqueous solution, paraffin-embedded sections must be cleared and hydrated. Indeed, dyes cannot react on tissue saturated with paraffin. Dewaxing is done with a solvent. Then, hydration is done by putting slides in baths containing decreasing degrees of ethanol and finally water.

1.3.6.2   Protocol
- Put slides on a flame (facultative), which promotes the elimination of air bubbles.
- Cyclohexane: 2 × 10 min
- Ethanol 95%, 5 min
- Ethanol 70%, 1 min
- Tap water, 3 s

1.3.7 COLLODIONING

1.3.7.1 Principle

During certain histological staining or histochemical reactions, sections can be unstuck. They must be protected by a celloidin film, by placing slides with sections into a celloidin solution. After solidification, celloidin will hold sections in position, even if the sticking agent fails.

1.3.7.2 Protocol

- Celloidin preparation
  - Collodion (celloidin), 1 g
  - Ethanol 100%, 50 mL
  - Ether, 50 mL
- Collodioning
  - Dewax
  - Collodion, 1 mL
  - Ethanol 95% (if albumin sticking) or ethanol–formalin (if gelatin sticking)
    - Ethanol, 90 mL
    - Formalin, 10 mL
  - Wash with tap water
  - Stain

1.3.8 CONFECTION OF SMEARS

1.3.8.1 Definitions

1.3.8.1.1 Smears

A smear is defined as isolated cells plated on a slide. Cells to be plated can be in a liquid medium, after centrifugation, for example, or in a blood sample. They can also be obtained by dilaceration of a compact tissue.

1.3.8.1.2 Imprint

An imprint is done by placing a slice of an organ against a slide or another support. The operation is done several times. To do an imprint, the organ must not be laterally displaced during the operation.

1.3.8.1.3 Squash

A squash is done by compressing a small fragment of an organ between two slides.

1.3.8.2 Making a Smear

1.3.8.2.1 Dry Blood Smear on Slide

- Put a drop of blood near a side of a slide.
- With another slide or a lamella, push the suspension to the other side. As for water adhesion, it is necessary to use perfectly cleaned slides. For that, slides are washed with ethanol and hydrochloric acid (1 vol/1 vol). Then they must be rinsed with distilled water, or purchase washed and oil-free slides.
- Let dry. Drying must be immediate. To accelerate it, the technician can wave the slide in the air with fan-shaped movements.
- Preserve the smear by immersing it in a mixing of ethanol and acetic acid. In certain cases, preservation can be done by other fixatives.

1.3.8.2.2 Dry Blood Smear on Coverslip

- Put a drop of blood near an edge of a slide. As for water adhesion, it is necessary to use perfectly cleaned slides. For that, slides are washed with ethanol and hydrochloric acid (1 vol/1 vol). Then they must be rinsed with distilled water, or purchase washed and oil-free slides.
- Cover the plate with another plate.
- Let the blood stem itself.
- Separate the plates, placing a smear on each. To avoid difficulties when separating the plates, hold them at a 45° angle.

1.3.8.2.3 Making a Wet Smear

Smears are done between a slide and a coverslip.

- Put a drop of blood near the edge of a slide. As for water adhesion, it is necessary to use perfectly cleaned slides. For that, slides are washed with ethanol and hydrochloric acid (1 vol/1 vol). Then they must be rinsed with distilled water, or purchase washed and oil-free slides.
- With another slide or a coverslip, push the suspension to the other side.
- Do not let dry.
- Preserve the smear by immersing it in fixative, which can be a mixture of ethanol and acetic acid. Preservation can also be done using osmium tetroxide steam.

1.3.9 CELL CULTURES

1.3.9.1 Monolayer Cell Culture

- Develop a cell layer on Leighton's tube lamella or at the bottom of the flask
- Preserve the cell layer
- Stain directly on the slide or on the bottom of the flask

1.3.9.2 Suspension Cell Culture

A preserved block that has been obtained after centrifugation can be embedded in paraffin or resin, then cut and placed on a slide, like a classical organ (Figure 1.5).

1.4 METHODS OF STAINING

1.4.1 DYES

1.4.1.1 Definition

A dye is a chemical substance that is able to permanently stain a cellular or tissue component.

1.4.1.2 Phenomenon

Staining is a very complex phenomenon. It involves chemical factors such as acidic–basic liaisons. It is also affected by physical factors such as dye diffusion and capillarity. In complex staining, competition between dyes and blinding of one dye by another can also be observed. The phenomenon is not always precisely known.

1.4.2 MECHANISM

1.4.2.1 Dyes

In histological staining, the result is known but is obtained from empirical methods because the underlying mechanisms have not been elucidated. Conversely, in histochemical methods, the effects

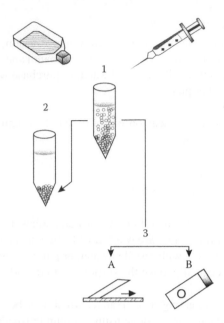

FIGURE 1.5 Treatment of cell cultures. (1) Cells are obtained from a culture, a suspension, or directly from the tissue. (2) The cell suspension is centrifuged. (A) Cells can be used to do a smear. (B) Or treated as a compact tissue. (From Morel, G., 1998, *Hybridation in situ*, Polytechnica, Paris. With permission.)

and molecular mechanisms regarding the action of the dye on the tissue are known, along with an understanding of the parameters, such as temperature or pH, that are necessary.

A dye is a molecule that empirically possesses two particular chemical groups: the chromophore, which gives the color, and another atomic group, the auxochrome, which is required to fix the dye molecules on the tissue-specific molecules (Figure 1.6). Certain substances are colored because they possess a chromophore group. However, they do not possess an auxochrome and they cannot fix themselves on the tissues. These substances are not dyes but chromogens. Certain chemicals are not chemically fixed on tissues, but they are dissolved into them. These substances are not dyes but lysochromes. They are used to stain lipids.

Most dyes are organic synthetic products; a few have a natural origin, essentially vegetal.

1.4.2.2 Chromophores

The main chromophore groups are

- Azoic
- Azine
- Indamine or thiazine
- Nitro
- Kinonic form of aromatic molecules and naphtokinones

The higher the number of chromophoric groups, the more intense the coloration will be.

1.4.2.3 Acidic and Basic Dyes

In classic histology, it is recognized that acidic dyes are cytoplasmic and basic dyes are nuclear. However, this acid/base terminology is not linked to the pH of the dye solution, and it is independent of the acidic or basic nature of this solution. The distinction refers to the auxochrome, which is the part of the dye molecule that is required for tissue fixation. Eosin sodium salts, which are acidic dyes, generally have a basic pH.

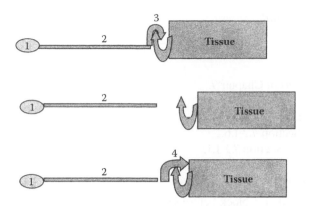

FIGURE 1.6 Structure of dyes. 1, Chromophores; 2, Neutral molecular part; 3, Auxochrome; 1 + 2, Chromogen; 4, Mordant.

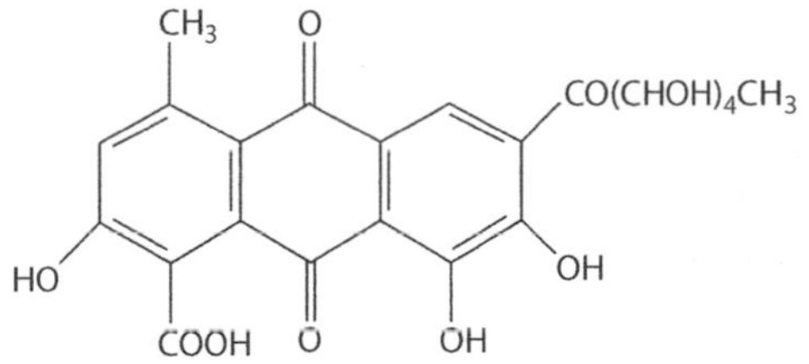

FIGURE 1.7 Carmine.

Basic dyes possess a $-NH_3^+$ cationic auxochrome. In contrast, acidic dyes possess an anionic auxochrome, such as $-COO^-$ or even SO_3^-. When a SO_3^- group is fixed on a basic dye, the latter is transformed in acidic dye (for example, acidic and basic fuchsin).

1.4.2.4 Mordancy

When a staining substance does not have an auxochromic group, a mordant is useful. A mordant is a molecule that allows the fixation of the staining substance on the tissue. Two methods can be used. The tissue can be first submitted to the action of this molecule and then to the staining substance. Or, the mordant can be added to the staining solution. Iron and ammonium alum are often used as mordant. Such a solution is called a lac. Staining with mordancy is used especially to visualize cell nuclei with hematoxylin or nuclear fast red.

1.4.3 NUCLEUS STAINING

1.4.3.1 Principle

Nuclear staining uses classic histological dyes that have been used by histologists for many years. These dyes have an animal or vegetal origin, and they can be obtained by organic synthesis. They are not used as histochemical reagents because their mode of action is not known and so cannot be controlled. Their use is essentially of descriptive interest. These dyes can be used in a progressive or regressive mode.

1.4.3.2 Carmine

Carmine (Figure 1.7) was the first dye used in histology. Today it is rarely used. It is the only dye with an animal origin. The staining molecule is carminic acid, which is characterized by kinonic groups (they are chromophoric groups). Carmine is obtained by treating dry body extracts of female cochineal, a homopterous Mexican insect, with iron alum. Carmine can be replaced with azocarmine, a synthetic molecule. See Section 1.4.7.1.

1.4.3.2.1   Chemical Formula

See Figure 1.7.

1.4.3.2.2   Preparation

See the following sections in Chapter 7.

- Borated carmine: See Section 7.2.1.5.
- Carmalum: See Section 7.2.1.6.
- Acetocarmine: See Section 7.2.1.1.

1.4.3.2.3   Staining with Borated Carmine

This staining is performed on a block before sectioning.

- Fixative—Recommended fixatives contain mercuric chloride (sublimate), and they are treated during 3 days with 70% ethanol.
- Reagents
  - Borated carmine
  - Hydrochloric ethanol
    - Ethanol 70%, 100 mL
    - Hydrochloric acid, 0.30 mL
- Protocol
  - Borated carmine, 3 days
  - Hydrochloric ethanol, 3 days. Hydrochloric ethanol is used for differentiation. It must be used to the point that the excess red dye is eliminated.
  - Ethanol 70%, 2 days
  - Dehydrate
  - Mount *in toto* or embed. After staining, the piece can be encased in paraffin or celloidin. It is then sectioned and slices are mounted between a slide and a coverslip after dewaxing, similar to a classic section.
- Results—Cell nuclei are red stained.

1.4.3.2.4   Carmalum Staining

Carmalum staining is a staining done on block before sectioning.

- Fixatives—Recommended fixatives contain mercuric chloride (sublimate) and they are treated during 3 days by ethanol 70%.
- Reagents
  - Carmalum
  - Potassium alum 1%
- Protocol
  - Carmalum, 2 days
  - Potassium alum, 1 day
  - Tap water, 1 day
  - Mount *in toto* or embed. After staining, the piece can be encased in paraffin or celloidin. It is then sectioned and slices are mounted between a slide and a coverslip after dewaxing, similar to a classic section.
- Results—Cell nuclei are red stained.

1.4.3.2.5   Chromosome Staining

- Fixative—Use a karyotype preparation.

- Reagents
  - Acetocarmine
- Protocol
  - Acetocarmine, a few drops
  - Cover with a coverslip
  - Push slightly
  - Dry with paper filter
  - Seal
- Results—Chromosomes are red stained. Chromosomes can be violet stained by adding ferric chloride into the acetocarmine solution.

1.4.3.3 Hematoxylin Staining

1.4.3.3.1 Principle

Hematoxylin (Figure 1.8) is a dye that is extracted from *Haematoxylum campechianum*, a South and Central American tree. Hematoxylin is not a dye by itself. It is hematein, an oxidation product, with a kinonic group that possesses chromophoric qualities (Figure 1.9). Therefore, before it is used, hematoxylin must be oxidized. This oxidation can be done in several ways. The classic manner is slow and consists of letting the hematoxylin solution be oxidized in air for 6 to 8 months. The other methods employ an oxidant.

Hematein, which does not possess an auxochrome group, has no affinity for tissues. It is necessary to use a mordant, generally an aluminum salt (iron and ammonium alum, potassium alum, etc.). Mordants can act on the tissue before the hematoxylin action. They can also act directly during the hematoxylin action. The hematein lacs that are obtained are called hemalum. (A lac is obtained

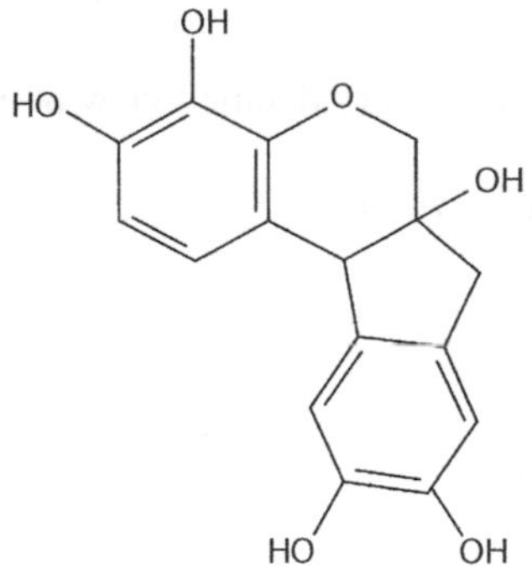

FIGURE 1.8 Hematoxylin.

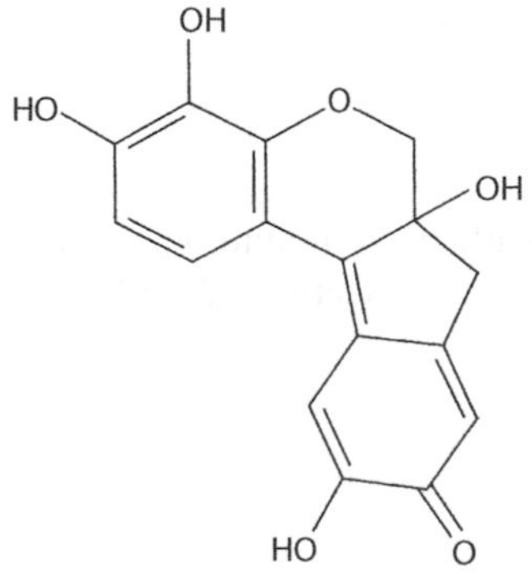

FIGURE 1.9 Hematein.

by mordant action on the dye.) Often the terms *hemalum*, *hematein*, and *hematoxylin* are used interchangeably. Stains that are obtained can be preserved for a long time without alteration, but the stain can be affected by the action of certain acidic molecules.

1.4.3.3.2   Formula
See Figure 1.8 and Figure 1.9.

1.4.3.3.3   Different Staining Types
- Staining preceded by the action of a mordant—These staining methods can be used for sections, smears, or cell cultures. Techniques are long and difficult. Nuclei are uniformly dark stained with a few chromatin details.
  - Regaud's hematoxylin
  - Heidenhain's hematoxylin
  - Mallory's phosphotungstic hematoxylin
- Hemalums—These dyes are lacs obtained by a mixing of hemalum and hematein. They are prepared by means of potassium alum. Methods using these dyes are the most classic. They are often used with a regressive mode of staining.
  - Hansen's hemalum
  - Masson's hemalum
  - Mayer's hemalum
  - Harris's hemalum
  - Ehrlich's hemalum
  - Delafield's hematoxylin
- Progressive iron hematoxylin lacs
  - Wegert's hematoxylin
  - Iron hematoxylin
  - Groat's hematoxylin (permits a very detailed view of nuclei with chromatin details)
- Other hematoxylin staining
  - Progressive chromic hematoxylin lacs
  - Hansen dihematein
  - Mallory phosphotungstic hematoxylin

1.4.3.3.4   Preparative Protocols
See the following sections in Chapter 7.

- Masson's hemalum: See Section 7.2.1.10.
- Hematoxylin: See Section 7.2.1.9.
- Groat's hematoxylin: See Section 7.2.1.7.
- Regaud's hematoxylin: See Section 7.2.1.13.

1.4.3.4   Nuclear Fast Red

1.4.3.4.1   Definition
Nuclear fast red is a synthetic dye with an anthraquinonic nature (Figure 1.10). It can be used alone or associated to other dyes. See Chapter 7, Section 7.2.1.12.

1.4.3.4.2   Formula
See Figure 1.10.

1.4.3.4.3   Staining
- Reagent
  - Nuclear fast red: See Chapter 7, Section 7.2.1.12.

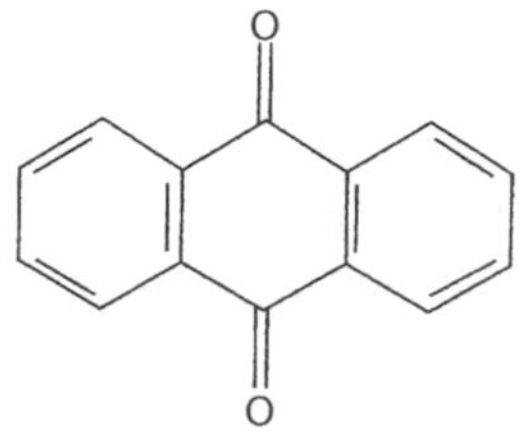

FIGURE 1.10   Nuclear fast red.

- Protocol
  - Dewax
  - Hydrate
  - Nuclear fast red, 2 min
  - Rinse with distilled water
  - Dehydrate, mount
- Results—Nuclei are red stained. This method is useful for obtaining a quick staining when another tissue component is visualized with a dark color.

1.4.4   STAINING OF CONNECTIVE TISSUE

1.4.4.1   Introduction

The connective tissue originates from the embryonic mesenchyme. It is characterized by several cells formed in a matrix, synthesized by the cells themselves. Several types of connective tissue can be distinguished: loose connective tissue, cartilage, and bone. Connective tissue is constituted with a fundamental substance, fibers of different natures, and cells. The homogenous and amorphous fundamental substance contains several molecules synthesized by cells, such as proteoglycans and exogenic substances. Specific molecules are also found in cartilage and bones. The fundamental substance of cartilage contains chondroitin sulfuric acid and chondrotin sulfate salts. In bone, fundamental substance contains calcium salts and more particularly hydroxy-apatite give the hardness to the bone.

The fibers found in connective tissue are collagen, reticulin, and elastic fibers. Reticulin fibers are single fibers of collagen. In some organs, such as large arteries, elastic blades are observed with elastic fibers. In loose connective tissue, fibroblasts and fibrocytes synthesize both fundamental substance and fibers. Several macrophage types are also observed. Mastocytes synthesize some proteoglycans. Plasmocytes are migrating cells belonging to the immunity system; these cells are characterized by the disposition of chromatin in a "wheel-rays" fashion. Leucocytes are also observed. Adipocytes stock fats; they are generally grouped together but separated from each other with fundamental substance and fibers. In cartilage, chondroblasts and chondrocytes synthesize the fundamental substance. In bone, osteoblasts and osteocytes are implicated in bone dynamics; osteoclasts are cells with several nuclei, devoted to the resorption of bony matrix.

Specific methods to visualize connective tissue are based on the detection of the fundamental substance or fibers.

1.4.4.2   Visualization of Fundamental Substance

Fundamental substance is rarely stained. Yet, it is possible to stain it with several dyes usually used in bichromic or trichromic reactions, giving a staining reaction with connective tissue. In hemalum–eosin staining, connective tissue is stained by eosin.

1.4.4.3   Collagen Visualization

1.4.4.3.1   Main Methods

Collagen is mainly composed with proteins. It appears as fibers found in the whole organism, except in nervous tissue. Collagen fibers are naturally birefringent, consequently to their molecular structure. They also present a birefringence due to the parallel disposition of underunits.

Collagen can be visualized in several manners. Fibers of collagen are acidophilic with current staining methods. There are not peculiar histochemical reactions but it is possible to use both acidophilic character and permeability of fibers to visualize them on histological sections. Collagen acidophily is related to its high isoelectric point.

Three kinds of methods allow visualization of collagen:

- Common methods as hemalum–phloxine–saffron in which saffron dissolved in ethanol gives a yellow coloration to collagen fibers.
- Methods using phosphomolybdic acid and/or aniline based dyes as fast green or aniline blue. Phosphomolybdic acid (PMA) can be replaced with phosphotungstic acid (PTA).
- Methods based upon competition between two acidic dyes, for example, Masson's trichroma and variations, azan staining, or Van Gieson's method.

1.4.4.3.2 Van Gieson's Method

- Fixative—All fixative agents are convenient. Fixatives with mercury chloride are also particularly recommended.
- Reagents
  - Hematoxylin: See Chapter 7, Section 7.2.1.9.
  - Van Gieson's picrofuchsin: See Chapter 7, Section 7.2.2.20.
- Protocol
  - Dewax, hydrate
  - Stain with an acidoresistant nucleus dye for 5 min. A dye based upon hematoxylin can be used. The optimal duration of staining is indicative and must be determined according to the dye used and organ studied.
  - Tap water, 10 to 30 min
  - Rinse with distilled water
  - Picrofuchsin, 2 to 5 min
  - Rinse quickly with distilled water or with ethanol 95%
  - Finish dehydration, mount
- Results—Nuclei are brown or black; cytoplasm, muscle fibers, red blood cells, and fibrin are yellow; collagen is red.

1.4.4.3.3 Detection of Reticulin by Use of Argyrophilic Reaction

The staining is based upon argyrophyly of reticulin fibers. See Chapter 2, Section 2.1.3.3.2 devoted to argyrophyly and argentaffinity.

- Fixatives
  - Bouin's fluid. Other fixatives such as Carnoy's fluid or ethanol can also be used.
  - pH 7 buffered formalin 10%
- Reagents
  - Potassium metabisulfite 3%
  - Periodic acid 0.5%
  - Distilled water neutralized with bromothymol blue. Bromothymol blue must be added drop by drop to water to obtain yellow reflects.
  - Silver ammonium carbonate
  - Ammoniacal water
    - Ammoniac, 2 drops
    - Distilled water, 100 mL
  - Formalin 2%
  - Gold chloride 0.2%
  - Oxalic acid 5%

- Sodium thiosulfate 5% or sodium hyposulfite
- Nuclear fast red: See Chapter 7, Section 7.2.1.12.
- Protocol
  - Dewax, hydrate
  - Potassium metabisulfite, 2 min
  - Rinse with distilled water
  - Periodic acid, 10 min
  - Rinse with neutral distilled water
  - Silver ammonium carbonate, 10 min at 56°C
  - Gold chloride, 10 to 15 min. Sections become gray.
  - Oxalic acid, 10 min. Use slides at level. Optimal duration of staining must be determined according to the tissue.
  - Sodium thiosulfate (or sodium hyposulfite), 10 min
  - Rinse with tap water
  - Nuclear fast red, 5 min
  - Rinse
  - Dehydrate
  - Mount
- Results—Reticulin fibers are black stained. Nuclei and cytoplasms are red and pink.

1.4.4.4 Visualization of Elastic Fibers

Generally, elastic fibers are not visualized with classic staining. It is necessary to use specific staining. The dyes frequently used are orcein or paraldehyde fuchsin. These staining methods give specific visualizations, but they are signaletical not histochemical.

1.4.4.4.1 Visualization of Elastic Fibers and Blades with Paraldehyde Fuchsin

Paraldehyde fuchsin was used for visualization of elastic fiber by Gomori (1950). It stains numerous secretions if it is preceded with an oxidation.

- Fixative—All fixatives are convenient but avoid prolonged fixation with bichromate.
- Reagent
  - Paraldehyde fuchsin: See Chapter 7, Section 7.2.2.17.
  - Groat's hematoxylin. Groat's hematoxylin can be replaced with nuclear fast red; nuclei will be red stained.
  - Picroindigocarmine: See Chapter 7, Section 7.2.2.7.
  - Hydrochloric acid 0.5% in absolute ethanol
- Protocol
  - Dewax, hydrate
  - Paraldehyde, 5 min
  - Wash with tap water
  - Absolute ethanol
  - Differentiate with hydrochloric ethanol
  - Wash with tap water
  - Groat's hematoxylin, 5 min
  - Wash with tap water. Continue to wash until the end of departure of dye.
  - Picroindigocarmine, 30 s
  - Dehydrate
  - Mount
- Results—Nuclei are dark brown or black, cytoplasms are green, elastic fibers are purple, and collagen fibers are blue stained. Groat's hematoxylin can be replaced with nuclear fast red; nuclei will appear red stained.

1.4.4.4.2 Visualization of Elastic Fibers and Blades with Paraldehyde Fuchsin with Oxidation

- Fixative—All fixatives are convenient but avoid prolonged fixation with bichromate.
- Reagent
  - Paraldehyde fuchsin: See Chapter 7, Section 7.2.2.17.
  - Groat's hematoxylin: See Chapter 7, Section 7.2.1.7.
  - Picroindigocarmine: See Chapter 7, Section 7.2.2.7.
  - Gomori's oxidant
    - $KMnO_4$ 2.5%, 15 mL
    - H_2SO_4 5%, 15 mL
    - Distilled water, 90 mL
- Bisulfite 2% in water
- Protocol
  - Dewax, hydrate
  - Gomori's oxidant, 20 to 30 s
  - Rinse with tap water
  - Sodium bisulfate (or sodium metabisulfite), passage
  - Tap water, 5 min. This operation is used to whiten the sections. It must be quick. Sodium metabisulfite can also be used.
  - Paraldehyde-fuchsin, 2 min
  - Wash with tap water
  - Groat's hematoxylin, 5 min
  - Tap water
  - Picroindigocarmine, 30 s
  - Dehydrate, mount
- Results—Elastic fibers and some secretions are purple, nuclei are dark brown, cytoplasms are green, collagen fibers are blue stained.

1.4.4.4.3 Visualization of Elastic Fibers with Orcein and Picroindigocarmine (Method of Unna–Tänzer)

- Fixative—All fixatives are convenient.
- Reagents
  - Orcein—Preparation
    - Orcein, 1 mL
    - Ethanol 100%, 100 mL
    - Hydrochloric acid, 0.7 mL
  - Picroindigocarmine: See Chapter 7, Section 7.2.2.7.
- Protocol
  - Dewax, hydrate. Collodion if necessary.
  - Orcein, 10 to 15 min
  - Rinse with distilled water. This operation is used to eliminate excess orcein.
  - Picroindigocarmine, 20 s
  - Differentiate with ethanol 100%
  - Mount
- Results—Elastic fibers are brown, collagen is blue-green.

1.4.4.4.4 Visualization of Elastic Fibers and Blades with Polychrome Blue (Method of Unna)

- Fixative—All fixatives are convenient.
- Reagents
  - Orcein—Preparation
    - Orcein, 0.25 mL
    - Ethanol 70%, 100 mL
  - Polychrome blue or methylene blue: See Chapter 7, Section 7.2.2.15.

- Protocol
  - Dewax, hydrate
  - Polychrome blue, 5 to 15 min
  - Tap water
  - Orcein, differentiate
  - Dehydrate
  - Mount
- Results—Elastic fibers are brown stained. Differentiation must be done under microscope control. Be careful; the differentiation can be extremely fast and provoke the elimination of nuclear staining. In this case, the only thing to do is another staining!

1.4.5 BICHROMIC STAINING

1.4.5.1 Hemalum Eosin

This staining method uses a nuclear dye (hemalum or Groat hematoxylin) and a cytoplasmic dye, eosin (which can be replaced with phloxin or erythrosine). See Figure 1.11.

- Fixative—All fixative agents are convenient.
- Reagents
  - Groat's hematoxylin: See Chapter 7, Section 7.2.1.7.
  - Eosin 1%: See Chapter 7, Section 7.2.2.8.
- Protocol
  - Dewax, hydrate
  - Groat's hematoxylin, 5 min
  - Wash with tap water to obtain a blue hematoxylin
  - Eosin, 30 s
  - Wash with tap water. Rinsing duration after adding Groat's hematoxylin is equal to the dye bath duration. When Groat's hematoxylin becomes too old, its staining qualities are weakened; nuclei are brownish colored, not dark blue, and it is necessary to use a new dye solution. However, to prolong the staining qualities of hematoxylin, it is possible to prolong the bath duration to 10 or 15 min and rinse for the same amount of time.
  - Dehydrate, mount
- Results—Nuclei are dark blue stained (they are brown if hematoxylin is too old), acidophilic cytoplasm are pink. Certain secretions remain uncolored. Hemalum eosin staining yields quick results, but it is not sufficiently selective to visualize tissue components with a great precision. Yet, it is very often used.

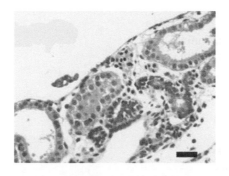

FIGURE 1.11 (See color insert.) Method: Hematoxylin–eosine. Tissue: Young amphibian kidney. Preparation: Paraffin section. Fixative: Bouin's fluid. Observations: Nuclei are blue stained, and cytoplasm, connective tissue, and smooth muscle are pink stained. Bar = 90 µm.

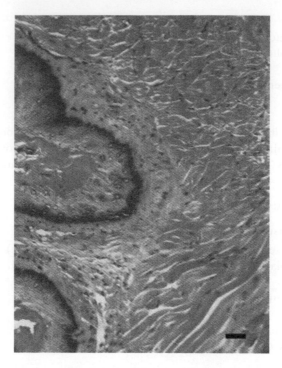

FIGURE 1.12   (See color insert.) Method: Hematoxylin–phloxine–saffron. Tissue: Mouse esophagus. Preparation: Paraffin section. Fixative: Bouin's fluid. Observations: Nuclei are dark stained, cytoplasm is pink, connective tissue is yellow, muscle are pink stained. Bar = 100 μm.

1.4.5.2   Hemalum Phloxine Saffron

This staining method uses hemalum as a nuclear dye, and phloxin as a cytoplasm dye (Figure 1.12). Saffron is collagen specific. This staining is used as a standard staining method in pathologic anatomy. It comes from Masson's trichroma. Hemalum can be replaced with hematoxylin.

- Fixative—All classic fixative are convenient.
- Reagent
  - Groat's hematoxylin: See Chapter 7, Section 7.2.1.7.
  - Phloxin 0.5% or 1% in distilled water: See Chapter 7, Section 7.2.2.18.
  - Saffron obtained by distillation in ethanol: See Chapter 7, Section 7.2.2.19.
- Protocol
  - Dewax, hydrate
  - Groat's hematoxylin, 5 min
  - Wash with tap water, 5 min
  - Phloxin, 3 min
  - Rinse
  - Ethanol 95%, 2 min
  - Ethanol 100%, 2 min
  - Saffron, 10 min
  - Ethanol 100%, quick
  - Butanol, cyclohexane
  - Mount
- Results—Nuclei are blue stained; cytoplasm, muscle fibers, and red blood cells are red; and collagen is yellow.

1.4.5.3 Hemalum Picroindigocarmine

This staining method uses hemalum or Groat's hematoxylin as a nuclear dye and a mix of dyes, and picroindigocarmine as a cytoplasmic dye.

- Fixative
  - Masson's hemalum: See Chapter 7, Section 7.2.1.10.
  - Calleja's picroindigocarmine: See Chapter 7, Section 7.2.2.7.
- Protocol
  - Dewax, hydrate
  - Hemalum, 3 min
  - Wash with tap water to obtain a dark brown staining
  - Picroindigocarmine, 30 s
  - Ethanol 100%, 10 min
  - Dehydrate
  - Mount
- Results—Nuclei and basophilic cytoplasm are brown colored. Acidophilic cytoplasm and nucleoli are yellow or green stained. Collagen fibers are blue and red blood cells are yellow. Glycoproteins are brown. Secretions are yellow or green stained. This staining method is useful for obtaining very quick results for a morphological appreciation. Variations of stain can be observed on the sections.

1.4.6 TRICHROMA

1.4.6.1 Mallory's Trichroma

The interest in Mallory's trichroma is essentially historical, because the trichroma used today are based upon Mallory's one. This method of staining remains useful for certain studies.

- Fixative—Sublimate or Zencker's fluid. The sublimate or Zencker's fluid are fixatives allowing a selective staining of connective tissue with methylene blue. The use of other fixatives can delete this selectivity.
- Reagents
  - Acidic fuchsin 1%
  - Phosphomolybdic acid 1%
  - Methylene blue–orange G—Preparation:
    - Methylene blue, 0.50 g
    - Orange G, 2 g
    - Oxalic acid, 2 g
    - Distilled water, 100 mL
- Protocol
  - Dewax, hydrate
  - Acidic fuchsin, 1 to 3 min. Avoid blue staining of other components than connective tissue.
  - Distilled water
  - Phosphomolybdic acid, 1 min. Phosphomolybdic acid (PMA) intensifies blue staining of connective tissue.
  - Rinse with distilled water
  - Methylene blue–orange G, 10 min. Time of staining is variable. Previous operations are useful. It is also possible to go directly from acidic fuchsin to the methylene blue mix according to this formula:
    - Methylene blue, 0.50 g
    - Orange G, 2 g
    - Distilled water, 100 mL

- Tap water
- Dehydrate, mount
- Methylene blue, 0.50 g
- Orange G, 2 g
- Distilled water, 100 mL
- Results—Collagen fibers and mucus are blue stained. Nuclei, elastic fibers, neuroglia are red. Red blood cells and myelinic fibers are yellow stained.

1.4.6.2 Masson's Trichroma (First Variant)

The nuclear dye can be Masson's hemalum or several hematoxylin solutions. The cytoplasmic dye is fuchsin. An aniline-derived dye allows differentiation of collagen fibers. This technique has numerous variants (hemalum, phloxine, and saffron or even Masson–Goldner's trichroma).

- Fixative—All fixative agents are convenient.
- Reagent
    - Regaud's hematoxylin: See Chapter 7, Section 7.2.1.13.
    - Acidic fuchsin culvert: See Chapter 7, Section 7.2.2.3.
    - Fast green: See Chapter 7, Section 7.2.2.12.
    - Iron and ammonium alum 5%
    - Picric acid saturated in ethanol
    - Acetic water 0.5%
    - Phosphomolybdic acid 1%
- Protocol
    - Dewax, hydrate
    - Iron and ammonium alum, 15 min at 50°C
    - Regaud's hematoxylin, 15 min at 50°C
    - Differentiate by picric ethanol to obtain gray reflection on the sections. Monitor differentiation duration carefully. If phosphomolybdic acid action is prolonged, all the nuclear dye can be extracted from the section. Conversely, if differentiation is too fast, an excess of dye can give a wrong staining of tissues. The entire section is "clogged" by hematoxylin; nuclei appear to be uniformly dark and chromatin cannot be seen.
    - Tap water, 10 min
    - Acid fuchsin culvert, 1 min at 50°C
    - Differentiate with phosphomolybdic acid, 5 to 10 min at 50°C
    - Rinse with acetic acid
    - Fast green, 10 min
    - Rinse with acetic water
    - Dehydrate, mount
- Results—Nuclei are brownish blue; cytoplasm is red; secretions are red or green; and muscles and collagen fibers are green stained. This staining method yields good results for morphological studies because the different parts of the tissue are very concentrated.

1.4.6.3 Masson's Trichroma (Second Variant)

The nuclear dye can be Masson's hemalum or several hematoxylin solutions. The cytoplasmic dye is fuchsin. An aniline-derived dye allows differentiation of collagen fibers. (See Figure 1.13.)

- Fixative—All fixative agents are convenient.
- Reagent
    - Regaud's hematoxylin: See Chapter 7, Section 7.2.1.13.
    - Aniline blue saturated in acetic water 2.5%
    - Acidic fuchsin: See Chapter 7, Section 7.2.2.2.

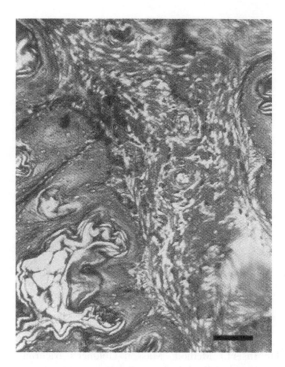

FIGURE 1.13 **(See color insert.)** Method: Masson's trichroma. Tissue: Mouse esophagus. Preparation: Paraffin section. Fixative: Bouin's fluid. Observations: Nuclei are brown stained, cytoplasm is pink, connective tissue is blue, muscles are pink stained. Bar = 200 μm.

- Iron and ammonium alum 5%
- Phosphomolybidic acid 1%
- Acetic water 1%
- Acetic ethanol 100%
- Protocol
  - Dewax, hydrate
  - Iron and ammonium alum, 24 h
  - Regaud's hematoxylin, 30 min at 50°C
  - Wash by distilled water
  - Let sections drain
  - Differentiate by iron and ammonium alum (or by picric acid saturated in ethanol 95%) to obtain a pure nuclear staining (dark brown)
  - Wash with distilled water
  - Acidic fuchsin, 5 min
  - Rinse with distilled water
  - Differentiate with phosphomolybdic acid, 5 min
  - Aniline blue, 5 min. Do not wash sections between phosphomolybdic acid and aniline blue.
  - Differentiate in acetic water, 5 to 30 min
  - Acetic ethanol 100%, passage. A passage consists of quickly dipping a slide into the mixture. An increased stay can have the same effect as a prolonged differentiation.
  - Dehydrate, mount
- Results—Nuclei are black; cytoplasm is red; and muscles, collagen fibers, and mucus are blue stained. This method yields good results for morphological studies. However, the blue staining of connective tissues provides a less contrasted picture of the sections than does the first variant.

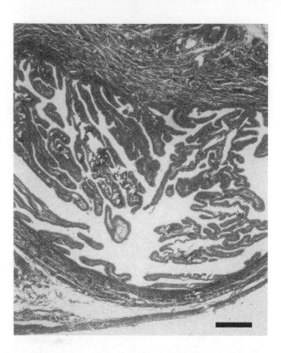

FIGURE 1.14 **(See color insert.)** Method: Masson–Goldner's trichroma. Tissue: Mouse oviduct. Preparation: Paraffin section. Fixative: Bouin's fluid. Observations: Nuclei are dark stained, cytoplasm is pink. Bar = 10 µm.

1.4.6.4 Masson–Goldner's Trichroma

The principle is the same as with Masson's trichroma. Nuclear dye is Groat's hematoxylin. (See Figure 1.14.)

- Fixative—Avoid fixative with osmium tetroxide.
- Reagents
  - Groat's hematoxylin. Groat's hematoxylin gives a particularly precise staining to nuclei. It is possible to see all the details of chromatin repartition. See preparative modes.
  - Acidic fuchsin culvert: See Chapter 7, Section 7.2.2.3.
  - Molybdic G orange: See Chapter 7, Section 7.2.3.24.
  - Acetic sulfo green, still called "acidic light green." See Chapter 7, Section 7.2.2.1.
  - Acetic water 1%
- Protocol
  - Dewax, hydrate
  - Groat's hematoxylin, 5 min. Groat's hematoxylin must be exclusively used.
  - Tap water to obtain a blue staining of sections
  - Fuchsin culvert, 5 min
  - Rinse with acetic water. It is often necessary to change acetic waters after each slide passage.
  - Molybdic G orange, passage
  - Acetic sulfo green, 10 min
  - Rinse with acetic acid
  - Dehydrate
  - Mount
- Results—Nuclei are black or dark blue. The bottom of cells is gray, acidophilic cytoplasm is pink; secretions are red or green stained. Muscles are red and collagen fibers green. This trichrome gives very good results for morphological studies. Nuclei are very detailed, and the different parts of tissues are well visualized.

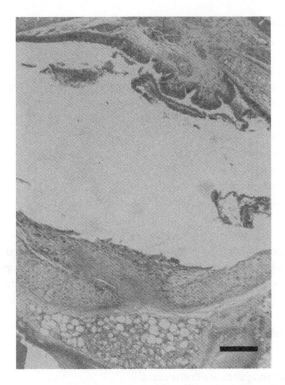

FIGURE 1.15 **(See color insert.)** Method: Prenant's triple staining. Tissue: Mouse trachea. Preparation: Paraffin section. Fixative: Bouin's fluid. Observations: Nuclei are brown stained, cytoplasm is pink, connective tissue is green. Bar = 200 µm.

1.4.6.5 Prenant's Triple Staining

This staining technique is derived from the old method called "iron hematoxylin." Its results are dependent upon preservation and protocol of staining. (See Figure 1.15.)

- Fixative—Avoid fixatives with osmium tetroxide.
- Reagent
  - Eosin 1%: See Chapter 7, Section 7.2.2.8.
  - Regaud's hematoxylin. See Chapter 7, Section 7.2.1.13.
  - Aqueous solution of sulfo green 0.5%. Still called "acidic light green." See Chapter 7, Section 7.2.2.1.
  - Iron and ammonium alum 0.5%
- Protocol
  - Dewax, hydrate
  - Eosin, 10 min
  - Iron and ammonium alum, 14 h
  - Regaud's hematoxylin, 24 h
  - Differentiate by iron and ammonium alum. Stop the differentiation when the general coloration of slide is gray.
  - Sulfo green, 20 to 30 s
  - Dehydrate by ethanol 100%
  - Mount
- Results—Nuclei and basophilic cytoplasm are dark brown stained, collagen is green. Cytoplasm, nucleoli, and erythrophilic secretions are green, as are cyanophilic secretions. This can be a good method for a morphological study. However, it is lengthy.

1.4.6.6   Prenant's Triple Staining (Method of Gabe)

- Fixative—Avoid fixatives with osmium tetroxide.
- Reagent
  - Groat's hematoxylin: See Chapter 7, Section 7.2.1.7.
  - Eosin sulfo green. Still called "acidic light green." See Chapter 7, Section 7.2.2.9.
  - Acetic water 0.5%
- Protocol
  - Dewax, hydrate
  - Hematoxylin, 5 min
  - Wash with tap water
  - Eosin sulfo green, 10 min
  - Rinse quickly with distilled water or hold in acetic water
  - Dehydrate
  - Mount
- Results—Nuclei and basophilic cytoplasm are dark brown stained; collagen is green. Cytoplasm, nucleoli, and erythrophilic secretions are pink. Cyanophilic secretions are green. This is a good method for morphological studies, and it is quicker than the original method.

1.4.6.7   Ramon y Cajal's Trichroma

See Figure 1.16.

- Fixative—All fixative agents are convenient.
- Reagents
  - Ziehl's fuchsin diluted at 1/5 with distilled water: See Chapter 7, Section 7.2.2.21.
  - Picroindigocarmine: See Chapter 7, Section 7.2.2.7.
  - Acetic water 0.2%

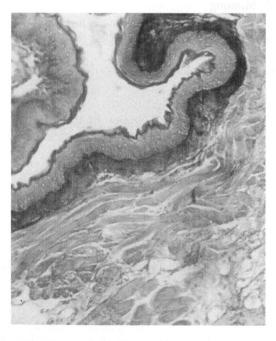

FIGURE 1.16   **(See color insert.)** Method: Ramon y Cajal's trichroma. Tissue: Mouse esophagus. Preparation: Paraffin section. Fixative: Bouin's fluid. Observations: Nuclei are red, cytoplasm are green or yellow, keratin is red stained, connective tissue is purple, muscles are yellow stained. Bar = 200 μm.

- Protocol
- After fixative without osmium tetroxide
  - Dewax, hydrate
  - Ziehl's fuchsin, 10 min
  - Wash with tap water
  - Wash with acetic water to start of fuchsin excess
  - Picroindigocarmine, 5 min
  - Acetic water, 10 min
  - Ethanol 100% as far as departure of red is stopped
  - Continue dehydration
  - Mount
- After a preservation with osmium tetroxide
  - Ziehl's fuchsin, 45 min at 60°C
  - Picroindigocarmine, 30 s
- Results—Nuclei and basophilic cytoplasm are red colored. Acidophilic cytoplasm is green or gray; collagen is blue. Secretions are green, blue, or red. Mucus is orange or purple colored. This method provides very well-contrasted results.

1.4.6.8 One-Time Trichroma

This is a quick method. Nuclei and cytoplasm are stained by azorubine. Phosphomolybdic acid is the mordant. Solid green stains connective fibers. Naphthol yellow (or Martius's yellow) stains red blood cells.

- Fixative—Avoid, if possible, mixtures with dichromate or osmium tetroxide.
- Reagent—One-time trichroma: See Chapter 7, Section 7.2.2.16.
- Protocol
  - Dewax, hydrate
  - One-time trichroma, 10 min

If the fixative contains osmium tetroxide:

- One-time trichroma, 45 min at 60°C
- Rinse
- Dehydrate, rinse. Sections can be stored in acetic water 1% after rinsing. If sections have an excess of red, treat them with yellow naphthol (or Martius's yellow) in saturated solution.
- Mount
- Results—Nuclei and cytoplasm are red stained, connective fibers are green, and red blood cells are yellow. Mucopolysaccharides are more often green. This method is quick, but it is often unsuccessful and results are sometimes deceiving.

1.4.6.9 Cleveland and Wolfe's Trichroma

Cleveland and Wolfe's technique is used to distinguish different secretory cell types belonging to adenohypophysis, which are characterized by their tinctorial properties (Figure 1.17). The original technique developed by Cleveland and Wolfe (1932) is not often used today. However, Herlant's modified method (1956) is used for studying cell types, along with other techniques, and more particularly for immunocytochemical detection of hormones. Cleveland and Wolfe's technique is essentially descriptive because it does not give any indication concerning the chemical structure of cell content. To study the pituitary, it is important to validate the method with a precise immunocytochemical study that will provide precise hormonal information for each cell category. Cleveland and Wolfe's trichroma can be used to visualize each tissue type.

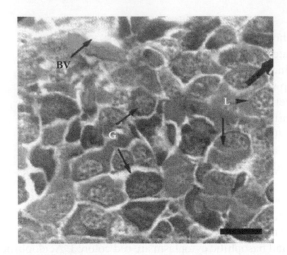

FIGURE 1.17   (See color insert.) Method: Cleveland and Wolfe's staining. Tissue: Amphibian adenohyopo-hysis. Preparation: Paraffin section. Fixative: Bouin's fluid. Observations: Nuclei are red stained, cytoplasm is blue, pink, purple or orange stained depending on cell type, connective tissue is blue, blood cells are orange stained. BV: blood vessel; G: gonadotrophic cell; L: lactotrophic cell. Bar = 10 μm.

- Fixative—Each classic fixative can be used. Halmi's liquid is recommended.
- Reagents
  - Erythrosin 1%: See Chapter 7, Section 7.2.2.10.
  - G orange 2% into phosphotungstic acid 1%: See Chapter 7, Section 7.2.3.24.
  - Aniline blue 1%: See Chapter 7, Section 7.2.2.6.
- Protocol
  - Dewax, hydrate
  - Erythrosine, 3 min
  - Rinse with distilled water
  - Phosphotungstic G orange, 30 s
  - Rinse with distilled water
  - Aniline blue, 1 to 2 min
  - Rinse with distilled water
  - Dehydrate
  - Mount
- Results—Nuclei are blue stained, nucleoli are pinkish, somatotropic cells are pinkish, thy-reotropic cells are purple, corticotropic cells are pale purple, lactotropic cells are orange stained, and gonadotropic cells are blue with pinkish granulations. This method is quick and provides very good visualizations of nuclei and the different parts of tissue. It is used for morphological studies.

1.4.6.10   Herlant's Tetrachroma

This method was initially used to stain the pituitary gland, but it can be used for all types of tissues and gives well contrasted results. (See Figure 1.18.)

- Fixative—The best fixative contains mercuric chloride (Susa, Halmi). Mixtures containing potassium dichromate are not recommended. Carnoy's fluid also gives good results.
- Reagents
  - Erythrosin (acetic solution): See Chapter 7, Section 7.2.2.10.
  - Diluted Heidenhain's blue: See Chapter 7, Section 7.2.2.13.

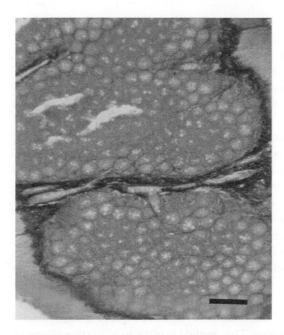

FIGURE 1.18 (See color insert.) Method: Herlant's trichroma. Tissue: Mouse intestine (caecum). Preparation: Paraffin section. Fixative: Bouin's fluid. Observations: Nuclei are red, cytoplasm is pink or non-stained, connective tissue is blue. Bar = 30 μm.

- Acidic alizarin blue: See Chapter 7, Section 7.2.2.4.
- Phosphomolybdic acid 5%
- Protocol
  - Dewax, hydrate
  - Erythrosin, 10 min. Erythrosine can be differentiated by ethanol 70%.
  - Rinse with distilled water
  - Heidenhain's blue, 5 min
  - Rinse with distilled water
  - Alizarin blue, 10 min
  - Rinse with distilled water
  - Phosphomolybdic acid, 10 min
  - Dehydrate
  - Mount
- Results—Chromatin is blue colored, nucleoli are red; cytoplasm and secretions are red or purple; mucus is blue; and cartilage and collagen are dark blue. This method is quick and provides very good visualization of nuclei and the different parts of tissue. It is used for morphological studies.

1.4.6.11 Paraldehyde Fuchsin

This technique has been used to stain elastic fibers (Gomori, 1950). It stains most secretions if oxidization has previously been performed.

- Fixative—All classic fixative agents are convenient but prolonged preservation by dichromate should be avoided.
- First variant: Without oxidization

- Reagents
  - Paraldehyde fuchsin: See Chapter 7, Section 7.2.2.17.
  - Groat's hematoxylin. Groat's hematoxylin can be replaced with nuclear fast red; nuclei will be red stained: See Chapter 7, Section 7.2.1.7.
  - Picroindigocarmine: See Chapter 7, Section 7.2.2.7.
  - Hydrochloric acid 0.5% in ethanol 100%
- Protocol
  - Dewax, hydrate
  - Paraldehyde fuchsin, 5 min
  - Wash in tap water
  - Dehydrate with ethanol 100%
  - Differentiate with hydrochloride ethanol as far as dye stop to go off
  - Wash with tap water
  - Groat's hematoxylin, 5 min
  - Wash with tap water
  - Picroindigocarmine, 30 s
  - Dehydrate
  - Mount
- Results—Nuclei are dark brown, cytoplasm is green stained, elastic fibers are purple, and collagen fibers are blue. This method is often used to stain elastic fibers on a tissue.
- Second variant with oxidization
- Reagents
  - Paraldehyde fuchsin: See Chapter 7, Section 7.2.2.17.
  - Groat's hematoxylin: See Chapter 7, Section 7.2.1.7.
  - Picroindigocarmine: See Chapter 7, Section 7.2.2.7.
  - Gomori oxidizing
    - $KMnO_4$ 2.5%, 15 mL
    - H_2SO_4 5%, 15 mL
    - Distilled water, 90 mL
  - Aqueous solution of sodium bisulfite or metabisulfite
- Protocol
  - Dewax, hydrate
  - Gomori oxidizer, 20 to 30 s
  - Rinse with distilled water
  - Whiten sections by a quick treatment with sodium bisulfate
  - Tap water, 5 min
  - Paraldehyde fuchsin, 2 min
  - Wash with tap water, 2 min
  - Groat's hematoxylin, 5 min
  - Tap water, 5 min
  - Picroindigocarmine, 30 s
  - Dehydrate
  - Mount
- Results—Elastic fibers and certain secretions are purple stained. Nuclei are brown-black. Cytoplasm is green and collagen fibers are blue stained. This method is often used to stain elastic fibers and secretions on a tissue.

1.4.6.12 Pappenheim Panoptic Staining

This staining method, first used for smears, is also used for sections. It allows one to visualize hematopoietic organs.

- Fixative—All the classic fixatives are convenient.
- Reagents
  - May–Grünwald—Preparation:
    - May–Grünwald, 10 mL
    - Distilled water, 80 mL
  - Giemsa—Preparation:
    - Giemsa, 10 mL
    - Distilled water, 750 mL
  - Acetic water 0.15%
- Protocol
  - Dewax, hydrate
  - May–Grünwald, 20 min at 35°C
  - Rinse
  - Giemsa, 40 min at 35°C
  - Differentiate with acetic water
  - Wash with tap water
  - Dehydrate by acetone
  - Mount
- Results—Nuclei are dark purple, basophilic cytoplasm is blue, acidophilic cytoplasm is red, collagen is pale blue, mucus is blue or purple, muscles are pinkish, and cartilage is blue. In blood cells, granulations of lymphoid cells are purple and that of myeloid cells are violet. Neutrophilic granulations are brownish or bluish. Erythrosinophilic granulations are brick red and basophilic granulations are blue colored. This relatively quick and easy method yields very good pictures of tissue. It permits visualization of the different blood cells on sections.

1.4.6.13 May–Giemsa Grünwald for Smears

- Fixative. See Chapter 1, Section 1.3.8.
- Reagents
  - May–Grünwald
  - PBS, pH 6.7—Preparation:
    - Crystallized disodium phosphate, 433 mL
    - Mono sodium phosphate (13.805 g/l), 567 mL. If crystallized disodium phosphate is not available, it can be prepared in the following manner:
      - Anhydrous phosphate, 14.8 g
      - Distilled water, 1000 mL
- Giemsa
  - Staining solution, 3 drops
  - PBS, 2 mL
- Protocol
  - May–Grünwald, 3 min
  - PBS, 3 min
  - Giemsa, 20 min
  - Rinse with water
  - Let dry
  - Mount with a hydrophobic wax
- Results—In blood cells, granulations of lymphoid cells are purple, that of myeloid cells are violet. Neutrophilic granulations are brownish or bluish. Erythrosinophilic granulations are brick red and basophilic granulations are blue colored. This relatively quick and easy method yields very good pictures of tissue. It permits visualization of the different blood cells on sections.

1.4.7 AZOIC DYES

1.4.7.1 Azan Staining

Azocarmine and aniline are the base of these types of staining. They allow several histological structures, particularly the chromatin repartition, to be visualized with great precision.

1.4.7.2 Heidenhain's Azan

- Fixative—All fixatives can be used, although fixatives with chromium or osmium should be avoided.
- Reagents
  - G or B azocarmine: See Chapter 7, Section 7.2.1.4.
  - Diluted Heidenhain's blue: See Chapter 7, Section 7.2.2.13.
  - Aniline 1% in ethanol 70%
  - Acetic acid 1% in ethanol 95%
  - Phosphotungstic acid 5% in distilled water
- Protocol
  - Dewax. Collodion if necessary.
  - Hydrate
  - If sections have been provided from tissue preserved with a fixative containing picric acid, the latter can be eliminated by 30 min in ethanol/aniline mixture. The step of picric acid elimination is optional.
  - Incubate in G azocarmine, 1 h at 60°C; or incubate in B azocarmine, 1 h at RT
  - Rinse in distilled water
  - Differentiate with aniline ethanol until an almost pure nuclear staining is obtained. Differentiation must be performed under microscope control. Caution: Differentiation can be extremely fast and can provoke the elimination of nuclear staining. In this case, the only thing to do is restain!
  - Acetic ethanol, 30 s. Acetic ethanol stops azocarmine differentiation. Stay of sections can be prolonged.
  - Wash with distilled water
  - Phosphotungstic acid, 30 min. Phosphotungstic acid acts as a mordant and makes Heidenhain's blue staining possible. It also prolongs azocarmine differentiation.
  - Wash with distilled water
  - Heidenhain blue, 1 h
  - Differentiate blue with ethanol 95%
  - Dehydrate directly with ethanol 100%
  - Mount
- Results—Nuclei and certain cytoplasms are red stained; other cytoplasms are yellow or gray. Collagen is blue stained. Secretions can be different in function of their nature. Acid mucopolysaccharides are blue stained. Nuclei are very well stained. If the protocol is well done, all details of chromatin can be observed.

1.4.7.3 Romeis's Azan

See Figure 1.19.

- Fixative—All the fixatives can be used, although fixatives with chromium or osmium should be avoided.
- Reagents
  - G or B azocarmine: See Chapter 7, Section 7.2.1.4.
  - Aniline blue: See Chapter 7, Section 7.2.2.6.

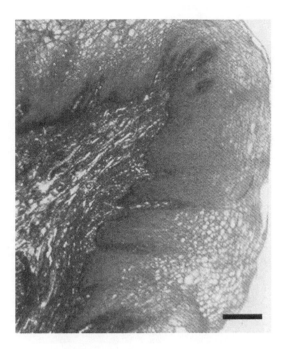

FIGURE 1.19 **(See color insert.)** Method: Romeis's azan. Tissue: Mouse cardia. Preparation: Paraffin section. Fixative: Bouin's fluid. Observations: Nuclei are red stained, cytoplasm is pink, connective tissue is blue. Bar = 200 μm.

- Aniline 1% in ethanol 70%
- Acetic acid 1% in ethanol 95%
- Protocol
  - Dewax. Collodion if necessary.
  - Hydrate
  - If sections from the tissue are preserved with a fixative containing picric acid, the latter can be eliminated by 30 min in ethanol/aniline mixture. The stage of picric acid elimination is optional.
  - Incubate in G azocarmine, 1 h at 60°C; or incubate in B azocarmine, 1 h at RT
  - Rinse in distilled water
  - Differentiate by aniline ethanol until an almost pure nuclear staining is obtained. Differentiation must be done under microscope control. Caution: Differentiation can be extremely fast and can provoke the elimination of nuclear staining. In this case, the only thing to do is restain!
  - Acetic ethanol, 30 s. Acetic ethanol stops azocarmine differentiation. Stay of sections can be prolonged.
  - Distilled water, 30 s
  - Phosphomolybdic G orange, 5 min
  - Wash with distilled water
  - Aniline blue, 10 min
  - Differentiate by ethanol 95%
  - Dehydrate
  - Mount
- Results—Nuclei and certain cytoplasms are red stained; other cytoplasms are yellow or gray. Collagen is blue stained. Secretions can be different depending on their nature. Acid mucopolysaccharides are blue stained. Nuclei are well stained with Romeis's azan. If the protocol is well done, all details of chromatin can be observed.

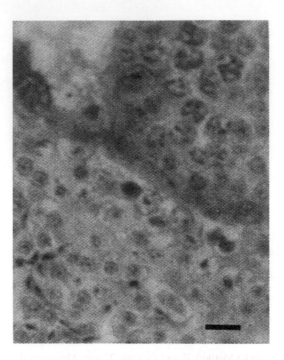

FIGURE 1.20   (See color insert.) Method: Modified azan. Tissue: Mammal testis. Preparation: Paraffin section. Fixative: Bouin's fluid. Observations: Nuclei are red stained with details, cytoplasm is pink, connective tissue is blue. Bar = 30 μm.

1.4.7.4   Modified Azan

See Figure 1.20. Nuclei are stained with nuclear fast red.

- Reagents
  - Nuclear fast red: See Chapter 7, Section 7.2.1.12.
  - Aniline blue: See Chapter 7, Section 7.2.2.6.
  - Molybdic G orange: See Chapter 7, Section 7.2.3.24.
- Protocol
  - Dewax, hydrate
  - Nuclear fast red, 15 min
  - Rinse
  - Molybdic G orange, 5 min
  - Rinse with water
  - Aniline blue, 2 to 5 min
  - Wash with distilled water. Washing with distilled water eliminates aniline blue in excess.
  - Ethanol 95%
  - Dehydrate, mount
- Results—They look like staining by Heidenhain or Romeis's azan, but details of nuclei are not observable as with a true azan (Romeis or Heidenhain).

1.4.8   STAINING OF SEMITHIN SECTIONS

1.4.8.1   Staining with Toluidine Blue

See Trump et al. (1961) and Figure 1.21.

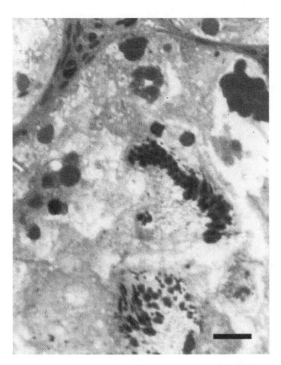

FIGURE 1.21 **(See color insert.)** Method: Toluidine blue (metachromasy). Tissue: Amphibian testis. Preparation: Inclusion in Epon, semithin sections. Fixative: Glutaraldehyde–paraformaldehyde. Observations: Nuclei are purple, cytoplasms are pink. Bar = 30 μm.

- Reagents
  - Toluidine blue (pH 11): See Chapter 7, Section 7.2.3.2.
    - Toluidine blue, 0.5 g
    - Sodium carbonate 2%, 100 mL
    - Boil
    - Filter
- Protocol
  - Put sections on a plate, 80°C
  - Toluidine blue, 1 drop
  - Evaporate
  - Distilled water
  - Evaporate
- Results—Tissues are stained in several blue colors or purple. Different staining are obtained according to metachromatic properties of tissues.

1.4.8.2 Staining with Toluidine Blue–PAS

- Reagents
  - Periodic acid 1%
  - Schiff's reagent
  - Sulfurous water—Preparation:
    - Sodium metabisulfite 10%, 1 mL
    - Distilled water, 20 mL
    - Hydrochloric acid 1 M, 1 mL
  - Toluidine blue pH 11: See Chapter 7, Section 7.2.3.2.
    - Toluidine blue, 0.5 g
    - Sodium carbonate 2%, 100 mL

- – Boil
- – Filter
- Protocol
  - Periodic acid, 15 min
  - Rinse with distilled water, 2 times
  - Schiff's reagent, 30 min
  - Sulfurous water, 2 × 2 min
  - Rinse
  - Toluidine blue, 1 min
  - Rinse
  - Mount sections
  - Blotting
  - Dry
- Results—Glycogen is pink; cytoplasm and nuclei are blue stained. This reaction is based upon histochemical reactions (see Chapter 2). Association of a metachromatic dye and PAS reaction gives several staining on the same semithin section.

1.4.8.3 Staining with Paraphenyldiamine

- Reagents
  - Paraphenyldiamine—Preparation:
    - – Paraphenyldiamine, 1 g
    - – Ethanol 95%, 100 mL
- Protocol
  - Put sections on a plate, 80°C
  - Reagent, 1 drop
  - Evaporate
  - Rinse
  - Evaporate
- Results—Tissues are pink stained.

1.5 MOUNTING

1.5.1 PRINCIPLE

After staining, sections must be mounted in a permanent manner so that they may be observed. The mounting mode depends on the section type: embedding with paraffin, celloidin wax, or frozen sections. It is also dependent both on the staining method itself, because certain methods do not support dehydration, and on the chemical reaction, because the results of certain histochemical reactions cannot be preserved for a long time.

1.5.2 MOUNTING AFTER DEHYDRATION

After staining or histochemical reaction, sections must be dehydrated. They then must be mounted with a medium that will prevent them from becoming aqueous. The medium for mounting, generally hydrophobic, is used after dehydration. In several cases, the staining can disappear after dehydration, so a hydrophilic medium is used, without dehydration.

1.5.2.1 Dehydration and Canada Balm Mounting

- Ethanol 70%, 2 min
- Ethanol 100%, 5 to 10 min
- Butanol, 5 min

- Cyclohexane, 10 min
- Place a drop of Canada balm on a coverslip
- Place the coverslip on the sections avoiding air bubbles. Mounting between slide and coverslip is a critical step. Be careful to avoid air bubbles, which will interfere with a correct observation.
- Let dry for 24 h at 60°C

1.5.2.2 Dehydration and Mounting with Eukitt

Eukitt is a synthetic medium that quickly dries in the air. Other synthetic media can be used, such as DPX, Permount, HSR, and Clarite, which are available from commercial distributors of histological products.

- Ethanol 70%, 2 min
- Ethanol 100%, 5 to 10 min
- Butanol, 5 min
- Cyclohexane, 10 min
- Put a drop of medium on a coverslip
- Place the coverslip on the sections avoiding air bubbles. It is common for air bubbles to be trapped between the slide and coverslip after mounting, which obscures a correct observation.
- Let dry for 4 h at RT

1.5.3 Aqueous Medium Mounting

1.5.3.1 Kaiser's Syrup Mounting

- Kaiser's syrup
  - Gelatin, 7 g
  - Distilled water, 42 mL at 50°C
  - Glycerin, 50 g
  - Phenol, 1 g
- Protocol
  - Place a drop of Kaiser's syrup on the slide. Kaiser's syrup must be stored at 60°C to avoid solidification.
  - Place a coverslip
  - Seal with paraffin or nail varnish
  - Let rest for 24 h at 4°C

1.5.3.2 "Crystal Mount" Mounting

"Crystal Mount" allows mounting in a hydrosoluble medium without seating. Crystal Mount medium has several advantages. It is a substance that polymerizes very quickly without bubble formation. When it is polymerized, the surface is level, smooth, and hard, and it is not necessary to use a coverslip (but it is possible). Crystal Mount also supports immersion oil and can be dissolved with tepid water without affecting the sections.

- Place a drop of Crystal Mount on the sections
- Let polymerize for 2 h at 37°C
- A coverslip is not necessary

1.5.4 Mounting for Celloidin Sections

- If the celloidin pellicle has been eliminated (i.e., the sections have been adhered with gelatin), mounting is done in the classic manner using Canada balm or a hydrophobic medium, such as Eukitt, after dehydration.

- If the celloidin pellicle is still present after staining:
  - Ethanol 95%. Never use ethanol 100% because it dissolves the celloidin pellicle.
  - Phenol
  - Xylene

1.5.5 Mounting for Plastic Wax Sections

Wax sections can be directly observed without mounting. They can also be covered directly with a drop of Canada balm or mounting medium and a coverslip. After drying, they can be observed.

1.5.6 Mounting for Fluorescent Preparations

Preparations that are to be observed with a fluorescent microscope must be mounted with a medium without its own fluorescence. Buffered glycerin is often used, such as Apathy's syrup, or other synthetic substances, for example "Fluoprep."

- Place a drop of mounting medium on the slide
- Cover with a coverslip
- Seal the coverslip with paraffin or nail varnish
- Let rest for 2 h at 4°C

REFERENCES

Cleveland, R., and J.M. Wolfe. 1932. A differential stain for the anterior lobe of the hypophysis. *The Anatomical Record* 51:409–413.

Exbrayat, J.M. 2001. *Genome visualization by classic methods in light microscopy*. CRC Press, Boca Raton, FL.

Gabe, M. 1968. *Techniques histologiques*. Masson et Cie, Paris.

Ganter, P., and G. Jolles. 1969. *Histochimie normale et pathologique*, 2 vol. Gauthier-Villars, Paris.

Lison, L. 1960. *Histochimie et cytochimie animales*. Gauthier-Villars, Paris.

Martoja, R., and M. Martoja-Pierson. 1967. *Initiation aux techniques de l'histologie animale*. Masson, Paris.

Morel, G. 1998. *Hybridation in situ*. Polytechnica, Paris.

Trump, B.F., E.A. Smuckler, and E.P. Benditt. 1961. A method or staining epoxy section for light microscopy. *Journal of Ultrastructural Research* 5:343–348.

2 Histochemical Methods

Jean-Marie Exbrayat

CONTENTS

2.1　GENERAL METHODS FOR HISTOCHEMISTRY

2.1.1　Purpose of Histochemistry

The purpose of histochemistry is to visualize *in situ* the chemical nature of cell and tissue structures. Histochemistry is used to study animals, vegetables, and humans in good health as well as pathology. To be visualized, cell and tissues must be first preserved in a state as near as possible to the living state. Then they will be visualized with several reactions based upon physical and histochemical properties. The use of control methods allows one to be precise if the staining obtained is specific. Contrary to biochemistry, substances stay in place during the histochemical investigation. Biochemistry also allows precise quantitative studies. For a long time, histochemistry was only qualitative, but today histochemistry allows quantification of results using image analysis.

2.1.2　Fixation and Characterization

2.1.2.1　Fixation

Fixatives used for histochemical investigations are the classic fixatives described in the first chapter of this section. It is indispensable to know if the fixative does not prevent the visualization of the researched molecule. In some cases, specific fixatives must be used.

2.1.2.2 Characterization of Chemical Nature of Structures

Visualization of the chemical nature of cell and tissue substances requires the use of chemical or physical methods.

2.1.2.2.1 Chemical Methods

Two types of chemical methods can be used. The first allows visualization of a peculiar atome or atomic group. In this category, the example is given by reaction of the guanidyl group belonging exclusively to arginine. The second is used to visualize an atomic group belonging to several different substances. In this category are the reactions used to detect carbonyl groups that allow visualization of carbohydrate or nucleic acid according to the method chosen to prepare carbonyl. In this case it is necessary to be precise with the method used to obtain carbohydrate and to use control methods.

2.1.2.2.2 Physical Methods

Several methods based upon physical properties are also useful. They can be used alone or combined with chemical methods. Among these methods, some of them are relatively simple, like solubility, extraction with solvents, microincineration, the observation with polarized light, and autofluorescence. Other methods are more complex and need a peculiar engineering: histophotometry, microspectrophotometry, and x-ray microanalysis.

2.1.3 GENERAL METHODS FOR HISTOCHEMICAL STAINING

2.1.3.1 Basophilic Staining

2.1.3.1.1 Principle

Basophilic components of tissues, such as nucleic acids, certain glucides, and proteins, react with basic dyes. Basophilic staining is in the category of histochemical methods in which the reaction between tissue or cell molecule and dye is known and controlled. Acidophilic staining occurs when tissues react with an acidic dye. The principles are the same as in basophilic staining.

2.1.3.1.2 Definition

A basic dye is a salt in which the cation is colored and the anion is not. An acidic dye is a salt in which the anion is colored and the cation is not.

2.1.3.1.3 Mechanism

A basophilic staining is characterized by the fixation of a colored cation on the tissue or cell element, which is colored by the cation. An acidophilic staining is characterized by the fixation of a colored anion on the tissue or cell element, which is stained with the anion.

2.1.3.1.4 Factors Acting on Basophilic Reaction

Basophilic reactions are subject to external factors that must be considered during the interpretation phase. Competition between the dye cation and other cations can exist. These latter cations can be uncolored or colored, and belong to the staining solution, to the tissue, or both. On the other hand, the staining reaction can be attenuated or suppressed if the penetration of stain molecules into the tissue is insufficient. Certain heavy dye molecules cannot access tissue sites to stain.

It is always possible that staining molecules are fixed to other tissue groups by different binding mechanisms (adsorption, hydrogen binding, etc.).

To be valuable, a basophilic reaction must follow certain rules:

- Cell and tissue acidic groups must be ionized
- Tissue and cell anions and cations must not be bound to one another
- Tissue anionic groups must be in sufficient quantity to be visualized

Numerous tissues and cell substances have acidic groups: nucleic acids, proteins, glucides, and lipids. Methyl green and pyronine are basic dyes with high specificity for nucleic acids. Pyronine is used to visualize RNAs and methyl green reacts on DNA. Both are used for methyl green–pyronine staining.

Basophilic reactions must be controlled to be certain of the nature of stained elements. Not all basic dyes react in the same manner on all basophilic structures. Some dyes are highly specific of nucleic acids. Some basic dyes, like toluidine blue, jointly visualize nucleic acids and other substances (see Section 2.1.3.2).

2.1.3.1.5   Precautions

To succeed in staining with a basic dye, some precautions must be taken:

* The dye must be as pure as possible and must not be in a mixture.
* The staining method must be progressive. Treatments after the staining are also important.
* It is advised to wash sections as few times as possible with a buffered solution at the pH of the stain.
* Dehydration with 2-methyl butan-2-ol is preferred to dehydration with ethanol.

This type of staining is well used for visualizing nucleic acids and proteins.

2.1.3.2   Metachromasy

2.1.3.2.1   Phenomenon

Metachromasy is a physical quality of staining solutions. It is used to visualize nucleic acids. Nucleic acids, and more especially DNA, and acidic carbohydrates can be stained with toluidine blue or thionine. It also permits one to differentiate, by different color, DNA and other substances during the staining of semithin sections of epoxy resin with toluidine blue. This staining is a method used to obtain quick results.

A metachromatic dye visualizes a tissue or cell structure by staining it with a color that is different from the color of the diluted dye solution. The color change is called metachromasy; the cell or tissue element is called the chromotrope. Most metachromatic stains are also basic dyes (toluidine blue, for example), but this is not an absolute rule because some metachromatic dyes are acidic. (However, they are rarely used.) Mann–Dominici staining is used to reveal metachromatic qualities of several cellular components. Some staining solutions are mixing solutions, for example, the trichroma composed of a mixture of azorubine (a basic dye), solid green, and naphthol yellow. These solutions are not metachromatic dyes.

2.1.3.2.2   Mechanism

A nonaqueous solution of metachromatic dye preserves the same color whatever its concentration or the temperature. In aqueous solution, variations of temperature or concentration modify the color solution. For example, solutions of toluidine blue are blue for the concentrations usually used. They become purple then red if the concentrations are increased. For the same concentration, the coloration of the solution is blue at high temperature and red at low temperature. A diluted solution of toluidine blue is blue colored. After staining, nuclei (DNA) are purple colored and cartilage (chondroitin-sulfate) is colored red.

A spectrometric study of these substances has been performed by Lison and Michaelis (Figure 2.1). When the concentration of aqueous solution increases or when temperature decreases, the wavelength for the maximal absorption (λ_{max} or α band) lowers to the red wavelength and a new absorption wavelength is observed (λ'_{max} or β band). Consequently, the color of solution is displaced from blue to purple or even red. Passage is observed from orthochromatic form (blue) to the metachromatic form (purple or red) for this dye. The explanation is based on physics and has supplanted other theories based on the presence of several molecular forms into the same dye solution.

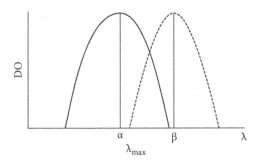

FIGURE 2.1   Spectrometric curves of a metachromatic substance showing the displacement of λ_{max} (α band) when a solution is diluted, or when the temperature is high to λ'_{max} (β band) when a solution is concentrated, or the temperature is low.

2.1.3.2.3   Important Factors

Tissue reaction depends upon negative charges (or positive in certain cases) that are present on the chromotrope. To obtain a metachromatic reaction, the tissue charges must be relatively dense (0.5 nm apart). The more this distance decreases, the more the metachromasy intensity increases. Conversely, if the distance increases, the tissue color is orthochromatic. On the other hand, the increase of charge alignment can increase the metachromatic reaction. This alignment can be increased by hydrophilic or hydrophobic groups. Water molecules, which allow polymerization of the dye, can also be used to increase the metachromatic reaction.

The molecular weight of tissue elements is also involved; chromotropic substances are generally molecules with a high molecular weight or molecules can aggregate themselves to obtain a high molecular weight overall.

External factors can modify the metachromatic qualities of a dye. Cations can be competitive to the basic dye (or anions if the dye is acidic), and they can lower metachromasy or even stop it. The higher the valence of competitive ions, the higher is the inhibitor effect. An addition of soluble proteins decreases or stops this phenomenon. Use of proteolytic enzymes can stop this negative effect.

2.1.3.2.4   Chromotropic Substances

Chromotropic substances are acidic mucopolysaccharides and nucleic acids (especially DNA). Certain lipids and silica particles also are metachromatic.

2.1.3.2.5   Choice of Dyes

The most used metachromatic dyes are toluidine blue, azure blue, thionine, and cresyl blue. All are thiazinic. They are used in aqueous solution. Concentrations are 1% (1g/L) for less chromotropic substances, and 0.1%, 0.01%, or 0.001% for very chromotropic substances. Generally, metachromatic dyes are also basic. However, some acidic dyes also are metachromatic.

2.1.3.2.6   Mounting the Section

Mounting is also an important factor because the mounting medium can inhibit staining. For nucleic acids, mounting with a hydrophobic medium after dehydration or with a hydrophilic medium without dehydration has no consequence on staining; however, this is not the case for cartilaginous tissue.

2.1.3.3   Visualization of Reductive Groups

2.1.3.3.1   Reductive Substances

Numerous substances possess reductive properties, which are used for their histochemical detection. Among these substances are cysteine, certain pigments, phenolic molecules such as catecholamine or serotonin, unsaturated lipids, vitamin C, and aldehydes that are produced by action of an oxidant with certain chemical groups belonging to glucides or DNA. The reductive properties of aldehydes are used in the Feulgen and Rossenbeck's reaction to visualize DNA, or in PAS reaction to visualize carbohydrates with photonic microscopy as well as with electron microscopy.

2.1.3.3.2 Reactions

Four types of reactions are useful to visualize the reductive properties of tissues and cell elements.

Reduction by a silver salt: argentaffinity or argyrophily. Reductive substances act on a silver salt to yield a black precipitate of reduced silver. For that, ammonium silver nitrate can be used.

$$2AgNO_3 + 2NaOH \rightarrow Ag_2O + 2NaNO_3 + H_2O$$

$$Ag_2O + 4NH_3 + H_2O \rightarrow 2(Ag(NH_3)_2OH)$$

This is a histochemical reaction in which the cell or tissue substance is the reactant. This substance is called argentaffine or reductive silver. It is visualized as a black deposit of metallic silver.

Reaction of silver nitrate with sodium hydroxide yields silver oxide. This reacts with ammonium to yield ammonium silver hydroxide, which is characterized by a diamine ion that will react with the tissue component.

$$R–CH=O + 2(Ag(NH_3)OH) \qquad R–COOH + 2Ag° + 4NH_3 + H_2O$$

Other reactions are based on the visualization of a tissue component by silver deposition, but these reactions are not histochemical. They are silver impregnations, which are often used in histology. For that, a silver salt is constituted; then it is reduced by formaldehyde. Certain cell or tissue substances fix the reduced silver by electrostatic binding, adsorption, and so on. But these substances are not used for the reaction. Silver impregnation is used to visualize the nucleolar organizer (AgNOR method). These cell or tissue components are called argentophilic. These reactions are not of histochemical nature

$$2AgNO_3 + 2NaOH \rightarrow Ag_2O + 2NaNO_3 + H_2O$$

$$Ag_2O + 4NH_3 + H_2O \rightarrow 2(Ag(NH_3)_2OH)$$

or

$$2AgNO_3 + Li_2CO_3 \rightarrow Ag_2CO_3 + 2LiNO_3$$

$$Ag_2CO_3 + 4NH_3 \rightarrow Ag(NH_3)_2CO_3$$

The silver salt can be prepared from sodium hydroxide or lithium carbonate. After the salt is obtained, it is reduced by formaldehyde.

$$H_2C=O + 2(Ag(NH_3)_2OH) \rightarrow HCOOH + 2Ag° + 4NH_3 + H_2O$$

or

$$2H_2C=O + 2(Ag(NH_3)_2CO_3) \rightarrow 2HCOOH + 2Ag° + 4NH_3 + 2CO_2$$

Reduction of ferric ferricyanide. Reductive groups can be visualized by reducing ferric ferricyanide to ferric ferrocyanide, also called Prussian blue. Reductive substances are blue stained. This reaction is used to visualize reductive –SH groups belonging to sulfhydryl proteins, such as keratin.

Other reductive methods (reduction of osmium tetroxide, tetrazolium salts) can also be used to visualize reductive groups.

2.1.3.4 Visualization of Carbonyl Groups

Carbonyl groups are characteristic of aldehyde and ketone, and their visualization is often used in histochemistry. This visualization can concern native groups or groups that have been created by certain reagents. This last type of reaction is useful to visualize DNA or carbohydrates. Among these reactions, PAS permits visualization of polysaccharides. The Feulgen and Rossenbeck reaction permits visualization of DNA in cell nuclei.

Visualization can be done in several manners: by reduction of a silver salt, which is linked to the reductive properties of aldehydes and whose reactions are argentaffine like; or by use of the Schiff reagent, which reacts with the carbonyl part of a molecular group. Other methods such as diamine phenylene or hydrazone can also be used for DNA detection, but they are less interesting. The use of Schiff reagent or a similar reagent allows visualization of aldehydes and ketones. In contrast, the use of a method based on reductive properties only allows visualization of aldehydes.

2.2 HISTOCHEMISTRY OF CARBOHYDRATES

2.2.1 Structure of Carbohydrates

Carbohydrates are organic molecules constituted with carbone, several hydroxyl chemical groups, and aldehyde or ketone groups. These molecules are used for storage and distribution of energy in the cell. Storage molecules are starch in vegetables and glycogen in animals. They also can be narrowly implied in the structure of cells. Ribose and deoxyribose enter the constitution of nucleic acids. Chitin is a part of arthropod exoskeleton; cellulose is a constituent of cell walls in plants. Chondroinin sulfate belongs to cartilaginous tissue.

2.2.1.1 Monosaccharides

Monosaccharides are the simplest carbohydrates (Figure 2.2). The smallest monosaccharide contains only three carbons. Among monosaccharides with five carbons, ribose and deoxyribose constitute a part of nucleic acids. Glucose is a six-carbon monosaccharide. By hydrolysis, these molecules are decomposed but never give smaller carbohydrates. Glucose is the most important energetic source for the cell in healthy organisms.

2.2.1.2 Disaccharides, Oligosaccharides, and Polysaccharides

These molecules can be hydrolyzed giving two or more monosaccharides. So, disaccharides can be hydrolyzed to give two monosaccharides, oligosaccharides give three to nine monosaccharides. Polysaccharides contain at least 10 monosaccharides. Classification into oligosaccharides and polysaccharides is often subjective when the number of monosaccharides is low.

Polysaccharides can be linked to other molecules such as proteins, giving mucopolysaccharides or proteoglycans. These molecules are used as energetic sources by the cells. Several of them are linked to cell membranes where they are implied in cellular knowledge.

2.2.2 Fixation and Preparation of Tissues

Fixation of tissues depends on the type of carbohydrate researched.

2.2.2.1 Fixation of Monosaccharides

Monosaccharides are very diffuse, so it is difficult to preserve them on tissues. It is possible to precipitate them with a solution of saturated barium hydroxide in methanol. Consequently, monosaccharides and barium combine themselves to give insoluble molecules in methanol that precipitate into the cells. This method allows the conservation of glucose and several hexoses on sections. It is possible that these molecules displace themselves into the tissues.

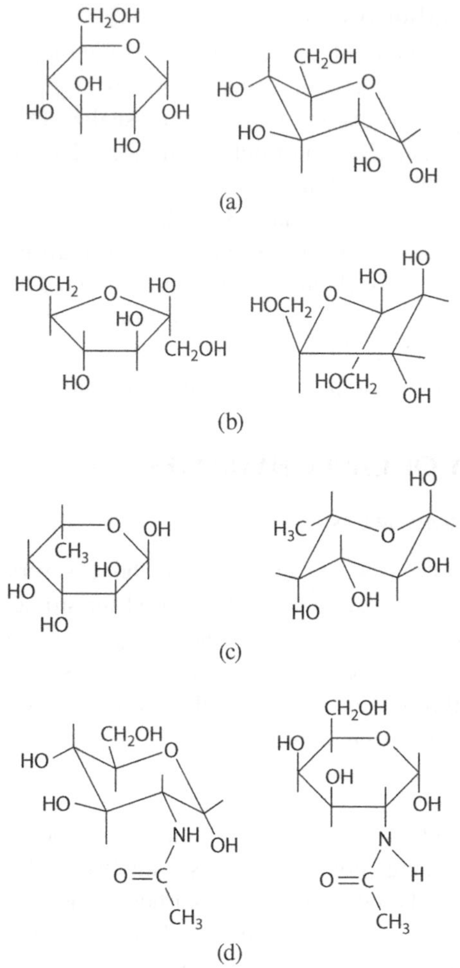

(a)

(b)

(c)

(d)

FIGURE 2.2 Several examples of monosaccharides. (a) α D glucose. (b) α D *N* acetylglycosamine. (c) β D fructose. (d) α L fucose.

2.2.2.2 Fixation of Glycogen

- Mechanism—Glycogen is a mixture of several molecules that are more or less polymerized. So, fixation is not easy because each molecule possesses its own solubility.
- Fixative—Several fixative agents can be used
  - Fixatives with ethanol—Carnoy's fluid, ethanol–formalin
  - Fixatives with picric acid—Bouin's fluid, Bouin-Allen's fluid, ethanol saturated with picric acid
  - Fixatives with formalin—Neutral formalin, salted formalin, buffered formalin
  - Avoid fixatives with mercury
  - Freezing
- Results—The best fixations give a homogenous repartition of glycogen in cells. But artifacts are often observed:
  - Leaky images—Glycogen is accumulated at the pole of the cell by where the fixative entered.
  - Granulous images—The aspect of glycogen granulations is not regular.
  - Glycogen displacement—Glycogen can displace into blood vessels or connective tissue.

2.2.2.3   Other Polysaccharides

Among polysaccharides, starch, cellulose, and chitin can be correctly preserved by use of the usual fixatives: formalin, Bouin's fluid, and Carnoy's fluid.

2.2.2.4   Fixation of Heterosides

Usual fixatives are generally sufficient to preserve complex carbohydrates. Several heterosides can be difficult to preserve. Sometimes, fixation can interact with staining, modifying the results.

- Chemical fixatives
  - Formalin can modify the staining qualities of certain carbohydrates. It is recommended to use buffered formalin.
  - Bouin's fluid
  - Ethanol—Do not prolong fixation. The staining will be better with use of dyes in ethanolic solution to avoid rehydration.
  - Lead salts—Good conservation of complex carbohydrates (heterosides) but tissue and cell structures can be damaged.
  - Metal salts—Avoid chromic salts because their oxidizing action can modify the results of staining.
  - Osmium tetroxide—The action of this fixative is both oxidizing and polymerizing, which can modify the results of staining.
- Freezing gives good results.

2.2.3   PAS

2.2.3.1   PAS Method or Periodic Acid, Schiff

See Figure 2.3. The PAS method oxidizes carbohydrates with periodic acid to give aldehyde functions (–CHO). This method can also give positive results with lipids or proteins that are also able to give an aldehyde function with periodic acid. Consequently, several control reactions must be added to the PAS method, even though these controls are not often performed. These functions react with Schiff's reagent (commonly called faded fuchsin), which gives a red staining to the sections.

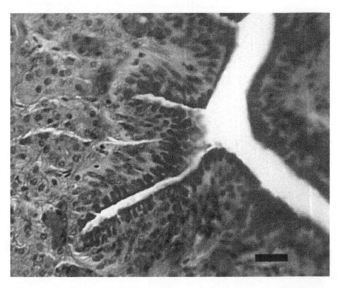

FIGURE 2.3   (See color insert.) Method: PAS. Tissue: Amphibian stomach. Preparation: Paraffin section. Fixative: Bouin's fluid. Observations: Nuclei are brown, carbohydrates are red, and collagen is red-purple. Bar = 100 μm.

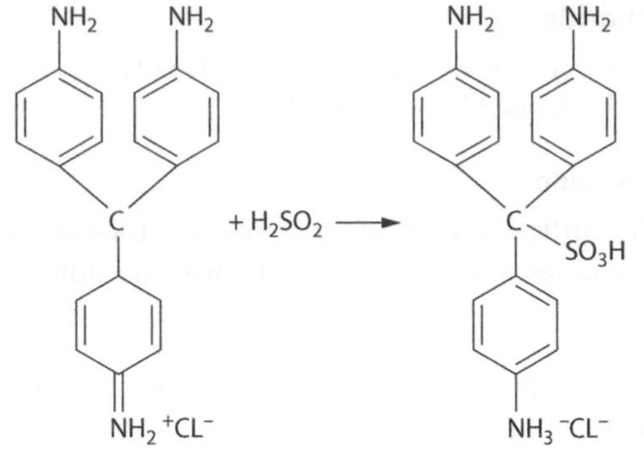

FIGURE 2.4 Schiff's reagent.

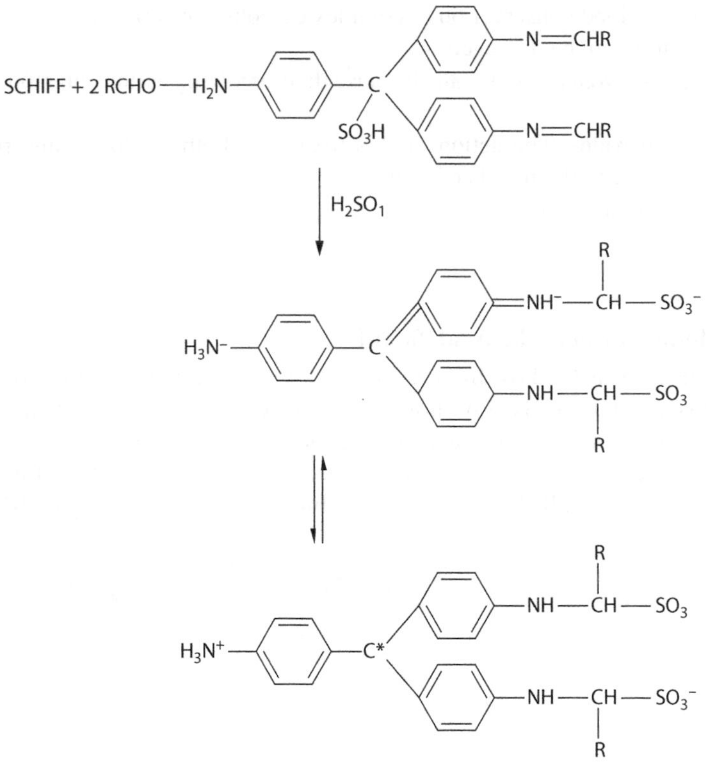

FIGURE 2.5 Reaction of aldehydes with Schiff's reagent.

2.2.3.1.1   Action of Periodic Acid

$$\text{R-CHOH-CHOH-R'} + \text{HIO}_4 \rightarrow \text{RCH} = \text{O} + \text{O} = \text{HC-R'}$$

2.2.3.1.2   Action of Schiff's Reagent

See Figure 2.4 and Figure 2.5.

- Fixatives—In certain cases, the oxidant power of fixative can modify the results. The current fixatives can be used for visualization of mucous carbohydrates.

- Reagents
  - Schiff's reagent: See Chapter 7, Section 7.2.3.28.
  - Periodic acid 1%
- Protocol
  - Dewax, hydrate
  - Periodic acid, 10 min
  - Tap water, 10 min
  - Schiff's reagent, 10 min
  - Tap water, 5 min
  - Dehydrate, mount
- Results—PAS positive substances are red stained. Glygogen is red; collagen is purple.

2.2.3.2   PAS–Hematoxylin–Picroindigocarmine

- Fixatives—In certain cases, the oxidant power of fixative can modify the results. The current fixatives can be used for visualization of mucous carbohydrates.
- Reagents
  - Schiff's reagent: See Chapter 7, Section 7.2.3.28.
  - Groat's hematoxylin: See Chapter 7, Section 7.2.1.7.
  - Picroindigocarmine: See Chapter 7, Section 7.2.2.7.
  - Periodic acid 1%
- Protocol
  - Dewax, hydrate
  - Periodic acid, 10 min
  - Tap water, 10 min
  - Schiff's reagent, 10 min
  - Tap water, 10 min
  - Groat's hematoxylin, 5 min
  - Tap water, 5 min
  - Picroindigocarmine, 30 s
  - Dehydrate, mount
- Results—PAS positive substances are red stained, nuclei are brown-blue, acidophilic cytoplasm and nucleoli are yellow or green, collagen is blue, and red blood cells are yellow.

2.2.3.3   PAS–Hematoxylin–Molybdic Orange G

- Fixatives—In certain cases, the oxidant power of fixative can modify the results. The current fixatives can be used for visualization of mucous carbohydrates.
- Reagents
  - Schiff's reagent: See Chapter 7, Section 7.2.3.28.
  - Groat's hematoxylin: See Chapter 7, Section 7.2.1.7.
  - Molybdic orange G: See Chapter 7, Section 7.2.3.24.
  - Periodic acid 1%
- Protocol
  - Dewax, hydrate
  - Periodic acid, 10 min
  - Tap water, 10 min
  - Schiff's reagent, 10 min
  - Tap water, 10 min
  - Groat's hematoxylin, 5 min
  - Tap water, 5 min
  - Molybdic orange G, 2 min

- Tap water
- Dehydrate, mount
- Results—PAS positive substances are red stained, nuclei are brown-blue, acidophilic cytoplasms are yellow, and collagen is purple.

2.2.3.4   Control for Glycogen Identification

- Principle—Glycogen identification needs comparison between sections with a normal reaction and sections that have been submitted to an enzymatic digestion. We describe here ptyalin digestions.
- Reagent—Ptyalin. A digestion with ptyalin or amylase contributes to glycogen hydrolysis.
- Protocol
  - Dewax, hydrate
  - Ptyalin, 30 min at 37°C
  - Tap water
  - PAS—Perform the reaction according to usual protocol
- Results—Glycogen disappears on the sections treated with ptyalin.

2.2.3.5   Control for Carbohydrate Nature of PAS Positive Substances

- Principle—1 to 2 Glycol functions (glucides) and primary hydroxylamine functions (proteins) are submitted to acetylation and, consequently, they lack PAS reactivity.
- After saponification, reactivity is again established only for 1 to 2 glycol functions (glucides).
- Reagents
  - Acetylation reagent
    - Acetic anhydride, 13 mL
    - Anhydrous pyridine, 20 mL
  - Saponification reagent
    - KOH 0.1 N in water
- Protocol—Two sets of sections are necessary
  - First set: Acetylation. During the reaction, acetic acid reacts with both glycol groups of carbohydrates and primary hydroxylamine functions of proteins. So the action of HIO_4 remains impossible.
    - Dewax
    - Ethanol 100%, 10 min
    - Let sections dry
    - Pyridine, 5 min
    - Acetylation reagent, 4 h at 37°C
    - Pyridine, 5 min
    - Ethanol 100%, 10 min
    - Ethanol 70%, 1 to 2 min
    - Tap water
    - Stain with PAS; perform the reaction according to the usual protocol
  - Second set: Saponification, just after acetylation. During the reaction, glycols are first blocked with acetylation reaction. After saponification, glycols are separated from acetic acid and they can be stained again with the PAS method. Primary hydroxylamine functions remain blocked and they cannot be visualized after PAS reaction.
    - Saponification reagent, 45 min
    - Rinse with tap water
    - Stain with PAS; perform the reaction according to the usual protocol.
- Results
  - For carbohydrates, the first set is stained but not the second one.
  - For proteins, both sets are not stained.

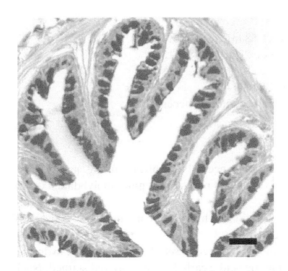

FIGURE 2.6 (See color insert.) Method: Alcian blue. Tissue: Amphibian esophagus. Preparation: Paraffin section. Fixative: Bouin's fluid. Observations: Nuclei are brown, acidic carbohydrates are blue. Bar = 100 µm.

2.2.4 Acidic Carbohydrates Detection

These molecules are nitrogenous carbohydrates with acidic functions: —COOH or —OSO3H.

2.2.4.1 Alcian Blue Method

See Figure 2.6. Dyes belonging to the alcian blue family are basic. A decreasing of pH increases their selectivity.

- Fixative—All current fixative agents can be used.
- Reagents
  - Alcian blue pH 2.5 and 4.5: See Chapter 7, Section 7.2.3.2.
  - Groat's hematoxylin: See Chapter 7, Section 7.2.1.7.
- Protocol
  - Dewax, hydrate
  - Alcian blue, 20 min
  - Tap water
  - Groat's hematoxylin, 5 min
  - Tap water, 5 min
  - Dehydrate, mount
- Results—Acidic mucous carbohydrates are blue stained; nuclei are brown-blue.

2.2.4.2 Metachromatic Reaction

- Principle—Metachromasy is the property of some dyes to stain cell or tissue substances in a different color than their proper one. These substances are chromotropic (chondroitin sulfuric acid of cartilage, granulations of mastocytes, acidic carbohydrates, nucleic acids). See Section 2.1.3.2.
- Fixative—Fixation must both preserve metachromatic structures and not provoke the apparition of abnormal structures. Avoid oxidizing fixatives. Fixatives with ethanol or Bouin's fluid can be used.
- Reactives
  - Toluidine blue: See Chapter 7, Section 7.2.3.31 or toluidine blue 0.2% in acetate/acetic acid buffer pH 4.2–4.6.

- Buffer:
    - Acetic acid 11.58 mL/L, 12 mL
    - Sodium acetate 27.21 g/L, 8 mL
- Acetic water 0.2%
- Ammonium molybdate 4%
- Protocol—Two sets of sections are necessary.
    - Dewax, hydrate
    - Toluidine blue, 5 min
    - Acetic water, 1 min
    - Observe. Observation is needed to control staining quality.
    - Ammonium molybdate, 10 min. Ammonium molybdate fixes the staining.
    - Tap water, 10 min
    - Mount one set of sections in hydrophilic medium. Use Kaiser's syrup or another hydrophilic mounting medium.
    - Dehydrate and mount the second set of sections. Use a hydrophobic resin.
- Results—Metachromatic acidic carbohydrates are red stained. Nuclei are blue.

2.2.4.3 Control for Nature of Metachromatic Acidic Substances

- Reagents
    - Methylation reagent
        - Methanol, 100 mL. Methanol blocks acidic functions. Only COOH can give a reverse reaction.
        - Hydrochloric acid, 0.8 mL
    - Saponification reagent
        - Ethanol 80%, 100 mL
        - Potassium hydroxide, 1 g
- Protocol—Two sets of sections are necessary.
    - First set: Methylation
        - Cyclohexane, 5 min
        - Ethanol 100%, 5 min
        - Methanol, 10 min
        - Methylation reagent, 2 h at 60°C
        - Tap water
        - Metachromatic reaction with first set. Perform the reaction according to the usual protocol.
    - Second set: Saponification
        - Saponification reagent, 30 min
        - Tap water
        - Metachromatic reaction. Perform the reaction according to the usual protocol.
- Results
    - First set: –COOH and –OSO$_3$H functions are not metachromatic
    - Second set: Only –COOH functions are metachromatic

2.2.5 COMBINED METHODS

2.2.5.1 Alcian Blue–PAS

See Figure 2.7. Simple monosaccharides and acidic carbohydrates can be visualized together with this method.

- Fixative—All current fixative agents can be used.

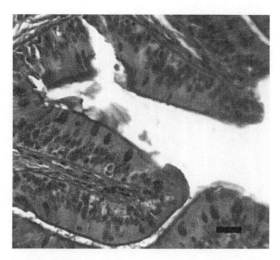

FIGURE 2.7   (See color insert.) Method: Alcian blue–PAS. Tissue: Amphibian intestine. Preparation: Paraffin section. Fixative: Bouin's fluid. Observations: Nuclei are brown, carbohydrates are red, and collagen is red-purple. Bar = 100 μm.

- Reagents
  - Alcian blue pH 2.5: See Chapter 7, Section 7.2.3.2.
  - Schiff's reagent: See Chapter 7, Section 7.2.3.28.
  - Groat's hematoxylin: See Chapter 7, Section 7.2.1.7.
  - Molybdic orange G: See Chapter 7, Section 7.2.3.24.
  - Periodic acid 1%
- Protocol
  - Dewax, hydrate
  - Alcian blue, 20 min
  - Tap water
  - Periodic acid, 10 min
  - Tap water, 10 min
  - Schiff's reagent, 10 min
  - Tap water, 10 min
  - Groat's hematoxylin, 5 min
  - Tap water, 5 min
  - Molybdic orange G, 2 min
  - Tap water
  - Dehydrate, mount
- Results—Carbohydrates are red, purple or blue stained, according to their chemical nature. Nuclei are brown stained, acidophilic cytoplasms are yellow, and collagen is purple.

2.2.5.2   Ravetto's Staining

See Figure 2.8 and Figure 2.9.

- Principle—At pH 0.5, alcian blue stains only sulfated carbohydrates. At pH 2.5, alcian yellow stains specifically carboxyled carbohydrates.
- Fixative—All current fixative agents can be used.
- Reagents
  - Buffer pH 0.5. Hydrochloric acid 0.5 N can also be used.
  - Alcian blue: See Chapter 7, Section 7.2.3.2.
  - Alcian yellow 0.5%, in pH 2.5 buffer. See Chapter 7, Section 7.2.3.2.
  - Nuclear fast red: See Chapter 7, Section 7.2.1.12.

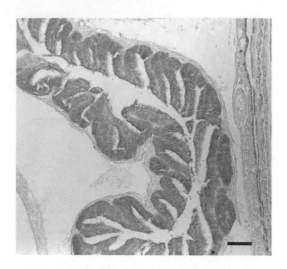

FIGURE 2.8   (See color insert.) Method: Ravetto's staining. Tissue: Amphibian oviduct. Preparation: Paraffin section. Fixative: Bouin's fluid. Observations: Nuclei are brown, acidic carboxyled carbohydrates are blue, sulfated carbohydrates are yellow, and mixtures of carboxyled and sulfated carbohydrates are green. Bar = 100 µm.

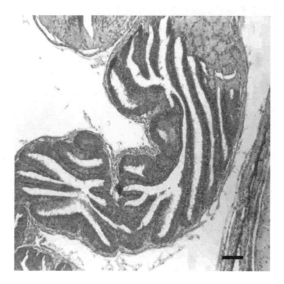

FIGURE 2.9   (See color insert.) Method: Ravetto's staining. Tissue: Amphibian oviduct. Preparation: Paraffin section. Fixative: Bouin's fluid. Observations: Nuclei are brown, acidic carboxyled carbohydrates are blue, sulfated carbohydrates are yellow, and mixtures of carboxyled and sulfated carbohydrates are green. Bar = 100 µm.

- Protocol
    - Dewax, hydrate. No collodioning.
    - Alcian blue, 30 min
    - Rinse with buffer pH 0.5, 10 s
    - Tap water
    - Alcian yellow, 30 min to 1 h. Duration of staining depends upon tissue and fixation.
    - Tap water
    - Nuclear fast red, 8 min
    - Dehydrate, mount

- Results—Carboxylic carbohydrates are yellow stained, sulfated carbohydrates are blue, and carbohydrate mixtures are green stained.

2.3 HISTOCHEMISTRY OF PROTEINS

2.3.1 STRUCTURE OF PROTEINS

Proteins are organic molecules constituted with a linear chain of amino acids bonded to each other with peptide bonds. According to the number of amino acids constituting the chain, the protein is also called peptide or polypeptide. A protein can also be composed with a single or several chains of polypeptides.

2.3.1.1 Amino Acids

Amino acids R-CH (COOH)NH$_2$ are organic molecules containing both acidic and amine chemical groups. The acidic group of a first amino acid can condense with the amine group of the second one, giving a peptide bond. Chaining of amino acids are peptides and polypeptides.

$$NH_2\text{-CHR-COOH} + HN_2 - CHR'\ COOH \rightarrow NH_2\text{-CHRCO-NH-CHR'-COOH}$$

Twenty standard amino acids, encoded by DNA, are found in the organism. But these amino acids can be modified after several chemical reactions. Reactions can be hydroxylation, phosphorylation, methylation, and so on. Some amino acids contain specific chemical groups that can be used for visualization.

2.3.1.2 Molecular Structure of Proteins

2.3.1.2.1 Primary Structure

The order with which amino acids are bonded gives specific properties at each protein. This chain order corresponds to the primary structure (Figure 2.10). This specific order partly explains why synthetic proteins do not have the same properties of proteins found in living organisms.

2.3.1.2.2 Secondary Structure

Long chains of amino acids are folded. Folds are maintained with hydrogen bonds. The main folds are α helix, β sheet, or turns. Super-secondary structures correspond to successions of secondary structures. They are intermediary between secondary and tertiary structures. Structural domains are parts of proteins resulting from local folding of proteins. They are larger than secondary structures (Figure 2.11).

2.3.1.2.3 Tertiary Structure

A tertiary structure corresponds to a general folding of the molecule. Stabilization of this structure is done with several interactions such as hydrogen bonds, ionic interactions, and disulfide bonds.

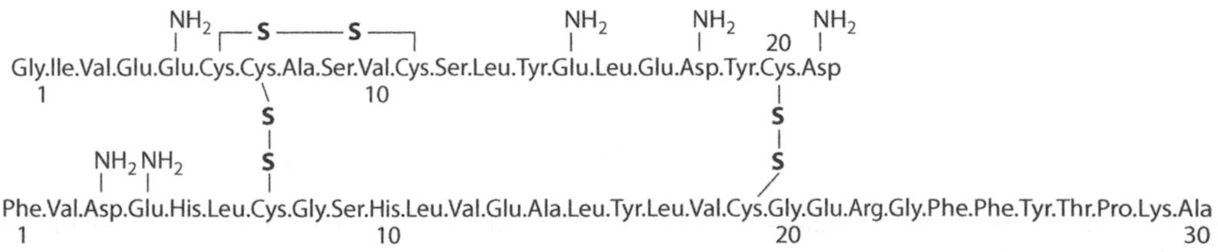

FIGURE 2.10 Primary structure of insulin. (From Smith, C.A., and E.J. Wood, 1997, *Les biomolécules*, Masson, Paris. With permission.)

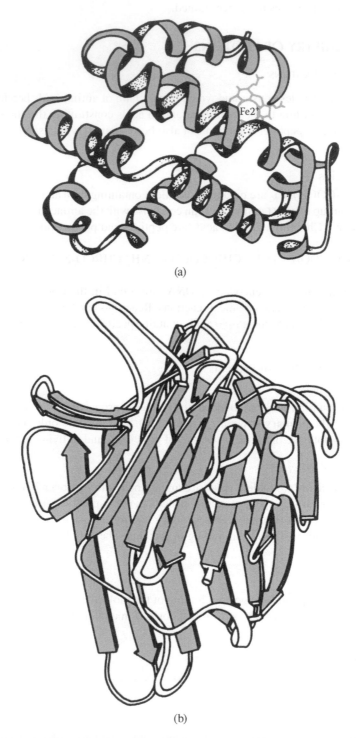

(a)

(b)

FIGURE 2.11 Examples of secondary structures: (a) Myoglobine. (b) Concanavaline A. (From Smith, C.A., and E.J. Wood, 1997, *Les biomolécules*, Masson, Paris. With permission.)

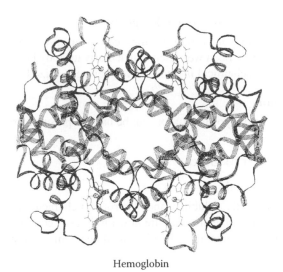

Hemoglobin

FIGURE 2.12 Quaternary structure of hemoglobin. (From Smith, C.A., and E.J. Wood, 1997, *Les bio-molécules*, Masson, Paris. With permission.)

2.3.1.2.4 Quaternary Structure

Quaternary structure is found in proteins containing several subunits, each one being a single poly-peptide chain; hemoglobin, for example (Figure 2.12).

2.3.1.3 Fibrillar and Globular Proteins

Fibrillar proteins (i.e., collagens, elastins, keratins, actin, myosin) are insoluble in water. They com-prise the main structural proteins. Globular proteins (i.e., hemoglobin, myoglobin) are water soluble.

2.3.1.4 Structural and Enzymatic Proteins

A lot of proteins constitute the structure of organisms, giving the shape of cells. Enzymatic pro-teins are implied in cell metabolism. Several proteins are also bonding ones. Several hormones are of proteic nature. Structural and enzymatic proteins can be visualized with staining meth-ods. Enzymatic activities can be visualized using specific techniques of histoenzymology (see Chapter 3).

2.3.2 Fixation and Preparation of Tissues

2.3.2.1 Effects of Fixatives

Even if generally all the current fixatives can be used to detect proteins, the reactive functions of pro-teins can be modified with fixative action. (See Chapter 1, Section 1.3.2.3.) These modifications are:

- Blocking of some reactive chemical groups used by the fixative to increase stability of cell and tissue structures. It is the consequence of polymerizing action of fixative.
- Deletion of some proteic groups. Globular proteins can be deleted.
- Stabilization of the molecular building in a natural state.
- Modification of molecular relationships between proteic chains. It is the denaturation (or coagulation) of proteins that are spread out with apparition of reactive chemical groups on the surface of the molecule. This phenomenon belongs to the fixative process.

2.3.2.2 Fixatives Used as Histochemical Reagents

Fixatives can modify solubility of protein, so they can be used as histochemical reagents. The physical effects of fixatives have been well used, particularly to study hormones of pituitary organs, before the use of immunocytochemical methods have been generalized.

2.3.2.3 Physical Fixation

Freezing, cryodessication, and cryosubstitution can be used to preserve proteins.

2.3.3 METHODS OF CHARACTERIZATION

2.3.3.1 Physicochemical Methods

These methods are based upon the solubility and electropolar character of proteins. They also comprise enzymatic digestions. These techniques are now rarely used consequently to the development of new methods, such as the immunocytochemical methods, which are more precise and specific than previously.

2.3.3.2 Histochemical Stainings

Only several histochemical stainings of proteins will be given here. These methods have been abandoned for specific methods, such as immunohistochemical ones based upon the use of antibodies. Among specific staining of protides, biuret reaction is directly inherited from the colorimetric assay methods used in biochemistry. Use of mercuric bromophenol blue is also a classic staining. A lot of dyes used in classic histology, like eosin, phloxin, and orange G can be considered as protein specific. Group reactions, such as xanthoproteic ones, are also issued from biochemistry.

Diazotation reactions and reactions specific to peculiar chemical groups are used to visualize peptides and proteins characterized by a lot of peculiar amino acids.

2.3.4 GENERAL HISTOCHEMICAL STAININGS

2.3.4.1 Hartig–Zacharias's Method

The Hartig–Zacharias's method is more signaling than histochemical.

- Fixative—All current fixative agents can be used. Avoid osmium tetroxide.
- Reagents
  - Nuclear fast red: See Chapter 7, Section 7.2.1.12.
  - Potassium ferrocyanure 2% in hydrochloric acid 1%
  - Ferric chloride (still called "iron perchloride") in distilled water
- Protocol
  - Dewax, hydrate
  - Ferrocyanure, 10 min
  - Tap water
  - Ferric chloride, 2 min
  - Tap water
  - Nuclear fast red, 1 min
  - Tap water
  - Dehydrate, mount
- Results—Proteins are blue stained. They are called "siderophilic."

2.3.4.2 Diazotation

In histochemistry, a diazotation is the reaction between a phenol and a diazonium salt at a basic pH. The result is a stained azoic molecule. In chemistry, the significance of "diazotation" is different:

it is the reaction transforming an aromatic amine (alanin, for example) into a diazoic molecule, characterized by a "–N = N–" radical. This reaction is performed by action of nitrous acid upon the amine. The substance obtained can combine itself with aromatic amines giving an azoic dye.

Diazotation reactions are used to visualize a lot of amino acids belonging to lateral chains of peptides and proteins. However, the intensity of this reaction was minuscule, and diazotation reactions have been replaced with tetrazoreactions, the results of which are well obvious.

2.3.4.3   Danielli's Tetrazoreaction

See Figure 2.13 and Figure 2.14.

- Principle—This staining method is based on a double coupling. The first reaction consists of reacting the tissue element weakly stained with a diazonium salt, which gives a first product. Then the product is combined with a naphthol, and the result is an intense reaction, the color of which varies with the salt used. The salt generally used in the reactions is orthodianisidine, also called fat blue B. Another salt that is often used is H acid, which is characterized by a β-naphthol function. Tissue and cell groups are brown stained. This method, first used to visualize proteins, has been adapted for nucleic acids. For nucleic acids, benzoylation or acetylation must be performed before staining. Chromosomes are intensely stained.
- Fixative—Avoid formalin
- Reagents
  - Veronal buffer pH 9.2
    - HCl 8.35 mL/L, 231 mL
    - Sodic veronal 20.618 g/L, 769 mL
  - Fast blue B (orthodianisidine) or 1-amino-8-naphthol-3,6-disulfonic acid (see Chapter 7, Section 7.2.3.11), 0.2% in buffer
  - Acid H (see Chapter 7, Section 7.2.3.1), 2% in buffer
- Protocol
  - Dewax, collodion, hydrate
  - Orthodianisidine: 5 min
  - Tap water
  - Wash in buffer, 3 × 2 min

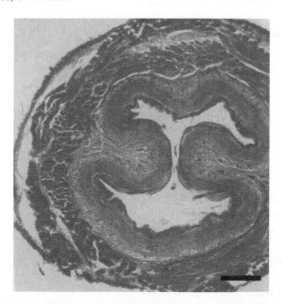

FIGURE 2.13   (See color insert.) Method: Danielli's tetrazoreaction. Tissue: Mouse esophagus. Preparation: Paraffin section. Fixative: Bouin's fluid. Observations: Proteins are brown stained. Bar = 100 μm.

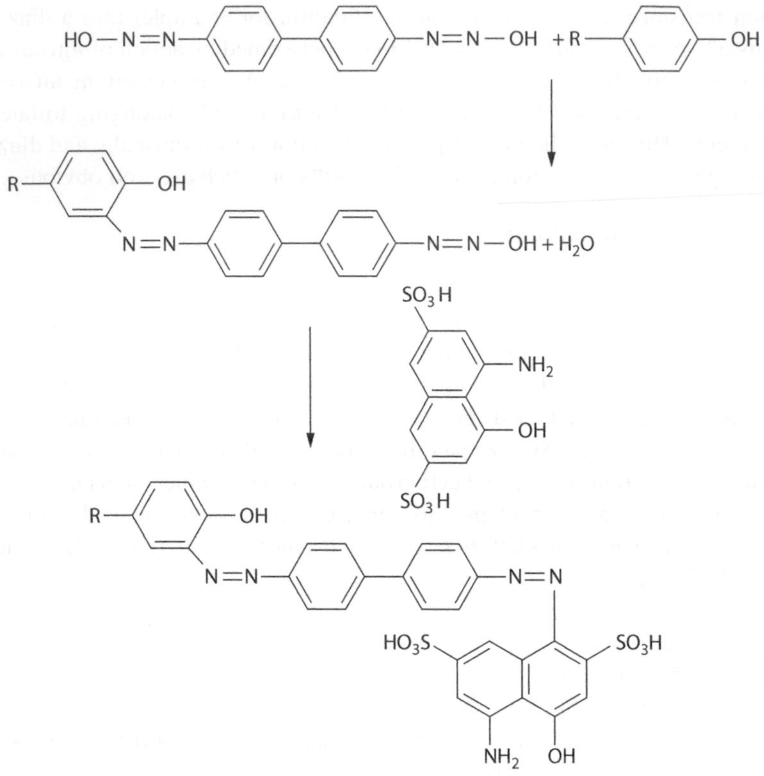

FIGURE 2.14   Danielli's tetrazoreaction.

- Acid H, 15 min
- Tap water
- Dehydrate, mount
- Results—Proteins are brown or blue stained.

2.3.4.4   Control Methods for Tetrazoreaction

Tetrazoreaction permits visualization of histidine, tryptophan, tyrosine, lysine, cysteine, and arginine. The use of clamping methods for some of these amino acids permits more precise characterization.

2.3.4.4.1   Clamping with Performic Acid

Performic acid is used to clamp tryptophan.

- Reagent
    - Performic acid. Performic acid is prepared just before use.
        - Formic acid 98%. 40 mL
        - Hydrogen peroxide 30%, 4 mL
        - Concentrated sulfuric acid, 0.5 mL
    - Let rest 1 h. This solution must be used within 24 hours.
- Protocol
    - Dewax, hydrate
    - Performic acid, 20 min
    - Wash with tap water
    - Tetrazoreaction. Perform reaction according to usual protocol.
- Results—Tetrazoreaction is positive for tyrosine, histidine, lysine, cysteine, and arginine.

2.3.4.4.2 Clamping with Dinitrofluorobenzene

Dinitrofluorobenzene is DNFB. This molecule clamps tyrosine, histidine, cysteine, and lysine.

- Reagents—DNFB saturated in sodium carbonate saturated in ethanol 90%.
- Protocol
  - Dewax
  - Ethanol 90%, 10 min
  - DNFB, 24 h
  - Wash with ethanol 90%
  - Wash with tap water
  - Tetrazoreaction. Perform reaction according to usual protocol.
- Results—Tetrazoreaction is positive for tryptophan and arginine. Used alone, DNFB permits one to visualize total proteins. See Section 2.3.4.8.

2.3.4.4.3 Clamping by Benzoylation

Hydroxyl and amine groups are clamped.

- Reagents—Benzoyl chloride 10% in anhydrous pyridine.
- Protocol
  - Dewax
  - Ethanol 100%, 10 min
  - Petrol ether, 3 min
  - Let air-dry
  - Benzoyl chloride, 12 h at 25°C
  - Acetone, 10 min
  - Ethanol 100%, 10 min
  - Hydrate
  - Tetrazoreaction. Perform reaction according to usual protocol.
- Results—Tetrazoreaction is positive for histidine and nucleic acids.

2.3.4.5 Staining of Section Bottom after Danielli's Tetrazoreaction

The tetrazoreaction alone provides sufficient staining. The bottom is generally weakly stained by the diazonium salt used. The main substances that are researched are specifically strongly stained.

2.3.4.6 Chloramine-T–Schiff Method

- Principle—This method is based upon the oxidative deamination with production of aldehydes that restained with Schiff's reagent.
- Fixative—All current fixative agents are convenient. Avoid osmium tetroxide.
- Reagents
  - Schiff's reagent: See Chapter 7, Section 7.2.3.28.
  - Phosphate buffer pH 7.5
    - Disodium phosphate (14.198 g/L), 841 mL
    - Monosodium phosphate (13.805 g/L), 159 mL
  - Chloramine-T 1% in buffer: See Chapter 7, Section 7.2.3.6.
  - Sodium thiosulfate or sodium hyposulfite 5%
  - Sulfurous water
    - Sodium metabisulfite 10%, 10 mL
    - Distilled water, 190 mL
- Protocol
  - Dewax, collodion, hydrate.

- Chloramine-T, 6 h at 37°C
- Rinse quickly to eliminate excess of chloramine-T
- Thiosulfate, 3 min
- Tap water
- Schiff's reagent, 30 min
- Rinse with sulfurous water
- Tap water
- Dehydrate, mount
- Results—Proteins with terminal –NH2 are pink stained.

2.3.4.7 Ninhydrin–Schiff Method

- Principle—Free primary amino acids are oxidized to give aldehydes that are visualized with Schiff's reagent.
- Fixative—All current fixative agents are convenient. Avoid osmium tetroxide.
- Reagents
  - Schiff's reagent: See Chapter 7, Section 7.2.3.28.
  - Ninhydrin solution. Preparation:
    - Ninhydrin, 0.5 g
    - Ethanol 100%, 100 mL
- Protocol
  - Dewax
  - Ethanol 100%
  - Ninhydrin solution, 16 to 20 h at 37°C
  - Tap water, 5 min
  - Distilled water
  - Schiff's reagent, 30 min
  - Tap water, 20 min
  - Distilled water
  - Dehydrate, mount
- Results—Proteins are red stained.

2.3.4.8 Staining with DNFB (2,4-Dinitro-1-Fluorobenzene)

- Principle. Free terminal amino acids and free lateral lysine, cysteine, tyrosine, and histidine are bound to DNFB giving a yellow staining for lysine and cysteine, but not for tyrosine and histidine.
- Fixative—All current fixative agents are convenient. Avoid osmium tetroxide.
- Reagents—DNFB: See Chapter 7, Section 7.2.3.10. This staining is used to visualize proteins by means of terminal and some lateral free amino acids staining. DNFB is also used to clamp Danielli's tetrazoreaction. See Section 2.3.4.4.2. In this case, tetrazoreaction is only possible with tryptophan and arginine.
- Protocol
  - Dewax
  - DNFB, 3 h at 65°C
  - Ethanol 50%, 2 × 1 min
  - Dry slides in air
  - Rinse in tertiary butanol
  - Rinse two times in xylene
  - Mount
- Results—Proteins are yellow stained.

2.3.4.9 Visualization of Proteins with Coomassie's Blue

- Reagents
  - Triton X 100 1%. Triton X 100 is a surfactant used to permeabilize the phospholipidic layers of membranes. This operation allows the penetration of dye.
  - Coomassie's blue: See Chapter 7, Section 7.2.3.7.
- Protocol
  - Dewax, hydrate
  - Triton X 100, 15 min
  - Coomassie's blue, 45 to 60 min. Duration of staining must be tested.
  - Rinse with distilled water or PBS
  - Mount in hydrophilic medium. Glycerol at 50% in distilled water or "Crystal Mount," for example.
- Results—Proteins are blue stained.

2.3.5 VISUALIZATION OF PECULIAR CHEMICAL GROUPS

2.3.5.1 Arginine Detection

Sakaguchi's technique (1925) modified by Baker (1947).

- Principle—Guanidyl radical of arginine is found in histones and protamines.
- Fixative—Bouin's fluid, salty formalin, Carnoy's fluid. Do not paste sections with albumin.
- Reagents
  - α-Naphthol in alcaline solution. Preparation:
    - – Sodium hydroxide 1%, 10 mL
    - – α-Naphthol 1% in ethanol 70%, 10 mL
  - Bleaching liquid (commercial solution), 20 mL
  - Chloroform–pyridine mixing
    - – Pyridine, 30 mL
    - – Chloroform, 10 mL
  - Pyridine
- Protocol
  - Dewax, hydrate
  - Dry with blotting paper. A supplementary step for collodioning can be useful between dewaxing and hydration to prevent ungluing of sections.
  - Alkaline solution of α-naphthol, 15 min
  - Dry with blotting paper
  - Chloroform–pyridine, 3 min
  - Mount in pyridine
  - Observe
- Results—Arginine is pink stained.

2.3.5.2 Tyrosine Detection

Millon's reaction (1849).

- Principle—This method is used to visualize phenol chemical groups, and more particularly tyrosine. In fact, only tyrosine gives a stained result.
- Fixatives—Bouin's fluid, salty formalin, Carnoy's fluid. Frozen sections are destroyed, so they must be avoided.

- Reagent
  - Millon's reagent
    - Solution A
      - $HgSO_4$, 10 g
      - H_2SO_4 10%, 100 mL
      - Heat until dissolution
      - Cool
      - Distilled water, 200 mL
    - Solution B
      - $NaNO_2$, 0.25g
      - Distilled water, 100 mL
    - Reagent. Mixture must be prepared just before use.
      - Solution A, 90 mL
      - Solution B, 10 mL
- Protocol
  - Dewax, hydrate
  - Millon's reagent and warm to simmering point, 2 min
  - Let cool
  - Distilled water, 3 × 2 min
  - Mount in hydrophilic medium. Glycerin can be used.
- Results—Tyrosine is reddish stained. Staining can also be purple or violet.

2.3.5.3 Tryptophan Detection
Adam's method (1957).

- Principle—This method is used to detect the indole groups of tryptophane, tryptamine, and serotonin. But the biogenic amines are in very small quantities, and only tryptophan gives a positive result.
- Fixative—Bouin's fluid, salty formalin, Carnoy's fluid, and other classic fixatives can be used. Frozen sections also give good results.
- Reagents
  - DMAB. Carbonyl group of DMAB (or p-methylaminobenzaldehyde) first reacts with tryptophan giving a red chemical component (β-carboline), then a pigment called carboline blue is formed by oxidation of β-carboline with nitrite.
    - p-Methylaminobenzaldehyde, 5 g
    - Concentrated HCl, 100 mL
  - Nitrite solution
    - $NaNO_2$, 1 g
    - Concentrated HCl, 100 mL
  - Acidic ethanol
    - Ethanol 70%, 99 mL
    - Concentrated HCl, 1 mL
- Protocol
  - Dewax, hydrate
  - Air-dry
  - DMAB, 30 s
  - Acidic ethanol
  - Dehydrate, mount
- Results—Tryptophan is blue stained.

2.3.5.4 Determination of Total Electropositive Groups; Deitch's Method

- Principle—In a strongly acidic solution, visualization of proteins is linked to the quantity of ionized acidic groups. The chemical groups concerned are essentially amine, guanidyl, and imidazole.
- Reagents
  - Naphthol yellow S. Use hydrosoluble naphthol yellow. See Chapter 7, Section 7.2.3.21.
  - Acetic acid 1%
- Protocol
  - Dewax, hydrate; do not collodion
  - Naphthol yellow, 15 min
  - Differentiate with acetic acid, 15 to 24 h
  - Dry with blotting paper.
  - Dehydrate, use tertiary butanol
  - Xylene
  - Mount
- Results—Proteins are yellow stained, with intensity proportional to the quantity of basic groups. This method can be useful for quantitative study.

2.3.6 VISUALIZATION OF SULFHYDRYLED PROTEINS

These methods are used to visualize thiols (R S H) and thioether (R S S R') in tissues. Methionin, cysteine, and cystin are particularly well distributed in skin, rodent esophagus, and cardia.

2.3.6.1 Chèvremont and Frédéric's Method

- Principle—Colorless ferricyanide is reduced with –SH groups in blue-colored ferric ferrocyanide. Ferric ferrocyanide is also called "Prussian blue."
- Fixative—Usual fixative agents can be used. Avoid osmium tetroxide.
- Reagents
  - Ferric ferricyanide: Prepare just before use.
    - Potassium ferricyanide 0.1%, 100 mL
    - Ferric sulfate 1%, 300 mL
  - Nuclear fast red. See Chapter 7, Section 7.2.1.12.
- Protocol
  - Dewax, hydrate
  - Ferric ferricyanide, 3 × 5 min
  - Tap water
  - Nuclear fast red, 30 s. Nuclear fast red is used to stain the bottom of sections. Hematoxylin can also be used but the contrast between nuclei, cytoplasm, and sulfhydryls will not be so obvious.
  - Wash
  - Dehydrate, mount
- Results: –SH groups are blue stained.

2.3.6.2 Bennet's Method, Pearse's Variant

- Principle—This staining is based upon the combination of proteins and mercury, giving mercaptide complexes.
- Fixative—Trichloracetic acid 5%, formalin, Carnoy's fluid, ethanol.
- Reagent—Red sulfhydryl reagent (R.S.R.) at saturation in ethanol 80%. Mercury orange can also be used.

- Protocol
  - Dewax, hydrate
  - R.S.R., 12 h
  - Ethanol 100%
  - Mount
- Results: –SH groups are orange stained. Staining is often weak.

2.3.6.3 Staining with D D D (2,2′ Dihydroxy-6,6′ Dinaphtyl Disulfur)

Still called Barnett and Seligman's (1952) method.

- Principle—D D D and –SH chemical groups react together to form a colorless thioether that will be stained with orthodianisidine.
- Fixatives—All usual fixatives can be used. Avoid osmium tetroxide.
- Reagents
  - D D D: See Chapter 7, Section 7.2.3.8.
  - Acetic water at 0.2%
  - Ether
  - Orthodianisidine. Or fast blue B. The solution must be prepared just before use: See Chapter 7, Section 7.2.3.11. Or
  - Orthodianisidine 0.1% in pH 7.4 phosphoric buffer.
    - Phosphoric buffer
      - Disodium phosphate (23.866 g/L of crystallized phosphate), 80 mL; monopotassium phosphate (9.066 g/L), 20 mL. Or
      - Anhydrous phosphate (9.465 g/L), 80 mL; monopotassium phosphate (9.066 g/L), 20 mL
- Protocol
  - Dewax, hydrate
  - D D D, 1 h at 60°C
  - Let sections cool 10 min
  - Acetic water, 2 times, 5 min each
  - Ethanol 100%
  - Ether, 2 × 5 min
  - Ethanol 100%
  - Tap water
  - Orthodianisidine, 2 min
  - Tap water
  - Dehydrate, mount
- Results: –SH chemical groups are blue stained.

2.3.6.4 Thioether Reduction

- Principle—Only thiols are detected with these methods of staining. So it is necessary to reduce thioether molecules to detect them.
- Reagents—Ammonium sulfur 5% (pH 9.5)
- Protocol
  - Dewax, hydrate
  - Ammonium sulfur, 1 h at 37°C
  - One effect of histochemical reactions
- Results: –SH and –S–S– chemical groups are stained.

2.3.6.5 Control of Thiol Presence

- Principle—Blocking methods are used to control the presence of thiols.
- Reagent
  - Lugol: See Chapter 7, Section 7.2.3.15.
- Protocol
  - Dewax, hydrate
  - Lugol, 4 h at 25°C
  - Wash with tap water
  - One effect of histochemical reactions
- Results: –SH chemical groups are not stained.

2.4 HISTOCHEMISTRY OF LIPIDS

2.4.1 STRUCTURE OF LIPIDS

2.4.1.1 General Aspects

Lipids are fat organic components with several biological functions. They enter in the composition of cell membranes and cytoplasm inclusions, as well as internal membranes. They have storage and structural functions, contributing to the shape of organism. Some of them are hormones. They are also used as thermic protectors; and they can prevent dehydration. Phospholipids are more particularly found in biological membranes. Lipids are stocked into specialized cells, the adipocytes. Sterols and steroidic hormones derive from lipids.

2.4.1.2 Chemical Structure

Lipids are esters obtained by reaction between a fatty acid on a fatty alcohol (Figure 2.15). Fatty acids possess a long chain of carbon and hydrogen. Alcohol can be glycerol or monohydroxyalcohol. The ester can result from the reaction of a single alcohol and a single fatty acid. It can also result from reaction between several hydrate functions belonging to a single alcohol and several single fatty acids. Lipids are complex when they are associated with a carbohydrate or a protein.

In waxes, only a single alcohol is combined with a single fatty acid. In triglycerides, three hydrate functions of glycerol are combined with three single fatty acids. In phospholipids, phosphoric acid is associated to the fatty alcohol, giving a complex lipid.

2.4.1.3 Classification

Several classifications are available. The classification given here is from Smith and Wood (1997).

2.4.1.3.1 Simple Lipids

- By hydrolysis, waxes give monohydroxy alcohol and fatty acids. Acylglycerides are also called neutral fats; triglycerides are acylglycerides.
- Acylglycerides are hydrolyzed in glycerol and fatty acids.

2.4.1.3.2 Complex Lipids

- Phospholipids can be divided into phosphoacylglycerides, giving glycerol, fatty acids, phosphate, and several other molecules; and sphingomyelines that can be decomposed into sphingoside, fatty acid, phosphate, and other small components. Sialic acid is found among simple carbohydrates of gangliosides.
- Glycolipides can be divided into cerebrosides decomposed into sphingol, fatty acid, and simple carbohydrates; gangliosides that can also be hydrolyzed into sphingol, fatty acid, and simple carbohydrates.
- Isoprenoids can be divided into steroids, carotenoids, or even liposoluble vitamins; it is not possible to hydrolyze these molecules.

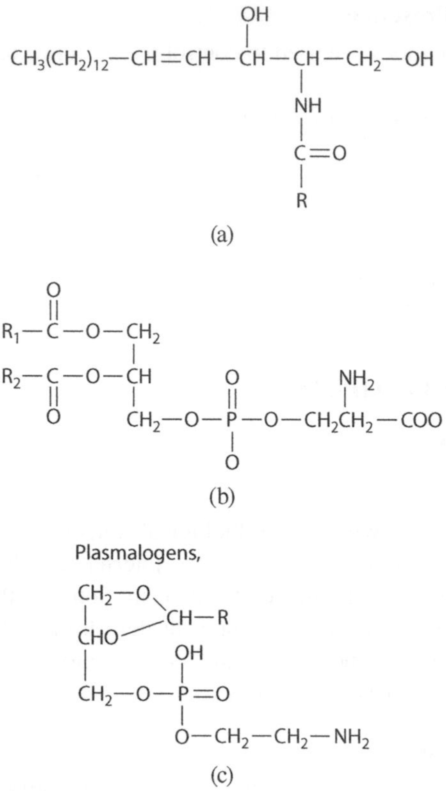

$$CH_3(CH_2)_{12}-CH=CH-\underset{\underset{NH}{|}}{\overset{\overset{OH}{|}}{CH}}-CH-CH_2-OH$$

with the NH bearing $C=O$ and R below.

(a)

(b)

Plasmalogens,

(c)

FIGURE 2.15 Examples of lipids. (a) Ceramide. (b) Plasmalogen. (c) Phosphatidyl serine.

2.4.2 FIXATION AND TISSUE PREPARATION

2.4.2.1 Lipid Types in Histology

Homophasic lipids are stable lipids physically isolated from the other components. Still called pure lipids, they look like droplets into cytoplasm of cells. They are hydrophobic. Heterophasic lipids are hydrophilic and associated to other components. Due to their hydrophily, heterophasic lipids are stable into organic solvents. But they are physiologically unstable because they are soluble in water. So both fixation and embedding must be considered according to the lipid type studied. Because a lot of lipids are soluble in organic solvents, the fixatives must not be of organic nature. Because of the organic solvent being used with paraffin embedding, lipids must be treated to become insoluble before embedding, or use frozen sections.

2.4.2.2 Fixation and Insolubilization

2.4.2.2.1 Ciaccio's Method

- Reagents
  - Mixture
    - Potassium bichromate 5%, 80 mL
    - Formalin, 15 mL
    - Acetic acid, 5 mL
  - Potassium bichromate 3%
- Protocol
  - Fixation into the mixture, 12 h

- Wash with potassium bichromate 3%, 5 min
- Potassium bichromate 3%, 2 days at 60°C
- Wash with tap water
- Frozen sections or paraffin embedding

2.4.2.2.2   Baker's Method
- Fix with formalin–calcium, 6 to 12 h
- Treat with potassium bichromate 3%, 2 days at 60°C

2.4.2.3   Demasking of Lipids

Several structures, in which biochemical analysis proved the presence of lipids, are not stained with lipids reactions. This can be due to two reasons: certain lipids are very dispersed and they cannot be detected with use of lysochromes; other lipids cannot be extracted with organic solvents because they are strongly bound to other substances. They are "masked lipids" and they are chemically linked to proteins or polysaccharides. Masking of lipids can be due to chemical links or to concentric layers of water molecules intercalated between lipids.

Some lipids are softly linked and they can be stained with proceedings using mordant or fixative; other lipids are strongly linked and must be demasked. For that there are organic solvents, acids, proteolytic enzymes, and salt solutions. Lipids are softly linked in myelin and red blood cells.

2.4.3   INCLUSIONS AND SECTIONS

Frozen sections of fixed or fresh tissues are often used. Rich-lipid tissues are often dilacerated and lipid droplets can be displaced in several parts of tissue.

Main embedding media used are

- Gelatin
- Carbowaxes without passing into solvents; protocol is complex
- Polyvinylglycols; protocol is complex
- Paraffine after insolubilized lipids

2.4.4   PHYSICAL METHODS

Among physical methods to study lipid histochemistry, we can use the polarizing microscope to detect isotropic, anisotropic, or birefringent character of lipids observed.

On frozen section of tissue fixed with formalin, homophasic lipids are characterized by "Malt Cross" if they are composed with cholesterides, phospholipids, or sphingolipides.

2.4.5   STAINING METHODS FOR LIPIDS WITH LYSOCHROMES

2.4.5.1   Staining with Black Sudan

See Figure 2.16.

- Principle—Dyes belonging to the "Sudan" family are lysochromes that dissolve into lipids giving them a staining.
- Fixative—Postchromisated material embedded into paraffin or frozen sections.
- Reagents
    - Black Sudan B: See Chapter 7, Section 7.2.3.5.
    - Nuclear fast red: See Chapter 7, Section 7.2.1.12.

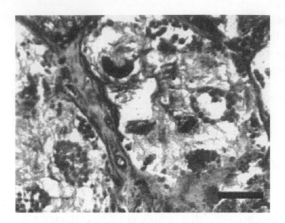

FIGURE 2.16   (See color insert.) Method: Sudan black and nuclear fast red. Tissue: Amphibian testis. Preparation: Frozen sections. Fixative: Buffered formalin. Observations: Nuclei are red stained and lipids are black. Bar = 30 μm.

- Protocol
    - Dewax for embedded sections
    - Ethanol 70%, 1 min. Use frozen or dewaxed sections.
    - Sudan black, 20 to 60 min. Use a blocked up bottle. Time must be tested according to the tissue.
    - Ethanol 70%, 1 min
    - Water
    - Nuclear fast red, 30 s. Nuclear fast red can be replaced by another nuclear dye.
    - Wash with tap water
    - Mount in hydrophilic medium, Kaiser's syrup or "Crystal Mount," for example
- Results—Lipids are black stained, nuclei are red.

2.4.5.2   Bromine–Black Sudan Staining

Bayliss and Adams's method (1972).

- Principle—Dyes belonging to the Sudan family are lysochromes that dissolve into lipids giving them a staining. Bromine is also used to stain lecithins, free fatty acids, and cholesterol.
- Fixative—Frozen sections postfixed in calcium–formalin during one hour or frozen fixed sections.
- Reagents
    - Black Sudan B: See Chapter 7, Section **7.2.3.5**.
    - Aquaeous bromine 2.5%
    - Sodium metabisulfite 0.5%
    - Nuclear fast red: See Chapter 7, Section 7.2.1.12.
- Protocol
    - Aquaeous bromine, 30 min.
    - Tap water
    - Sodium metabisulfite, 1 min. Metabisulfite eliminates excess bromine.
    - Wash in distilled water. After washing and drying, if sections are immersed into acetone, only phospholipids are stained.
    - Sudan black, 20 to 60 min
    - Ethanol 70%, 1 min
    - Water

- Nuclear fast red: 30 s. Nuclear fast red can be replaced by another nuclear dye. Nuclei will be stained according to the dye.
- Wash with tap water
- Mount in hydrophilic medium
- Results—Lipids are blue-black stained, nuclei are red.

2.4.5.3 Staining with Red Oil

Lillie and Ashburn's method (1943).

- Fixative—Frozen sections postfixed in calcium–formalin during one hour or frozen fixed sections. Postchromisated material embedded into paraffin can also be used.
- Reagents
  - Red oil O: See Chapter 7, Section 7.2.3.27.
  - Mayer's hemalum: See Chapter 7, Section 7.2.1.11.
- Protocol
  - Dewax (for embedded sections)
  - Isopropanol 60%, rinse
  - Red oil, 15 min
  - Isopropanol 60%, rinse
  - Tap water
  - Mayer's hemalum, 3 min
  - Tap water
  - Mount in hydrophilic medium, Kaiser's syrup or "Crystal Mount," for example
- Results—Hydrophobic lipids are red stained, nuclei are black blue.

2.4.5.4 Staining with Nile's Blue Sulfate

Method of Holczinger (1959) used to visualize free fatty acids.

- Principle—Nile's blue sulfate molecule is a blue colored oxazine. The detected lipids are unsaturated. They must not be soluble in the dye solution. In sulfuric solution, it is partially decomposed giving both the free base (oxazine) and an oxazone. Oxazone is obtained by methylation of oxazine. Nile's red, which is a lysochrome, stains in red or pinkish each lipid, whatever their chemical structure, and Nile's blue is a basic dye that stains each acidic lipid. The blue color masks the red one. So a sulfuric solution of Nile's sulfate stains neutral fat in red, acidic fats in dark blue, and nuclei and cytoplasm in blue. This staining is used to distinguish neutral from acidic lipids.
- Fixatives—Formalin or frozen sections. Paraffin embedded sections can be used but several lipids are lost.
- Reagents
- Nile's blue: See Chapter 7, Section 7.2.3.22.
- Acetic acid 1%
- Protocol
  - Dewax for embedded sections
  - Nile's blue, 5 min at 60°C
  - Wash with tap water
  - Differentiate with acetic acid, 30 s at 60°C
  - Wash with tap water
  - Mount in hydrophilic medium, Kaiser's syrup or "Crystal Mount," for example
- Results—Neutral lipids (glycerides, cholesterides) are pinkish stained, acidic lipids (serin-cephalines, sulfatides, gangliosides) are blue stained.

2.4.5.5   Acetone–Nile's Blue Sulfate Method

Method of Dunnigan (1968a,b) used to visualize phospholipids.

- Principle—This staining is used to distinguish phospholipids after dissolution of other lipids in acetone.
- Fixative—Frozen fixed sections with formalin. This technique can be used for tissue embedded in paraffin but several lipids are lost.
- Reagents
  - Nile's blue: See Chapter 7, Section 7.2.3.22.
  - Hydrochloric acid
  - Acetic acid 1%
  - Acetone
- Protocol
  - Hydrochloric acid, 1 h. This operation is used to desaponify calcium salts.
  - Tap water
  - Dry sections
  - Acetone, 20 min at 4°C. Extraction of all the lipids except phospholipids.
  - Dry sections
  - Nile's blue, 5 min at 60°C
  - Wash with tap water
  - Differentiate with acetic acid, 30 s at 60°C
  - Wash with tap water
  - Mount in hydrophilic medium, Kaiser's syrup or "Crystal Mount," for example
- Results—Only phospholipids are blue stained.

2.4.5.6   Copper–Rubeanic Acid Method

- Principle—This staining is used to distinguish free fatty acids after dissolution of other lipids in acetone.
- Fixative—Frozen fixed sections with formalin or calcium formalin.
- Reagents
  - Cupric acetate 0.005%
  - Rubeanic acid 0.1% in 70% ethanol. Rubeanic acid is dithiooxamide. Preparation:
    - Rubeanic acid, 0.1 g
    - Ethanol 70%, 100 mL
  - Hydrochloric acid 1M
  - EDTA. pH of EDTA must be adjusted at 7.0 with sodium hydroxide.
  - Acetone
  - Nuclear fast red: See Chapter 7, Section 7.2.1.12.
- Protocol
  - Hydrochloric acid, 1 h. This operation is used to desaponify calcium salts.
  - Distilled water
  - Dry sections
  - Acetone, 20 min at 4°C
  - Dry sections
  - Cupric acetate, 3 h
  - Sections can be treated with 0.025% dimethylaminobenzylidine-rhodanine in 70% ethanol. Free fatty acids will be red stained. In this case nuclei can be stained with a hemalum to give them a dark color.
  - EDTA, 2 × 10 s
  - Distilled water

- • Rubeanic acid, 10 min
- • Ethanol
- • Nuclear fast red, 30 s
- • Tap water
- • Mount in hydrophilic medium, Kaiser's syrup or "Crystal Mount," for example
- Results—Free fatty acids are dark green stained. Nuclei are red.

2.4.5.7 Control Reaction

Other structures being stained, controls must be prepared.

- • Compare sections stained with Sudan black with sections stained with Nile's blue.
- • In doubtful cases, examine a control slide on which lipids have been extracted.

Before staining, use the mixture:

- • Chloroform, 20 mL
- • Methanol, 20 mL for 2 to 12 h

2.4.6 VISUALIZATION OF STEROLS

2.4.6.1 Staining of Sterols with Liebermann and Schultz's Staining

- • Fixative—Frozen sections of material preserved with formalin.
- • Reagents
  - • Iron alum 2.5%
  - • Acidic mixture. Composition of acidic mixture:
    - – Acetic acid, 50 mL
    - – Concentrated sulfuric acid, 50 mL
- • Protocol
  - • Iron alum, 3 days
  - • Blot
  - • Cover slides with acidic mixture
  - • Cover with a coverslip and observe
- • Results—Sterols are blue or purple stained, then preparation becomes green. Stained preparation can be conserved for only a few hours. Take a photograph.

2.4.6.2 Lewis and Lobban Method

Coming from Liebermann and Schultz's method.

- • Fixative—Frozen sections of material preserved with formalin
- • Reagent—Sulfuric acid 80% with 0.5% iron alum
- • Protocol
  - • Dispose 1 drop of reagent on dry slides
  - • Observe immediately
- • Results—Testosterone and derivatives are blue-green stained. Estrogens and derivatives are pink or purple. Stained preparation can be conserved for only a few hours. Take a photograph.

2.4.6.3 Perchloric Acid–Naphthoquinone Method

Adams's (1961) PAN method.

- • Fixative—Frozen sections of material preserved with formalin.

- Reagents
  - Perchloric acid–naphthoquinone. See Chapter 7, Section 7.2.3.23.
  - Ferric chloride 1%
- Protocol
  - Ferric chloride, 4 h
  - Wash in distilled water
  - Dry sections
  - Dispose reagent with a brush
  - Heat slides, 1 to 2 min at 70°C. Stop the reaction when color is developed.
  - Mount in perchloric acid
- Results—Cholesterol and its esters are blue stained. Staining is stable for several hours. Take a photograph.

2.4.6.4   Digitonin–PAN Method

This method is used to visualize free cholesterol only.

- Fixative—Frozen sections of material preserved with formalin.
- Reagent
  - Digitonin 0.5% in 40% ethanol.
- Preparation:
  - Digitonin, 0.5 g
  - Ethanol, 100 mL
  - Perchloric acid–naphthoquinone
  - Ferric chloride 1%
  - Acetone
- Protocol
  - Digitonin, 2 h. Precipitates cholesterol.
  - Acetone, 1 h. Eliminates cholesterides.
  - Ferric chloride, 4 h
  - Wash in distilled water
  - Dry sections
  - Dispose reagent with a brush
  - Heat slides, 1 to 2 min at 70°C
  - Mount in perchloric acid
- Results—Free cholesterol is blue stained. Staining is stable for only a few hours. Take a photograph.

2.4.7   STAINING OF UNSATURATED LIPIDS

2.4.7.1   Osmium Tetroxide Method

- Fixative—Frozen sections of material preserved with formalin.
- Reagents
  - Osmium tetroxide 1%
  - Nuclear fast red. Optional: See Chapter 7, Section 7.2.1.12.
- Protocol
  - Osmium tetroxide, 1 h
  - Distilled water
  - Nuclear fast red, 30 s. Optional.
  - Mount section in hydrophilic medium, Kaiser's syrup or "Crystal Mount," for example
- Results—Unsaturated lipids are brown-black stained. Other lipids are not stained.

2.4.7.2   Osmium Tetroxide–α-Naphtylamine

Osmium tetroxide alpha naphtylamine (OTAN) reaction of Adams (1959).

- Fixation—Frozen sections of material preserved with formalin.
- Reagents
  - Osmium–chlorate solution
    - Osmium tetroxide 1%, 10 mL
    - Potassium chlorate 1%, 30 mL
  - α-Naphtylamine reagent
    - Some crystals of α-naphtylamine
    - Distilled water, 40 mL at 40°C. Use at 37°C.
    - Filter
  - Alcian blue 2% in acetic acid 5%: See also Chapter 7, Section 7.2.3.2.
- Protocol
  - Osmium–chlorate, 18 h
  - Distilled water, 10 min
  - α-naphtylamine, 20 min at 37°C
  - Distilled water
  - Alcian blue, 15 s
  - Rinse in distilled water
  - Mount section in hydrophilic medium
- Results—Cholesterol esters in degenerating myelin are black stained, phospholipids in normal myelin are orange stained.

2.4.8   Reactions for Other Lipids

In this part several reactions used to visualize several groups of lipids are given. Lipids considered here are triglycerides, phosphoglycerides, plasmalogens, lecithin, sphingomyeline, cerebrosides, and sulfatides.

2.4.8.1   Calcium–Lipase for Triglyceride Visualization

- Fixative—Frozen sections of material preserved with formalin. It is possible to stain floating sections that will be mounted on slides after staining.
- Reagents
  - Lipase medium
    - Tris buffer, 0.2 M, pH 8.0, 15 mL
    - Calcium chloride 2%, 10 mL
    - Distilled water, 25 mL
    - Pancreatic lipase, 50 mg
  - Lead nitrate 1%
  - Ammonium sulfide 1%
  - Mayer's hemalum: See Chapter 7, Section 7.2.1.11.
- Protocol
  - Lipase medium, 3 h at 37°C (filter before use and use at 37°C)
  - Tap water
  - Lead nitrate, 15 min
  - Ammonium sulfide, 20 s
  - Tap water
  - Mayer's hemalum, 3 min

- Tap water. Washing with water permits differentiation of Mayer's hemalum; stop the reaction when sections become blue stained.
- Mount section in hydrophilic medium, Kaiser's syrup or "Crystal Mount," for example.
- Results—Triglycerides are brown stained, cell nuclei are blue.

2.4.8.2   Gold Hydroxamic Acid Method for Phosphoglyceride Visualization

- Fixative—Frozen sections of material preserved with formalin. It is possible to stain floating sections that will be mounted on slides after staining.
- Reagents
  - Alkaline hydroxylamine
    - Hydroxylamine hydrochloride, 2.5 g
    - Sodium hydroxide, 6 g
    - Distilled water, 100 mL
  - Silver nitrate solution
    - Silver nitrate, 0.1 g
    - Ammonium nitrate, 0.2 g
    - Distilled water, 100 mL
  - Sodium hydroxide. Used to obtain a pH 7.8 solution.
  - Acetic acid 1%
  - Yellow gold chloride 0.2%
  - Sodium thiosulfate 5%
  - Methyl green 1%: See Chapter 7, Section 7.2.3.18.
- Protocol
  - Alkaline hydroxylamine, 20 min
  - Distilled water, 3 × 5 min
  - Silver solution, 2 h
  - Distilled water
  - Rinse in acetic acid
  - Distilled water
  - Yellow gold chloride, 10 min
  - Distilled water
  - Sodium thiosulfate, 5 min
  - Distilled water
  - Methyl green, 5 min
  - Mount section in hydrophilic medium, Kaiser's syrup or "Crystal Mount," for example
- Results—Phosphoglycerides are red or purple stained, cell nuclei are green.

2.4.8.3   Plasmal Reaction or "Feulgen's Plasma Reaction"

- Fixative—Unfixed frozen sections.
- Reagents
  - Mercuric chloride 2%
  - Schiff's reagent: See Chapter 7, Section 7.2.3.28.
  - Mayer's hemalum: See Chapter 7, Section 7.2.1.11.
- Protocol
  - Air-dry section
  - Mercuric chloride, 10 min
  - Rinse with distilled water, 3 times
  - Schiff's reagent, 10 min
  - Tap water, 10 min. Running water is used to develop the color.
  - Mayer's hemalum, 3 min

- Tap water
- Mount section in hydrophilic medium, Kaiser's syrup or "Crystal Mount," for example
- Results—Plasmalogen are magenta stained. Nuclei are blue.

2.4.8.4   Dichromate–Acid Hematein (DAH Staining)

- Fixative—Frozen sections of material preserved with formalin or calcium-formalin (see Chapter 7, Section 7.1.1.8).
- Reagents
  - Dichromate solution:
    - Potassium dichromate 5%, 100 mL
    - Calcium chloride, 1 g
  - Tetraborate-ferricyanide:
    - Sodium tetraborate, 0.25 g
    - Potassium ferricyanide, 0.25 g
    - Distilled water, 100 mL
  - Methyl green: See Chapter 7, Section 7.2.3.18.
  - Acid hematein: See Chapter 7, Section 7.2.1.2.
- Protocol
  - Dichromate solution, 18 h at room temperature (RT)
  - Dichromate solution, 2 h at 60°C
  - Distilled water, 30 min
  - Acid hematein, 2 h at 37°C
  - Sodium tetraborate and potassium ferricyanide, 2 h at 37°C
  - Distilled water
  - Methyl green, 5 min
  - Distilled water
  - Mount section in hydrophilic medium, Kaiser's syrup or "Crystal Mount," for example
- Results—Lacithin and sphingomyelin are blue stained, nuclei are green.

2.4.8.5   Sodium Hydroxide–Dichromate Acid Hematein

- Fixative—Frozen sections of material preserved with formalin or calcium–formalin.
- Reagents
  - Acid hematein: See Chapter 7, Section 7.2.1.2.
  - Dichromate solution
    - Potassium dichromate 5%, 100 mL
    - Calcium chloride, 1 g
  - Tetraborate–ferricyanide
    - Sodium tetraborate, 0.25 g
    - Potassium ferricyanide, 0.25 g
    - Distilled water, 100 mL
  - Sodium hydroxide, 2M
  - Acetic acid 1%
  - Methyl green: See Chapter 7, Section 7.2.3.18.
- Protocol
  - Sodium hydroxide, 1 h at 37°C
  - Distilled water, 30 min
  - Acetic acid, 5 s
  - Dichromate solution, 18 h at RT
  - Dichromate solution, 2 h at 60°C
  - Distilled water, 30 min

- • Acid hematein, 2 h at 37°C
- • Sodium tetraborate and potassium ferricyanide, 2 h at 37°C
- • Distilled water
- • Methyl green, 5 min
- • Distilled water
- • Mount section in hydrophilic medium, Kaiser's syrup or "Crystal Mount," for example.
- • Results—Sphingomyelin are blue stained, nuclei are green.

2.4.8.6   Ferric Hematoxylin for Phospholipids

- • Fixative—Unfixed frozen sections. It is also possible to use frozen sections of material preserved with formalin or calcium–formalin.
- • Reagents
  - • Solution A
    - – Hydrochloric acid, 2 mL
    - – $FeCl_3$, $6H_2O$, 2.5 g
    - – $FeSO_4$, 7 H_2O, 4.5 g
    - – Distilled water, 298 mL
    - – Acetic acid, 2 mL
  - • Solution B
    - – Hematoxylin, 1 g
    - – Distilled water, 100 mL
  - • Working solution (ferric hematoxylin)
    - – Solution A, 30 mL
    - – Solution B, 10 mL
  - • Hydrochloric acid 0.2%
- • Protocol
  - • Use two groups of dried sections
    - – Group A—Chloroform–methanol, 1 h. Treatment of sections eliminates lipids other than phospholipids.
    - – Group B—Acetone, 15 min at 4°C
  - • Fix sections with calcium–formalin, 30 min. Do not consider this step if material is already fixed.
  - • Distilled water
  - • Ferric hematoxylin, 7 min
  - • Distilled water
  - • Hydrochloric acid, several times.
  - • Tap water
  - • Dehydrate, mount
- • Results—Phospholipids are blue stained.

2.4.8.7   Sodium Hydroxide–Ferric Hematoxylin for Sphingomyelin

- • Fixative—Unfixed frozen sections.
- • Reagents
  - • Ferric hematoxylin: See Chapter 7, Section 7.2.1.8.
  - • Acetic acid 1%
  - • Hydrochloric acid 0.2%
- • Protocol
  - • Sodium hydroxide, 1 h
  - • Distilled water
  - • Acetic acid, 5 s

- Distilled water
- Ferric hematoxylin, 7 min
- Distilled water
- Hydrochloric acid, several times
- Tap water
- Dehydrate, mount
- Results—Sphingomyelin are blue stained.

2.4.8.8 PAS Reaction for Cerebrosides

- Principle—Cerebrosides are glycolipids. The glucidic part is stained with PAS reaction. Chloramine–T is used to convert amino acids in carbonyls. Two groups of sections are used: First group without any treatment on which PAS reaction visualizes both proteins and cerebrosides; the second group of sections is treated with a chloroform–methanol mixture to eliminate lipids. Cerebrosides are stained only on sections without control; other glucides and proteins are stained on both groups of sections.
- Fixative—Frozen sections of material preserved with formalin or calcium–formalin, postfixed frozen sections.
- Reagents
  - Chloramine-T 10% in water: See Chapter 7, Section 7.2.3.6.
  - Performic acid solution
    - Formic acid 98%, 45 mL
    - Hydrogen peroxide (100 vol), 4.5 mL
    - Sulfuric acid (concentrated), 0.5 mL
  - Dinitrophenylhydrazine solution
    - Hydrochloric acid 1M, 100 mL
    - 2-4 dinitrophenylhydrazine, saturated
  - Schiff's reagent: See Chapter 7, Section 7.2.3.28.
  - Mayer's hemalum: See Chapter 7, Section 7.2.1.11. Mayer's hemalum can be replaced by another nucleic dye.
- Protocol
  - Aqueous chloramine-T, 1 h at 37°C
  - Use duplicate sections on slides. One slide is stained according to the protocol. The second slide first treated with chloroform–methanol (50 mL/50 mL) mixture is used as a control.
  - Distilled water. Washing must be as quick as possible.
  - Performic acid, 10 min
  - Distilled water. Several baths of distilled water are used.
  - Solution of dinitrophenylhydrazine, 2 h at 4°C
  - Distilled water
  - Periodic acid, 10 min
  - Distilled water
  - Schiff's reagent, 15 min
  - Tap water, 15 min
  - Mayer's hemalum, 3 min
  - Tap water. Running water is used to develop the color.
  - Distilled water
  - Mount in hydrophilic medium, Kaiser's syrup or "Crystal Mount," for example
- Results—Cerebrosides are magenta stained. Lipids in control sections are unstained. Proteins and glucides are stained on both normal and stained sections.

2.4.8.9 Toluidine Blue–Acetone Reaction for Sulfatides

- Fixative—Frozen sections of material preserved with formalin or calcium–formalin, post-fixed frozen sections.
- Reagents
  - Toluidine blue
  - Acetone
- Protocol
  - Toluidine blue, 16 h
  - Tap water
  - Acetone, 5 min
  - Xylene, mount
- Results—Sulfatides are red-brown stained.

2.4.8.10 Diamine–Alcian Blue for Sulfatides

- Fixative—Frozen sections of material preserved with formalin or calcium–formalin, post-fixed frozen sections.
- Reagents
  - Diamine reagent: Diamine salts are toxic. They must be handled with precaution: See Chapter 7, Section 7.2.3.9.
  - Alcian blue: See Chapter 7, Section 7.2.3.2.
- Protocol
  - Diamine reagent, 18 h
  - Tap water
  - Alcian blue 1%, 5 min
  - Tap water
  - Dehydrate, mount
- Results—Sulfatides are purple stained.

2.4.9 STAINING OF PIGMENTS

2.4.9.1 Pigments in Histochemistry

Some pigments, such as melanine, are observed in healthy tissues; others are found in several pathologic cases. Melanines are found in several healthy organs (skin, liver) as well as in pathological ones. Hemozein is a pigment characteristic of paludism (malaria). Visualization can be done by direct observation of pigment: color and autofluorescence. Solubility in several solvents and whitening in oxidizing reagents are also used to characterize pigments. Several staining methods are also used.

2.4.9.2 Visualization of Melanin by Reduction of a Silver Salt

See Figure 2.17. Melanin reduces ammoniacal silver complexes.

- Fixative—Formalin, buffered formalin, Bouin's fluid
- Reagents
  - Fontana's fluid. Preparation:
    - Silver nitrate 10%, 20 mL
    - Ammoniac (NH_4OH) drop after drop until the dissolution of precipitate
    - Distilled water, 20 mL
    - Let rest
    - Decant (use a black bottle)
    - Filter
    - Storage: 1 month. Stock at low temperature and at darkness.
  - Nuclear fast red: See Chapter 7, Section 7.2.1.12.

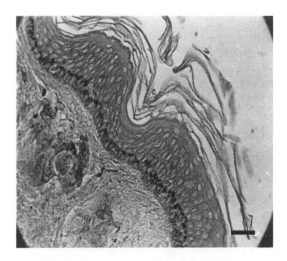

FIGURE 2.17 (See color insert.) Method: Argyrophily, hemalum. Tissue: Melanin in mouse skin. Preparation: Paraffin sections. Fixative: Bouin's fluid. Observations: Melanin is dark stained. Bar = 100 μm.

- Protocol
  - Dewax, hydrate
  - Distilled water, 2 h
  - Fontana's fluid (darkness), 36 h
  - Rinse with distilled water
  - Sodium hyposulfite, 30 s
  - Rinse
  - Nuclear fats red, 1 min
  - Rinse
  - Dehydrate, mount
- Results—Melanin and argyrophilic substances are black stained.

2.4.9.3 Staining of Lipofuscin with PAS
- Principle—The method consists of oxidizing carbohydrates with periodic acid to obtain aldehyde functions (–CHO). PAS reaction can be used to visualize lipofuscins (lipopigments). These functions react with Schiff's reagent giving a red staining to the sections. Schiff's reagent is commonly called "faded fuchsin."
- Fixative—Unfixed frozen sections.
- Reagents
  - Schiff's reagent: See Chapter 7, Section 7.2.3.28.
  - Periodic acid 1%
- Protocol
  - Periodic acid, 10 min
  - Tap water, 10 min
  - Schiff's reagent, 10 min
  - Tap water, 5 min
  - Dehydrate, mount
- Results—Lipofuscins are magenta stained.

2.4.9.4 Staining of Lipofuscin with Thionine
- Fixative—Unfixed frozen sections.
- Reagents—Thionine: See Chapter 7, Section 7.2.3.30.

- Protocol
  - Cover sections with reagent. No mounting medium can be used; staining remains unstable. Take a photograph.
  - Coverslip.
- Results—Lipofuscins are green stained. Other lipids, such as cerebrosides and gangliosides, are pinkish.

2.4.9.5 Staining of Lipofuscin with Luxol Fast Blue

- Fixative—Unfixed frozen sections.
- Reagents
  - Ethanol 95%
  - Lithium carbonate 0.05%
  - Luxol fast blue: See Chapter 7, Section 7.2.3.16.
  - Nuclear fast red: See Chapter 7, Section 7.2.1.12.
- Protocol
  - Luxol fast blue, 1 min
  - Rinse with ethanol 95%
  - Tap water
  - Lithium carbonate, differentiate. Control differentiation under microscope. An excess of differentiation can eliminate luxol blue.
  - Tap water
  - Nuclear fast red, 1 min
  - Tap water
  - Dehydrate, mount
- Results—Lipofuscins are blue stained. Nuclei are red.

2.4.9.6 Staining of Lipofuscin with Ziehl–Neelsen Method

- Fixative—Unfixed frozen sections.
- Reagents
  - Fuchsin: See Chapter 7, Section 7.2.3.13.
  - Methylene blue 2%: See Chapter 7, Section 7.2.2.15.
- Protocol
  - Fuchsin, 3 h at 60°C
  - Tap water
  - Acid ethanol, differentiate. Control the differentiation under microscope. An excess of differentiation can eliminate fuchsin.
  - Tap water.
  - Methylene blue, 1 min
  - Tap water
  - Dehydrate, mount
- Results—Lipofuscins are magenta stained. Nuclei are blue.

2.5 STAINING OF BIOGENIC AMINES

2.5.1 Biogenic Amines

2.5.1.1 General Considerations

Biogenic amines are implicated in several ways in the organism. They can be secreted by endocrine cells, and they have a hormonal function. Biogenic amines can also be used, such as neurotransmitters.

Biosynthesis of biogenic amines starts from the modification of the corresponding amino acid, mainly with decarboxylation. Biogenic amines react after several modifications, under enzymatic

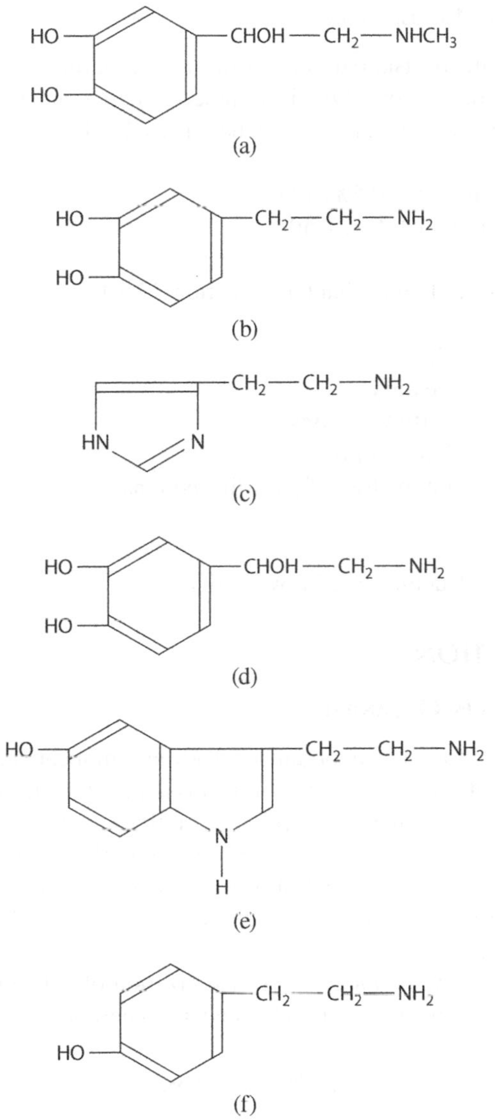

FIGURE 2.18 Several biogenic amines. (a) Adrenaline. (b) Dopamine. (c) Histamine. (d) Noradrenaline. (e) Serotonin. (f) Tyramine.

control. In this chapter we give only the chromaffin method of visualization, which can be used for all the biogenic amines.

2.5.1.2 Biogenic Amines

Adrenaline, noradrenaline, and dopamine are catecholamine because they are synthesized from catechol (Figure 2.18).

2.5.2 Fixation

Formalin 10%, buffered formalin, and neutral formalin are commonly used fixators. Frozen sections of fresh tissue preserved with formalin can also be used. In a general manner, classic fixative agents are available to preserve biogenic amines. When possible, it is useful to control the effects of each fixative.

2.5.3 STAINING METHOD: CHROMAFFINITY

- This reaction is useful to visualize adrenalin, noradrenalin, and serotonin. These components are oxidized on sections. Oxidized molecules are condensed and brown stained.
- Fixative—The fixative indicated here gives better results than other fixatives.
  - Mixture
    - Potassium bichromate 0.5%, 10 mL
    - Potassium chromate 0.5%, 1 mL
  - Formalin 10%
- Reagent—Nuclear fast red: See Chapter 7, Section 7.2.1.12.
- Protocol
  - Fix with mixture, 48 h
  - Formalin, 24 h, if necessary
  - Frozen sections or paraffin sections
  - Dewax, maximum 30 min, only for paraffin
  - Nuclear fast red, 1 min; hydrate if paraffin sections
  - Rinse
  - Dehydrate, mount
- Results—Chromaffin structures are brown stained.

2.6 MINERAL DETECTION

2.6.1 MINERAL ELEMENTS IN ORGANISM

Several mineral elements are found in an organism. They are implicated in several ways: membrane polarization and depolarization, nerve influx transmission, intercell cement, coenzyme factors, muscle contraction, and so on. So they are important for physiological activity of cells and tissues.

Mineral concentration is low in organisms; several mineral elements are important for physiological processes, but an increased concentration can become lethal for tissues and, consequently, the entire organism. The presence of an increased amount of mineral elements can be the sign of a pathological state of tissue.

In this part, we indicate only the method for visualization of calcium and iron, the most often researched, but the research of other mineral elements is sometimes useful.

2.6.2 GENERAL PRINCIPLES OF MINERAL VISUALIZATION

2.6.2.1 Chemical Fixative

- It does not bring mineral element
- It must not dissolve mineral elements out of sections
- The best fixative is ethanol 95%

2.6.2.2 Microincineration

Microincineration is a technique that burns the tissue. Organic elements are eliminated and mineral elements such as ashes are conserved. These ashes stay in place without any important displacement, so minerals can be observed. The preparations are called "spodograms." Mineral elements can be directly observed or they can be the subject of chemical reactions.

2.6.2.3 Staining Methods

Mineral elements are visualized with colored reactions, which are based on several principles:

- Metal substitution
- Formation of colored lacs

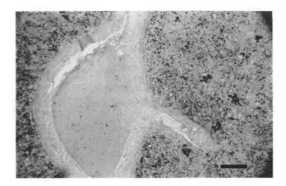

FIGURE 2.19 **(See color insert.)** Method: Perls's staining. Tissue: Spleen. Preparation: Paraffin sections. Fixative: Bouin's fluid. Observations: Iron is blue stained. Bar = 100 μm.

- Formation of complexes
- Reactions of conversion

2.6.2.4 Physical Methods

Several physical methods can be used to visualize mineral elements. But these are more used for analytical purpose than for visualization so they will not be detailed here.

2.6.3 Methods to Visualize Minerals

2.6.3.1 Visualization of Iron: Perls's Method

This method is used by pathologists to detect calcium in pathologic tissues (Figure 2.19).

- Principle—This reaction is used to visualize ionic ferric iron.
- Fixative
  - Formalin 10% buffer at pH 7
  - Bouin's fluid
- Reagents
  - Potassium ferrocyanide 5%
  - Hydrochloric ferrocyanide. Composition:
    - Potassium ferrocyanide 5%, 50 mL
    - Hydrochloric acid 5% in water, 50 mL
  - Nuclear fast red: See Chapter 7, Section 7.2.1.12.
- Protocol
  - Dewax, hydrate
  - Potassium ferrocyanide, 5 min
  - Hydrochloric ferrocyanide, 10 min
  - Tap water, 5 min
  - Nuclear fast red, 1 to 5 min
  - Rinse
  - Dehydrate, mount
- Results—Ferric iron is blue stained.

2.6.3.2 Visualization of Calcium: Stoelzner's Method

- Principle—A heavy metal is substituted to calcium, then it is precipitated to give a sulfide.
- Fixative
  - Ethanol

- • Formalin
- • Carnoy's fluid
- • Avoid aqueous fixatives because they dissolve calcium salts
- • Reagents
  - • Cobalt nitrate 2%, prepare just before use
  - • Ammonium sulfide
  - • Nuclear fast red: See Chapter 7, Section 7.2.1.12.
- • Protocol
  - • Dewax, hydrate
  - • Cobalt nitrate, 5 min
  - • Tap water
  - • Ammonium sulfide, 2 min
  - • Tap water
  - • Nuclear fast red, 1 min
  - • Tap water
  - • Dehydrate, mount
- • Results—Components containing calcium are black stained.

2.6.3.3   Control Method with Decalcification

- • Principle—It is necessary to compare results with decalcified sections.
- • Reagent—Ammonium citrate 10%. Ammonium citrate is used to decalcify the sections.
- • Protocol
  - • Dewax, hydrate
  - • Ammonium citrate, 10 min
  - • Wash with tap water
  - • Perform reaction of visualization. See Sections 2.6.3.1 and 2.6.3.2.
- • Results—Components with calcium are not stained.

2.7   NUCLEIC ACIDS HISTOCHEMISTRY

2.7.1   Structure of Nucleic Acids and Nucleoproteins

2.7.1.1   DNA

DNA is characterized by a sugar, deoxyribose, and more particularly the 2-D-deoxyfuranose. This molecule is associated to a nitrogenous base and a molecule of phosphoric acid (Figure 2.20). Groups belonging to deoxyribose–phosphate are bound by phosphodiester bonds between the 3′ and 5′ carbons belonging to two successive deoxyriboses. A deoxyribose and a nitrogenous base constitute a deoxynucleoside (or a nucleoside). Deoxyribose, nitrogenous base, and phosphate constitute a deoxyribonucleotide (a nucleotide). By convention, carbons belonging to pentoses are numbered from 1′ to 3′. Atoms belonging to bases are numbered from 1 to 5 or from 1 to 9.

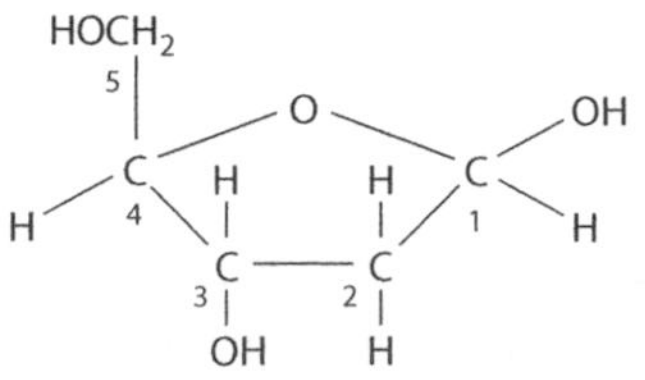

FIGURE 2.20   Ribose.

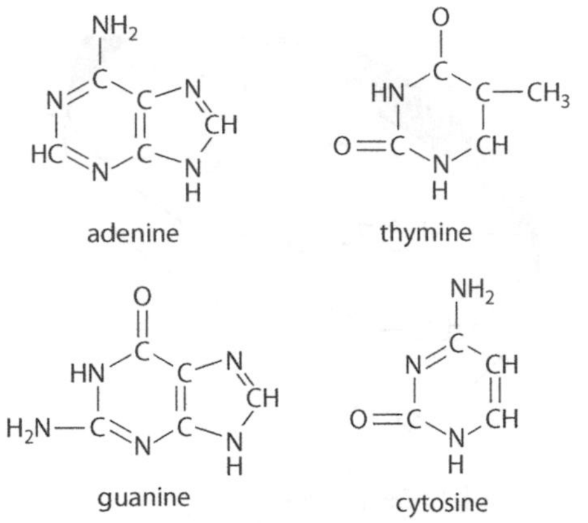

FIGURE 2.21 Puric and pyrimidic bases.

All the bases belong to the purine or pyrimidine family (Figure 2.21). Puric bases are adenine and guanine; pyrimidic bases are cytosine and thymine. (Thymine is the 5-methyl uracile.) Other bases can be observed in DNA composition, but they are very rarely found and their quantities are very small. Certain molecular sites in the bases can incur hydrogen binding and can allow binding with proteins.

The molecular structure of DNA is that of a double helix, which was discovered by Watson and Crick in 1953 (Figure 2.22). This double helix is composed of two molecular chains that are linked by nitrogenous bases. The structure corresponds to a stack of deoxynucleotides, linked to each other by diester bridges between the 3′ and 5′ carbons belonging to two adjacent deoxyriboses. These deoxyribose–phosphate bonds are oriented in such a way that gives each a helical aspect. These stacks form the primary structure of DNA.

The two molecular chains are oriented in the opposite direction along the helix axis: one from 5′ to 3′, the other from 3′ to 5′. These trends are a dyad and they are antiparallel. The entire molecule is ladderlike of which the bases are rungs.

Adenine is always bound to thymine by two hydrogen liaisons and the cytosine at guanine by three.

The two halves of the DNA double helix are bound by the pairing of two complementary nitrogenous bases. Binding two bases are hydrogen bonds, which allow the double helix to open at the moment of DNA replication or during copy of sequences in RNA. The described structure is that of β-DNA, which is the majority of *in vivo* DNA. Other molecular structures can be found. α-DNA is denser. It possesses 11 pairs of bases by helix turn. Z-DNA possesses 12 pairs of bases by turn and its general form is a zigzag.

Nitrogenous bases are complementary. A puric base is always linked to a pyrimidic base.

DNA geometry possesses a great and a small groove.

The DNA molecule is very stable. This stability is a consequence of hydrogen bonds, which are not very stable. However, when they are numerous, they possess hydrophobic interactions between two neighboring bases and they stabilize nucleoplasmic phosphoric groups. Mg^{++} ions also contribute to the stability of DNA molecule. The histones and basic proteins that are characteristic of eukaryotic organisms are bound to DNA to form nucleoproteins. The latter also contribute greatly to the stability of DNA.

2.7.1.2 RNA

RNA is characterized by a ribose (Figure 2.23). More precisely, the ribose is the 2-D-deoxyfuranose. As with DNA, this sugar is associated with one nitrogenous base and a phosphoric group. The molecule composed of ribose and one nitrogenous base is called ribonucleoside (or nucleoside). Binding with a phosphoric group is one ribonucleotide (nucleotide).

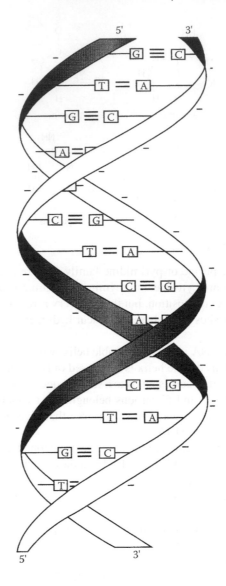

FIGURE 2.22   DNA double helix. (From Smith, C.A., and E.J. Wood, 1997, *Les biomolécules*, Masson, Paris. With permission.)

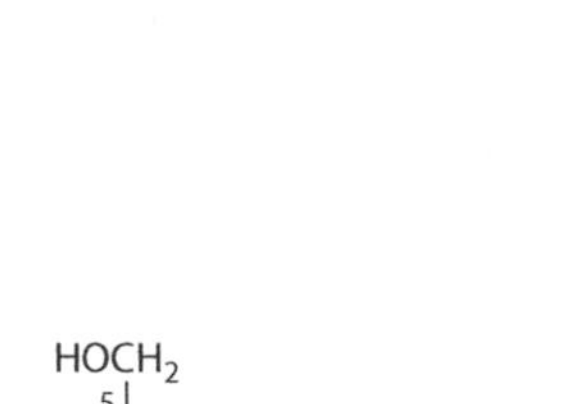

FIGURE 2.23   Ribose.

RNA molecules are generally single stranded. They are generated from one single DNA strand. However, free RNA bases can be paired if their structure is suitable. During copying of DNA in RNA, a DNA–RNA duplex is formed after complementary bases have been paired. Adenine is linked to uracil and guanine to cytosine.

At present, four RNA types are known. The four RNA types are independently transcribed from DNA. Messenger RNAs (mRNAs), which are copied from DNA, bring genetic information into the cytoplasm. This mRNA is then translated in a polypeptide into ribosomes, which are very small particles linked to one ribosomal RNA (rRNA). During the translation, mRNA–ribosome complexes are paired with amino acids that are brought into the cytoplasm by transfer RNAs (tRNAs). A fourth RNA class, the "small nuclear RNAs" (snRNAs), is combined with proteins to form ribonucleoprotein particles, which are required in nuclear metabolism of RNAs.

2.7.1.3   Nucleoproteins

Nucleoproteins are substances arising from proteins and nucleic acid combinations. Their liaisons are formed from basic groups of lateral chains linked to negative loads of phosphoric groups belonging to nucleotides. In the nucleus, histones are associated with DNA. Nucleoli are deprived of histones. Other deoxyribonucleic types can be observed into cell nucleus. Histones possess numerous basic amino acids, especially arginine and lysine.

2.7.2   General Principles

Histochemical methods permitting visualization of nucleic acids can be classified into four categories that are a function of the visualized molecular component:

- Puric and pyrimidic acid permit nucleic acid detection by use of ultraviolet spectrography.
- Presence of phosphoric acids gives an electronegative character to nucleic acids.
- Presence of pentose permits certain methods to be used for sugar detection.
- Presence of a proteinic part permits methods for protein detection to be used.

2.7.3   Detection of Puric and Pyrimidic Bases

2.7.3.1   Caspersson's Spectrophotometric Method

This spectrophotometric method is rarely used. Puric and pyrimidic bases are the only tissue molecules that strongly absorb ultraviolet wavelength at 260 nm. This method can be used on smears, cell cultures, sections, whatever the fixative used. RNA and DNA absorption curves are similar. Therefore, this method may only be used for normal sections and for sections from which one or the other nucleic acid has been extracted.

2.7.3.2   Danielli's Tetrazoreaction after Benzoylation or Acetylation

2.7.3.2.1   Original Method

(Also see Section 2.3.4.3.) The original method consists of combining tetrazoted benzidine and certain amino acids at an alkaline pH and at a low temperature. The resulting molecule is then coupled to an aromatic amine. It is better to replace tetrazoted benzidine by fast blue B. The salt generally used in the reactions is orthodianisidine, also called fast blue B. Another salt that is often used is H acid, which is characterized by a β-naphthol function. Tissue and cell groups are brown stained.

Some authors have supposed that nucleic acids were visualized because they were associated with proteins. Whatever the reaction, this method is useful to visualize nucleic acids.

- Fixative—Avoid mixtures with formalin.
- Reagent
  - Fast blue B (orthodianisidine) 0.2% in veronal buffer pH 9.2: See Chapter 7, Section 7.2.3.11.
  - Acid H 2% in veronal buffer pH 9.2: See Chapter 7, Section 7.2.3.1.
  - Veronal buffer. Preparation:
    - Hydrochloric acid 8.35 g/l, 231 mL
    - Sodium veronal 20.618 g/l, 769 mL
- Protocol
  - Dewax, hydrate
  - Orthodianisidine, 5 min
  - Wash with tap water
  - Veronal buffer, 3 × 2 min
  - Acid H, 5 min
  - Wash with tap water
  - Dehydrate
  - Mount
- Results—Proteins with associated nucleic acids are purple or brown stained.

2.7.3.2.2   Control Methods for Tetrazoreaction
See Section 2.3.4.4.

2.7.4   PENTOSE VISUALIZATION

2.7.4.1   Method of Turchini

2.7.4.1.1   Principle
Turchini's method is based on a condensation between pentoses and 9-phenyl, 2, 3, 7-tetrahydrofluo-rone. However, the specificity of this method has not been perfectly established.

Fluorone (or its derivative) is condensed with the pentose after an acidic hydrolysis. The product obtained yields, at a basic pH, a stained insoluble molecule. Methyl derivative gives a blue-purple staining of ribose (RNA) and red-orange with deoxyribose (DNA).

2.7.4.1.2   Protocol
- Fixative—All the classic fixatives can be used.
- Reagents
  - Hydrochloric acid M
  - Ethanol 80%
  - Sodium carbonate 1%
  - Fluorone. This solution must be used within 24 hours: See Chapter 7, Section 7.2.3.12.
- Protocol
  - Dewax, hydrate
  - Hydrochloric acid M, 10 min at 60°C. Duration of hydrolysis is varying according to the fixative. Time indicated here is given for fixation with Zenker's fluid.
  - Ethanol 80%, 15 sec
  - Fluorone, 12 h. Incubation time of fluorone can vary according to the studied tissue. In the original technique, the time varies from 4 to 14 h. It is necessary to test each material studied to determine the optimal time.
  - Sodium carbonate 1%, 2 min
  - Distilled water, 2 min
  - Acetone 50%, 5 min
  - Absolute acetone, 5 min

- Xylene–acetone (50/50), 5 min
- Xylene, 10 min
- Mount
- Results—DNA is violet-blue stained. RNA is orange red.

2.7.4.1.3 Hydrolysis Duration

The duration of hydrolysis varies according to the fixative.

- Bouin's fluid, 7 min
- Formalin, 7 min
- Carnoy's fluid, 8 min
- Helly's fluid, 25 min
- Zenker's fluid, 10 min

Durations given here are only indications. It is necessary to determine the hydrolysis duration each time a new material is studied.

2.7.4.2 Other Methods

Other methods permit visualization of deoxyribose but they are rarely used. The most used method is the Feulgen and Rossenbeck nucleal reaction, which is based on specific denaturation of deoxyribose and visualization of aldehyde groups by Schiff's reagent (see Section 2.7.7). Nucleic acids staining is not linked to nitrogenous bases; it is due to the histidine belonging to the proteinic part.

2.7.5 VISUALIZATION OF PROTEIN FRACTION

See Section 2.3.3.

2.7.6 BASOPHILIC REACTIONS

2.7.6.1 General Principles

Nucleic acids are linked to ionized acid groups that are present on phosphate parts. At a pH less than 2, electronegative charges appear. These charges can fix basic dyes. Negative phosphoric groups are also bound with proteins in live cells. This fact explains why nucleic acids cannot fix basic dyes in live cells. Preservation frees these groups, but there is always some competition between cations of dye and other cations belonging to the tissue, particularly amine groups belonging to the proteinic part of nucleoproteins. Basic dyes, as previously mentioned, can fix themselves onto other acidic groups that do not belong to nucleic acids. One often must use dyes with a certain specificity, for example, methyl green or pyronine for DNA and RNA, respectively. But use of dyes with a large spectrum of possibilities, such as toluidine blue, is not excluded. DNA visualization is relatively easy. For RNA, control reactions must be done.

2.7.6.2 Gallocyanine Method

2.7.6.2.1 Principle

At high temperature, gallocyanine forms three lac types with chrome alum. One of them is cationic and fixes itself onto phosphate groups belonging to nucleic acids. This forms a dark blue complex, which is obtained at a pH varying from 0.8 to 4.2. However, to prevent this substance from fixing on other molecules, operating at a pH ranging from 1.50 to 1.75 is recommended. In these conditions, the stain is highly selective for nucleic acids.

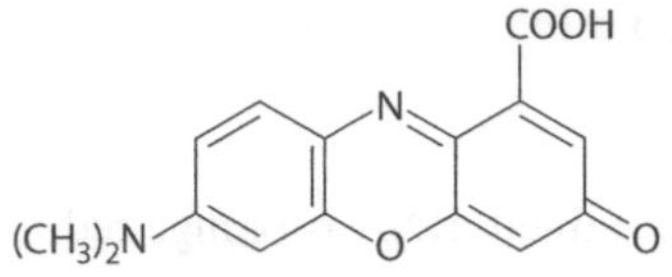

FIGURE 2.24   Gallocyanine.

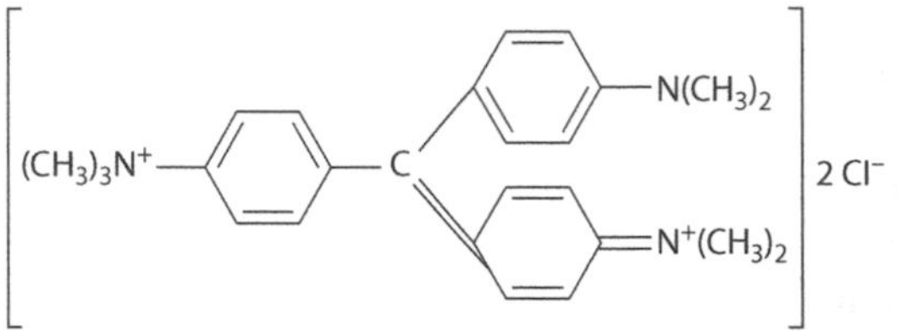

FIGURE 2.25   Methyl green.

2.7.6.2.2   Chemical Formula
See Figure 2.24.

2.7.6.2.3   Protocol
- Fixative—All classic fixatives can be used. If possible, avoid fixation with osmium tetroxide and fixing sections after postchromisation.
- Reagent—Gallocyanine chromic lac: See Chapter 7, Section 7.2.3.14.
- Protocol
  - Dewax, hydrate
  - Gallocyanine lac, 24 h. The gallocyanine immersion time can be increased.
  - Current tap water, 5 min
  - Dehydrate
  - Mount
- Results—DNA and structures with basophilic components are dark blue stained. Other dyes can be added to gallocyanine, for example, picrofuchsin.

2.7.6.3   Methyl Green Method

2.7.6.3.1   Principle
This method allows visualization of DNA alone. Methyl green stains DNA if the DNA is not depolymerized. Other basophilic substances, such as mucopolysaccharides or cartilaginous substances, are also stained.

2.7.6.3.2   Chemical Formula
See Figure 2.25.

2.7.6.3.3   Method
- Fixative—Carnoy's fluid is recommended but numerous other fixative fluids can be used. Practically all the classic fixative are convenient, but a short preservation to avoid depolymerization is recommended. Only a few hours in the fixative are necessary.
- Reagents—Methyl green: See Chapter 7, Sections 7.2.3.17 and 7.2.3.18.
- Protocol
  - Dewax, hydrate
  - Methyl green, 10 min
  - Dry the slide on paper filter

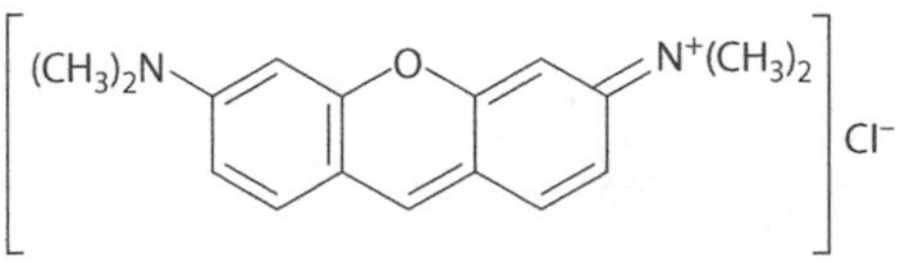

FIGURE 2.26 Pyronine.

- Dehydrate one slide at a time with two quick baths in butanol
- Cyclohexane, 10 min
- Mount
- Results—DNA is green stained. Mucopolysaccharides can also be green stained. Chromatin details are very well visualized. However, only nuclear staining permits the presence of DNA to be certified.

2.7.6.4 Pyronine Method

2.7.6.4.1 Principle

Pyronine is a basic dye that is highly RNA specific. It is generally used in Pappenheim–Unna's staining, associated to methyl green. It can also be used alone to visualize RNAs. The deletion of the staining after ribonuclease action shows that it has actual specificity.

2.7.6.4.2 Chemical Formula

See Figure 2.26.

2.7.6.4.3 Method

- Fixative—Among fixatives, Bouin's fluid can be avoided because it depolymerizes DNA. Practically all the classical fixatives are convenient, but a short preservation to avoid nucleic acid depolymerization is recommended. Only a few hours are necessary.
- Reagent—Pyronine: See Chapter 7, Section 7.2.3.25 or Section 7.2.3.26.
- Protocol
  - Dewax, hydrate
  - Pyronine, 10 min
  - Dry one slide at a time on paper filter
  - Dehydrate slide after slide by two quick passages in butanol
  - Cyclohexane, 10 min
  - Mount
- Results—RNAs are red stained. This stoichiometric method is very useful to appreciate all RNA molecules in cells. It can be useful in studying the evolution of RNA during embryonic development or as a pathological phenomenon.

2.7.6.5 Pappenheim–Unna's Staining

See Figure 2.27.

2.7.6.5.1 Principle

Pappenheim–Unna's staining technique is also called methyl green–pyronine staining. It permits one to stain simultaneously, and in a differentiated manner, RNA molecules in red and DNA in green. Caution! Although pyronine is an RNA specific dye, it differs from methyl green, which also reacts with mucus substances, granulations of mastocytes, and sulfate chondroitins in cartilage. However, nuclear staining will always permit one to verify the presence of DNA. Basic dyes, which are positively charged, mix with the negative components (acids) that are being investigated, in this case, nucleic acids. If the tissues are submitted to two basic dyes, the coloration obtained will

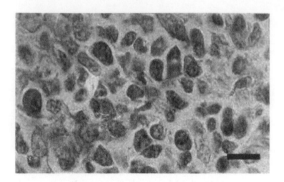

FIGURE 2.27 **(See color insert.)** Method: Pappenheim–Unna's staining or methyl green–pyronine. Tissue: Plasmocytes in connective tissue. Preparation: Paraffin. Fixative: Buffered formalin. Observations: DNA ion nuclei is purple stained and RNAs in cytoplasm are pink. Bar = 30 µm.

depend on the affinity of tissue or cell molecule for these dyes. These affinities will depend on pH. At pH 1.5, methyl green alone reacts and DNA alone is stained. At pH 9.3 and greater, pyronine alone reacts and RNA alone is stained.

In the present case, for a pH between 4 and 5, methyl green will stain DNA and pyronine will stain RNA molecules. The addition of both dyes into nuclei will stain them purple.

2.7.6.5.2   Protocol

- Fixative—Carnoy's fluid is recommended. Acidic fixatives must be avoided. However, such fixatives as formalin or Bouin's fluid can be used. Practically all the classic fixatives are convenient, but short preservation is recommended to avoid nucleic acid depolymerization. Only a few hours are necessary.
- Reagents—Methyl green–pyronine. See Chapter 7, Section 7.2.3.19 or Section 7.2.3.20.
- Protocol
  - Dewax, hydrate
  - Methyl green–pyronine, 10 min
  - Dry one slide at a time on paper filter
  - Dehydrate slide after slide by two quick passages in butanol
  - Cyclohexane, 10 min
  - Mount
- Results—Nuclei DNA and RNA are purple-blue stained. In cytoplasm, RNA is pinkish. Acidic mucopolysaccharides can be purple stained. The Pappenheim–Unna staining method is very useful to appreciate the evolution of nucleic acids during the differentiation of embryonic tissues or the evolution of certain cell types (sperm, for example).

2.7.6.6   Controls

2.7.6.6.1   Ribonuclease Brachet's Test

- Fixative—Carnoy's fluid is recommended. Acidic fixative must be avoided. However, such fixatives as formalin or Bouin's fluid can be used. Practically all the classical fixatives are convenient, but a short preservation is recommended to avoid nucleic acid depolymerization. Only a few hours are necessary.
- Reagents
  - Crystallized ribonuclease 0.01% in distilled water
  - Methyl green–pyronine: See Chapter 7, Section 7.2.3.19 or 7.2.3.20.
- Protocol
  - Prepare three groups of deparaffined slides, which are neither collodioned nor hydrated
  - One group is treated with ribonuclease 1 h at 37°C

- • Wash with tap water
- • A second group is treated with distilled water 1 h at 37°C
- • Stain the three groups with methyl green-pyronine. Perform reaction according to usual protocol.
- • Dry one slide at a time on paper filter
- • Dehydrate slide after slide by two quick passages in butanol
- • Cyclohexane, 10 min
- • Mount
- • Results—A pure green staining of DNA must be observed on the slide treated by ribonuclease.

2.7.6.6.2   RNA Extraction by Hydrochloric Acid

- • Fixative—Carnoy's fluid is recommended. Acidic fixative must be avoided. However, such fixatives as formalin or Bouin's fluid can be used. Practically all the classical fixatives are convenient, but a short preservation is recommended to avoid nucleic acid depolymerization. Only a few hours are necessary.
- • Reagents—Normal solution of hydrochloric acid.
- • Protocol
  - • Prepare three groups of deparaffined slides that are neither collodioned nor hydrated
  - • Treat one group with hydrochloride acid, 10 min at 60°C
  - • Treat the second group with distilled water, 0 min at 60°C
  - • Stain the three groups with methyl green–pyronine, 10 min. Perform reaction according to usual protocol.
  - • Dry one slide at a time on paper filter
  - • Dehydrate slide after slide by two quick passages in butanol
  - • Cyclohexane, 10 min
  - • Mount
- • Results—A pure green staining of DNA must be observed on the slide treated by ribonuclease.

2.7.6.7   Mann–Dominici's Staining

The Mann–Dominici's staining method is not generally used as a histochemical method; however, it is based on the use of a basic and metachromatic dye, toluidine blue. This method is useful to gauge specific histochemical reactions to analyze tissue components. This method can be useful to detect nucleic acids.

- • Fixative—All classical fixatives are convenient but oxidizing fixatives containing potassium dichromate or osmium tetroxide should be avoided.
- • Reagents
  - • Erythrosine–G orange: See Chapter 7, Section 7.2.2.11.
  - • Toluidine blue: See Chapter 7, Section 7.2.3.31.
  - • Potassium permanganate ($KMnO_4$) in aqueous solution 0.25%. Potassium permanganate solution is prepared from a stock solution at 2.5%.
  - • Sodium bisulfite (or metabisulfite) 2%
  - • Acetic water 0.25%
- • Protocol
  - • Dewax, hydrate
  - • Potassium permanganate, 30 s
  - • Rinse with distilled water
  - • Sodium bisulfate, 1 min
  - • Wash with tap water
  - • Rinse with distilled water. Immerse slides in acetic water to obtain generalized purple staining. Control under microscope.

- Toluidine blue, 1 min
- Rinse with distilled water
- Acetic water
- Ethanol 95%. At this stage, toluidine blue differentiation continues. Stop the reaction under microscope control.
- Ethanol 100%. Ethanol 100% stops the differentiation.
- Continue to dehydrate
- Mount
- Results—Nuclei, basophilic cytoplasm, and several secretions are purple-blue stained. Acidophilic cytoplasm, nucleoli, and several secretions are pinkish. Proteinic secretions and pigments are green-blue. Metachromatic mucus is purple stained.

2.7.6.8 Semithin Section Staining by Toluidine Blue
See Chapter 1, Section 1.4.8.

2.7.6.9 Love and Liles's, and Love and Suskind's Methods
2.7.6.9.1 Principle
Phosphoric groups that are bound to nucleoprotein amine groups are released by nitrous acid or formaldehyde. Toluidine blue is then fixed on these phosphoric groups. Toluidine blue then reacts with molybdate and yields a metachromatic reaction with color varying as a function of the nature of the nucleic acid nature (DNA or RNA).

2.7.6.9.2 Methods for Paraffin Sections
The reaction is based on comparing the results obtained with sections submitted to sublimate formalin and those not submitted to it (mordant) and the results to those submitted to nitrous deamination.

- Fixative—Sublimate–formalin is recommended.
- Reagents
  - Lugol: See Chapter 7, Section 7.2.3.15.
  - Sodium hyposulfite
  - Nitrous acid
  - Toluidine blue 0.01%: See Chapter 7, Section 7.2.3.31.
  - Ammonium molybdate 15%
  - Sublimate–formalin
- Protocol
  - First group of slides. The first group of sections gives results for tissues that are deaminated by nitrous acid.
    - Dewax, hydrate
    - Tap water, 5 min
    - Lugol, 5 min
    - Sodium hyposulfite, 5 min
    - Tap water, 5 min
    - Nitrous acid, 18 h
    - Tap water, 5 min
    - Toluidine blue 0.01%, 30 min
    - Ammonium molybdate 15%, 15 min
    - Tap water, 10 sec
    - Dehydrate. Dehydration must be done with tertiary butanol (2 methyl propane-2 ol).
    - Mount. Mounting with a "Permount" medium is recommended.

- Second group of sections. The second group of sections gives results for tissues that are deaminated by formaldehyde.
  - Sublimate–formalin, 2, 3, and 4 h
  - Dewax, hydrate
  - Tap water, 5 min
  - Lugol, 5 min
  - Sodium hyposulfite, 5 min
  - Tap water, 5 min
  - Nitrous acid, 18 h
  - Tap water: 5 min
  - Toluidine blue 0.01%, 30 min
  - Ammonium molybdate 15%, 15 min
  - Tap water, 10 s
  - Dehydrate. Dehydration must be done by tertiary butanol (2 methyl propane-2 ol).
  - Mount. Mounting by a "Permount" medium is recommended.
- Third group of sections. The third group of sections gives results for tissues that are not deaminated.
  - Dewax, hydrate
  - Tap water, 5 min
  - Lugol, 5 min
  - Sodium hyposulfite, 5 min
  - Tap water, 5 min
  - Toluidine blue 0.01%, 30 min
  - Ammonium molybdate 15%, 15 min
  - Tap water, 10 s
  - Dehydrate. Dehydration must be done by tertiary butanol (2 methyl propane-2 ol).
  - Mount. Mounting by a "Permount" medium is recommended.
- Results—DNA and RNA molecules are stained with several intensities with toluidine blue. Controls can be done by ribonuclease action or by comparison with other staining as the Feulgen reaction. In the original method, the authors distinguish three staining stages. Stage I staining is given without deamination. Stage II staining is given after sublimate–formalin deamination between 2, 3, or 4 hours. Stage III is given after nitrous acid action.

2.7.6.9.3 Method for Smears

- Fixative—Sublimate–formalin is recommended.
- Reagents
  - Lugol: See Chapter 7, Section 7.2.3.15.
  - Sodium hyposulfite
  - Nitrous acid
  - Toluidine blue 0.004% and 0.01%
  - Ammonium molybdate 15%
  - Sublimate–formalin
- Protocol
  - First smear. The first smear gives results for tissues that are deaminated by nitrous acid.
    - Sublimate–formalin, 1 h at 37°C
    - Tap water, 5 min
    - Lugol, 5 min
    - Sodium hyposulfite, 5 min
    - Tap water, 5 min
    - Nitrous acid, 18 h

- Tap water, 5 min
- Toluidine blue 0.004%, 30 min
- Ammonium molybdate 15%, 15 min
- Tap water, 10 s
- Dehydrate. Dehydration must be done by tertiary butanol (2 methyl propane-2 ol).
- Mount. Mounting by a "Permount" medium is recommended.
- Second smear. The second smear gives results for tissues that are deaminated by formaldehyde.
  - Sublimate–formalin, 5 and 10 min at 37°C
  - Tap water, 5 min
  - Lugol, 5 min
  - Sodium hyposulfite, 5 min
  - Tap water, 5 min
  - Nitrous acid, 18 h
  - Tap water, 5 min
  - Toluidine blue 0.01%, 30 min
  - Ammonium molybdate 15%, 15 min
  - Tap water, 10 s
  - Dehydrate. Dehydration must be done by tertiary butanol (2 methyl propane-2 ol).
  - Mount. Mounting by a "Permount" medium is recommended.
- Third smear. The third smear gives results for tissues that are not deaminated.
  - Sublimate formalin, 1 and 2 h at 37°C
  - Tap water, 5 min
  - Lugol, 5 min
  - Sodium hyposulfite, 5 min
  - Tap water, 5 min
  - Toluidine blue 0.01%, 30 min
  - Ammonium molybdate 15%, 15 min
  - Tap water, 10 s
  - Dehydrate. Dehydration must be done by tertiary butanol (2 methyl propane-2 ol).
  - Mount. Mounting by a "Permount" medium is recommended.
- Results—DNA and RNAs are stained with several intensities by toluidine blue. Controls can be done by ribonuclease action or by comparison with other staining as the Feulgen reaction.

2.7.7 Feulgen and Rossenbeck's Nuclear Reaction

2.7.7.1 Principle

2.7.7.1.1 General Principle

"Nuclear reaction" should be distinguished from "plasmal reaction." The Feulgen and Rossenbeck nuclear reaction was published in its definitive form in 1924. The method consists of submitting DNA to an acidic hydrolysis that exclusively reacts with puric bases and deoxyribose binding. This releases aldehydes, which are then stained by Feulgen's reagent or a similar substance. The reaction is done in two steps: first, hydrolysis of DNA to release an aldehyde and sugars and, second, detection of the aldehyde by a "Feulgen-Schiff" reagent type. This method is DNA specific and cannot be applied to RNA. The Feulgen and Rossenbeck plasmal reaction is a method used to visualize acetalphosphatide molecules (lipids) in the cytoplasm of some cell types. See Section 2.4.8.3.

2.7.7.1.2 Chemical Reaction of Acidic Hydrolysis

The chemical reactions are not precisely known but one mechanism is currently admitted. Hydrolysis only affects deoxyribose bound to a purine base, which yields a DNA-specific hydrolysis

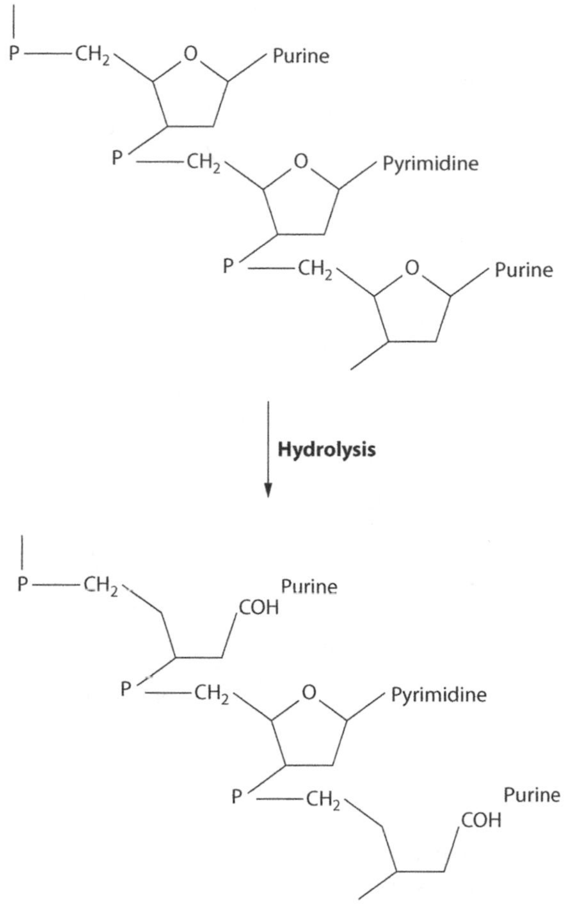

FIGURE 2.28 Feulgen hydrolysis reaction.

(Figure 2.28). Unstained or pale yellow-stained Schiff's reagent reacts with aldehyde to yield a red-stained product.

2.7.7.1.3   Schiff's Staining
See Figure 2.29.

2.7.7.2   Schiff's Reagent
Schiff's reagent, also called Schiff's leucofuchsin, is a mixture of basic fuchsin and sulfuric acid.

- Preparation: See Chapter 7, Section 7.2.3.28.
- Result—A pale yellow fluid must be obtained.

2.7.7.3   Histochemical Practice
2.7.7.3.1   Preservation
Tissue preservation for visualizing DNA by the Feulgen and Rossenbeck method is particularly important because the reaction protocol depends on the nature of the fixative. The fixative must provide the best morphological, cytological, or histological preservation possible. But on the other hand, the duration of the hydrolysis, which is fundamental for visualization of DNA, also depends on the nature of the fixative. Certain fixatives, such as acidic fluids, must be avoided. They induce the beginning of hydrolysis. If hydrolysis is too strong, DNA molecule degradation will continue

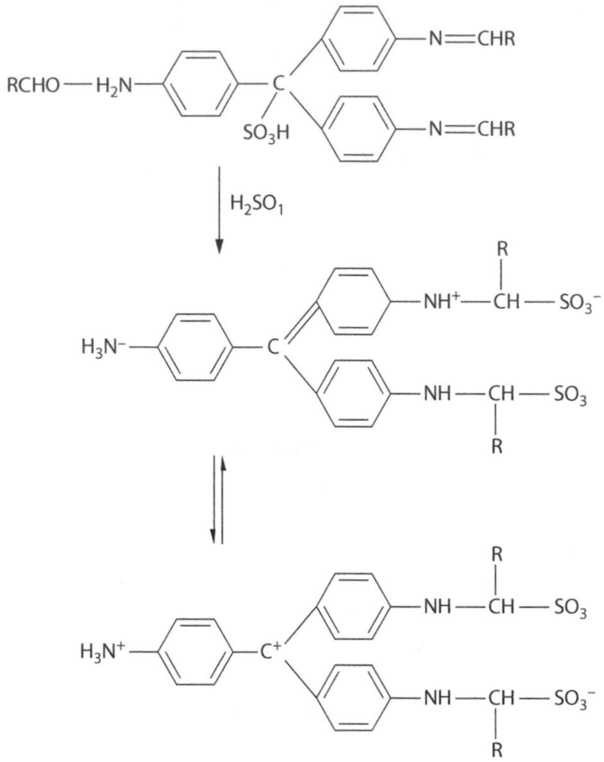

FIGURE 2.29   Feulgen's reaction.

beyond separation of puric bases and deoxyribose. That will induce total degradation of the DNA molecule, making it impossible to visualize.

Numerous studies concern the fixative and pH effects. It is now possible to realize a Feulgen and Rossenbeck reaction on tissues preserved with very different fixatives. Therefore, it is necessary to modulate the duration of the hydrochloric acid action.

Studies dealing with the effects of preservation have been done with Bouin's fluid. This fixative must be avoided because it is acidic and can inhibit Feulgen and Rossenbeck's reaction. However, the true effects of this fixative are not known. According to Gabe (1968), Bouin's fluid could induce a high level of DNA polymerization by binding with hydrolysis sites. This effect would prevent hydrochloric acid action from liberating aldehyde groups.

On the contrary, according to Ganter and Jolles (1969), Bouin's fluid begins to hydrolyze DNA molecules that would then be strongly degraded by a too lengthy acidic hydrolysis.

However, it is still possible to have good results with a Bouin's fluid preservation.

2.7.7.3.2   Hydrolysis

- Duration—In the classic method, acidic hydrolysis is done by hydrochloric acid M at 60°C. It is also possible to use hydrochloric acid 5 M at room temperature (20°C). This method can be very useful to obtain very precise visualization of chromatin details. The duration of hydrolysis is an essential factor in obtaining a good result. If the hydrolysis duration is very short, puric base and deoxyribose separation is incomplete and aldehydes are not released or they are released in very tiny quantities. If the hydrolysis duration is very long, DNA depolymerization continues. When an organ is studied for the first time, it is necessary to conduct a series of tests to determine the optimal duration of hydrolysis. When a Feulgen and Rossenbeck reaction is rigorously done, results can be studied by quantitative analysis, because the reaction is stoichiometric. Molecules with aldehyde functions are lost because, separated from the whole molecule, they go into solution and are weakly stained. In some cases, there is no reaction. An optimal duration must be strictly respected for the hydrolysis. Spectrophotometric studies have shown that hydrolysis reaction is different according to the studied tissue.

- Duration of hydrolysis in function of fixative
  - Bouin's fluid, 2 min
  - Champy's fluid, 25 min
  - Carnoy's fluid, 8 min
  - Formalin, 8 min
  - Sublimate–formalin, 8 min
  - Ethanol 100%, 5 min
  - Flemming's fluid, 16 min
  - Helly's fluid, 8 min
  - Heidenhain's Susa, 18 min
  - Zenker's fluid, 5 min
- Temperature—Acidic reaction is generally done at 60°C to accelerate the reaction. However, several authors have tried to modify this protocol to limit the effects of a high temperature on the sections. It is also possible to use low-temperature hydrolysis.
- Alternative methods—Hydrolysis can be done with hydrochloric acid in ethanol 100% solution and at 60°C. In this case, the small DNA molecules and their binding with proteins are preserved.
- In another method, hydrolysis is done at room temperature, with hydrochloric acid 5 M. It is also possible to proceed to a slow hydrolysis using hydrochloric acid at pH 1.2 and at 37°C.
- Other acids—Acids other than hydrochloric acid can be used: citric acid, perchloric acid, phosphoric acid, chromic acid (chrome trioxide), or sulfuric acid.
- Stopping reaction—Reaction can be stopped with cold water.

2.7.7.4 Protocol

See Figure 2.30.

2.7.7.4.1 General Protocol

- Fixative—Carnoy's or Flemming's fluids are often recommended, but it is possible to use numerous other fixatives. It is necessary to determine the optimal duration of hydrolysis for each. For determination of hydrolysis duration, see Section 2.7.7.3.2.
- Reagents
  - Schiff's reagent: See Chapter 7, Section 7.2.3.28.
  - Hydrochloric acid M

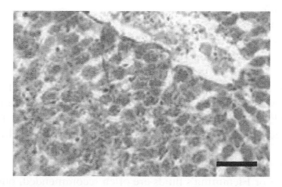

FIGURE 2.30 (See color insert.) Method: Feulgen and Rossenbeck's staining. Tissue: Mouse spleen. Preparation: Paraffin section. Fixative: Bouin's fluid. Observations: Nuclei are red stained. Bar = 30 μm.

- Sulfurous water or sodium metabisulfite 10%
  - Metabisulfite, 10 mL
  - Distilled water, 190 mL
- Protocol
  - Dewax, hydrate
  - Hydrochloric acid, 8 min at 60°C. This time given is for tissues preserved with Carnoy's fluid. The duration may differ with another fixative.
  - Tap water, 8 min
  - Schiff's reagent, 1 h
  - Tap water, 5 min
  - Rinse in sulfurous water, 3 × 1 min
  - Tap water, 5 min
  - Dehydrate
  - Mount
- Results—DNA is red stained. The Feulgen and Rossenbeck's reaction is stoichiometric and thus useful for DNA quantification by automatic methods.

2.7.7.4.2   First Variant

- Fixative—Carnoy's or Flemming's fluids are often recommended, but it is possible to use numerous other fixatives. It is necessary to determine the optimal duration of hydrolysis for each.
- Reagents
  - Schiff's reagent: See Chapter 7, Section 7.2.3.28.
  - Hydrochloric acid M
  - Picroindigocarmine: See Chapter 7, Section 7.2.2.7.
  - Sulfurous water or sodium metabisulfite 10%
    - Metabisulfite, 10 mL
    - Distilled water, 190 mL
- Protocol
  - Dewax, hydrate
  - Hydrochloric acid, 8 min at 60°C. This time given is for tissues preserved with Carnoy's fluid. The duration may differ with another fixative. For determination of hydrolysis duration, see Section 2.7.7.3.2.
  - Tap water, 8 min
  - Schiff's reagent, 1 h
  - Tap water, 5 min
  - Rinse in sulfurous water, 3 × 1 min
  - Tap water, 5 min
  - Picroindigocarmine, 30 s
  - Tap water, 5 min
  - Dehydrate directly by ethanol 100%
  - Mount
- Results—DNA is red stained. Acidophilic cytoplasm is yellow or green; collagen is blue; red blood cells are yellow stained; and glycoproteins are brown. Secretions can be yellow or green stained. Feulgen and Rossenbeck's reaction is stoichiometric and thus useful for DNA quantification by automatic methods. However, bottom staining can hinder quantification.

2.7.7.4.3   Second Variant

- Fixative—Carnoy's or Flemming's fluids are often recommended, but it is possible to use numerous other fixatives. It is necessary to determine the optimal duration of hydrolysis for each. For determination of hydrolysis duration, see Section 2.7.7.3.2.

- Reagents
  - Schiff's reagent: See Chapter 7, Section 7.2.3.28.
  - Hydrochloric acid 5 M
  - Picroindigocarmine: See Chapter 7, Section 7.2.2.7.
  - Sulfurous water or sodium metabisulfite 10%
    - Metabisulfite, 10 mL
    - Distilled water, 190 mL
- Protocol
  - Dewax, hydrate
  - Hydrochloric acid, 8 min. This time given is for tissues preserved with Carnoy's fluid. The duration may differ with another fixative.
  - Tap water, 5 min
  - Schiff's reagent, 1 h
  - Tap water, 5 min
  - Rinse in sulfurous water, 3 × 1 min
  - Tap water, 5 min
  - Picroindigocarmine, 30 s
  - Tap water, 5 min
  - Dehydrate directly by ethanol 100%
  - Mount
- Results—DNA is red stained. Acidophilic cytoplasm is yellow or green; collagen is blue; red blood cells are yellow stained; and glycoproteins are brown. Secretions can be yellow or green stained. Feulgen and Rossenbeck's reaction is stoichiometric and thus useful for DNA quantification by automatic methods. However, bottom staining can hinder quantification.

2.7.7.5 Alternative Methods to Feulgen and Rossenbeck's Reaction

2.7.7.5.1 General Principles

Some methods based on the same principle can replace the original Feulgen and Rossenbeck reaction. All these methods are based on an acidic hydrolysis that separates puric bases and deoxyribose, releasing aldehyde groups. The difference between the reactions is generally linked to the methods used to visualize aldehydes.

2.7.7.5.2 Thionine–SO_2 Method

This method is also called De Lamater's method. In this technique, a 0.25% thionine–SO_2 solution is used to replace Schiff's reagent. Thionine–SO_2 can be replaced with azure A–SO_2.

- Fixative—Carnoy's or Flemming's fluids are recommended.
- Reagents
  - Thionine: See Chapter 7, Section 7.2.3.30. Before use, add one drop of thionine chloride for 10 mL thionine or for 5 mL azure A.
  - Hydrochloric acid M
  - Sulfurous water or sodium metabisulfite 10%
    - Metabisulfite, 10 mL
    - Distilled water, 190 mL
  - Van Gieson's picrofuchsin: See Chapter 7, Section 7.2.2.20.
- Protocol
  - Dewax, hydrate
  - Hydrochloric acid, 4 min. This time given is for tissues preserved with Carnoy's fluid. The duration may differ with another fixative.
  - Tap water, 4 min

- Thionine–thionine chloride, 1 h
- Rinse with sulfurous water
- Van Gieson's picrofuchsin, 30 s
- Dehydrate directly with ethanol 100%
- Mount
- Results—DNA is deep blue stained. Acidic substances are metachromatic. This method is stoichiometric. The blue-stained DNA can be easily quantified by automatic image analysis.

2.7.7.5.3   Himes and Moriber's Method

In this method, nucleal reaction and PAS are associated.

- Fixative—Carnoy's, Flemming's.
- Reagents
  - Azure A–SO$_2$: See Chapter 7, Section 7.2.3.4.
  - Whiting fluid. Prepare at time of use:
    – Potassium (sodium) metabisulfite 5%, 5 mL
    – Hydrochloric acid M, 5 mL
    – Water, 90 mL
  - Hydrochloric acid M
  - Periodic acid
  - Schiff's reagent: See Chapter 7, Section 7.2.3.28.
  - Naphthol yellow S: See Chapter 7, Section 7.2.3.21.
- Protocol
  - Dewax, hydrate
  - Hydrochloric acid, 4 min at 60°C. This time given is for tissues preserved with Carnoy's fluid. The duration may differ with another fixative.
  - Wash with distilled water
  - Azure A–SO$_2$, 5 min
  - Rinse with distilled water
  - Whiting solution: 2 × 2 min
  - Periodic acid, 2 min
  - Rinse with distilled water
  - Schiff's reagent, 2 min
  - Whiting solution, 2 × 2 min
  - Naphthol yellow S, 2 min
  - Rinse with tap water
  - Dry with paper filter
  - Dehydrate. Tertiary butanol (2-methylpropane-2 ol) is recommended.
  - Mount
- Results—Nuclei are blue or green stained. PAS positive components are red and basic proteins are yellow.

2.7.7.5.4   Benson's Method

- Fixative—Carnoy's or Flemming's fluids are recommended. In the original method, preservation is done with formalin 10% added to 0.5% cetylpyridinium chloride.
- Reagents
  - Azure A–SO$_2$: See Chapter 7, Section 7.2.3.4.
  - Alcian blue pH 2: See Chapter 7, Section 7.2.3.2.
  - Hydrochloric acid 5 M and 0.01 M

- • Periodic acid
- • Schiff's reagent: See Chapter 7, Section 7.2.3.28.
- • Naphthol yellow S: See Chapter 7, Section 7.2.3.21.
- • Acetic acid 1%
- • Protocol
  - • Dewax, hydrate
  - • Hydrochloric acid, 9 min at RT
  - • Tap water, 1 min. Washing can be prolonged.
  - • Azure A–SO$_2$, 10 min
  - • Whiting solution, 2 × 2 min
  - • Tap water, 1 min. Washing can be prolonged.
  - • Alcian blue, 10 min.
  - • Hydrochloric acid 0.01 M, 1 min. Hydrochloric acid hydrolysis must be precisely adapted to avoid interferences between DNA blue staining by azure A and the blue alcian staining.
  - • Tap water, 1 min. Washing can be prolonged.
  - • Periodic acid, 5 min
  - • Tap water, 1 min. Washing can be prolonged.
  - • Whiting solution, 2 × 2 min
  - • Tap water, 1 min. Washing can be prolonged.
  - • Naphthol yellow S, 2 min
  - • Acetic acid 1%, 2 min
  - • Dry with paper filter
  - • Dehydrate
  - • Mount
- • Results—Nuclei are blue or green stained. PAS positive components are red, basic proteins are yellow, and acidic carbohydrates are blue.

2.7.7.5.5   Silver Methenamine Method

This method is also called Korson's method.

2.7.7.5.5.1   Principle
With a reducing substance, silver methenamine is reduced with metal silver deposit on the reducing group. For DNA visualization, the aldehyde group obtained after acidic hydrolysis is the reducing substance and it is visualized by a deposit of silver. Silver methenamine can be replaced by other silver salt as tetramine hexaethylene. These silver salts can also be used to visualize glucides by the PAS method during which aldehyde groups are formed from glycols belonging to the glucide molecule.

During the hydrochloride acid hydrolysis that is used for Feulgen and Rossenbeck's reaction, deoxyribose is diffused even before reducing reaction. It would be difficult to obtain a concluding result. To avoid this diffusion, the reaction is done with citric acid.

Aldehydes are visualized by silver and are black stained. The precision of the black staining is very strong. This method can be used to visualize DNA in transmission electron microscopy.

2.7.7.5.5.2   Protocol

- • Fixative—Neutral formalin and Carnoy's fluid fixed organs, freezing sections, smears, and chromosome preparations. All the usual fixatives are convenient.
- • Reagents
  - • Citric acid M. Warm citric acid before each use.
  - • Silver methenamine: See Chapter 7, Section 7.2.3.29.
  - • Gold chloride 0.2%

- Protocol
  - Dewax, hydrate
  - Citric acid 1 M, 30 min 60°C. For frozen section and smears, begin hydrolysis directly with citric acid. Reaction duration can vary as a function of the fixative. It is necessary to conduct a series of tests to determine the optimal duration.
  - Distilled water, 5 min
  - Silver methenamine, 1 h at 60°C
  - Rinse with distilled water
  - Gold chloride 0.2%, 5 min. The use of gold chloride is optional.
  - Rinse with distilled water
  - Dehydrate
  - Mount
- Results—DNA is black stained.

2.7.8 FLUORESCENT METHODS

2.7.8.1 Use of Fluorescent Dyes

Nucleic acids can be visualized by use of fluorochromes. Some of these methods have been perfected for examining nucleic acids with light microscopes. This chapter discusses a few of these staining methods. The development of flux cytometry necessitated the use of numerous fluorescent dyes. Their use has often been adapted for microscopic examination.

2.7.8.2 Fluorescent Dyes

Several types of fluorescent dyes can be defined as a function of their association with a nucleic acid.

2.7.8.2.1 Intercalating Fluorescent Dyes

These molecules are fixed between the two DNA strands or within an RNA loop. Acridine orange, ethidium chloride or bromide, and coriphosphine O belong in this category.

2.7.8.2.2 Feulgen–Schiff-Like Fluorescent Dyes

Several fluorescent dyes can be used, such as Schiff's reagent on aldehyde functions that are obtained after DNA acidic hydrolysis. Acriflavine or auromicine belong to that category.

2.7.8.2.3 Pair-Base Specific Fluorescent Dyes

Several fluorescent dyes react with specific bases pairs. Some of them are specifically intercalated between guanine and cytosine, for example, DAPI, (4′,6-diaminido-2-phenylindole), DIPI (4′,6-diaminido-2-imidazolinyl-^{4}H-^{5}H), and Hoechst 33258. The use of Hoescht dye permits visualization of nucleic acids belonging to mycoplasma when they infect cell cultures.

Others intercalate between adenine and thymine. They are fluorescent antibiotic molecules, such as chromomycin A3, mithramycin, or still olivemycin.

2.7.8.2.4 Interest to Use a Fluorescent Dye

One of the advantages of using a fluorescent dye is to achieve a stoichiometric staining. In certain cases, these dyes allow one to visualize nucleic acids that cannot be observed with other methods.

2.7.8.3 Acridine Orange Staining

2.7.8.3.1 Mechanism

Acridine orange is a basic dye belonging to the acridine family. It lends a green fluorescence to live cells and a red fluorescence to dead cells. Several studies have shown a variability of fluorescence that is associated with the fixative.

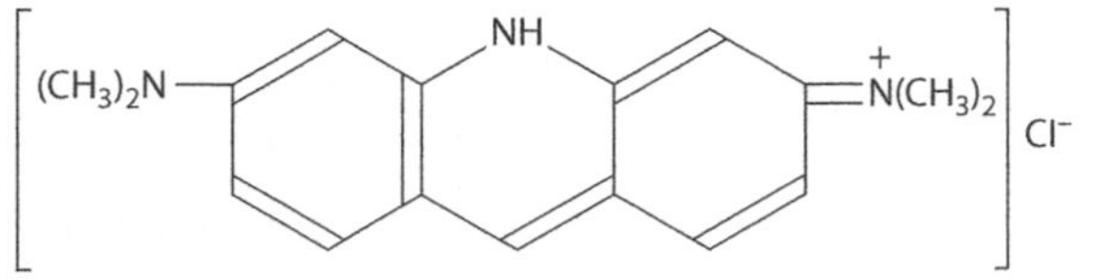

FIGURE 2.31 Acridine orange.

In preserved cells, the nucleus is orange-red stained, and the cytoplasm is slightly red. The other structures are green or yellow fluorescent.

On a fresh tissue, DNA is green fluorescent, and nucleus and cytoplasm RNA is red. Other structures can also be red stained. On a fresh tissue, RNase action prevents one to obtain a red fluorescence.

2.7.8.3.2 Formula
See Figure 2.31.

2.7.8.3.3 Effects of External Factors
The staining depends upon pH and dye solution concentration. In certain cases, nucleic acid fluorescence is obtained for a pH between 1.5 and 3.5. In other conditions, the pH must be between 3.5 and 5.

2.7.8.3.4 Action
Acridine orange reacts by intercalation between the two DNA strands or into a RNA loop, by salt binding or Van Der Waals forces. The dye absorbs at 520 nm wavelength and the emission is variable as a function of the fresh or preserved state of the tissue.

2.7.8.3.5 Protocol
- Preservation—All fixatives are convenient. It is also possible to use nonpreserved fresh tissues, smears, or cell cultures. As a function of the tissue preparation, the staining method varies.
  - For smears use ethanol–ether (1/1), 30 min
  - For paraffin section use a mixing solution with ethanol
- Reagents
  - Acetic acid 1%
  - Acridine orange 0.1%
  - Phosphate buffer M/15, pH 6
  - Calcium chloride M/10
- Method
  - Hydrate. For smears:
    - Ethanol 80%, 10 s
    - Ethanol 70%, 10 s
    - Ethanol 50%, 10 s
    - For sections hydrate as usual
  - Acetic acid 1%, 6 s
  - Distilled water, 2 × 3 s
  - Acridine orange 1%, 3 min
  - Phosphate buffer M/15, 1 min
  - Calcium chloride, 30 s
  - Mount with phosphate buffer pH 6. Preparation mounting can also be done in a mounting medium without fluorescence.
- Results—DNA emits a green fluorescence, RNA, a red one.

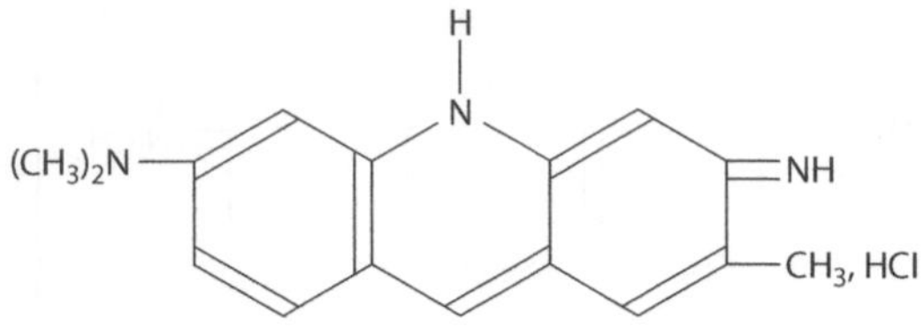

FIGURE 2.32 Coriphosphine.

2.7.8.4 Coriphosphine O Staining

2.7.8.4.1 Mechanism

Coriphosphine O is an acidic dye belonging to the acridine family. Under certain conditions, this dye yields results that are comparable with those given by acridine orange. It is not sensitive to an increased action of ultraviolet light. This dye gives a green fluorescence to DNA. Cytoplasmic RNAs are copper colored and nucleus RNA is orange fluorescent.

2.7.8.4.2 Formula

See Figure 2.32.

2.7.8.4.3 Action

Coriphosphine O is an intercalating dye that is fixed between two DNA strands or into RNA loops by salt binding or van der Waals forces. The absorption wavelength is 520 nm.

2.7.8.4.4 Protocol

- Preservation—All classic fixatives are convenient. It is also possible to use smears and cell cultures. In the original method, Carnoy's fluid is recommended for smears and also pieces to be embedded.
- Reagents
  - Phosphate buffer M/15, pH 7.1 with 0.2% phenol. Buffer can be replaced by PBS from a commercial source.
  - Coriphosphine O
- Method
  - Hydrate
  - Phosphate buffer pH 7.1, 5 min
  - Coriphosphine O, 5 min
  - Phosphate buffer pH 7.1, 10 s
  - Dry into glycerin
  - Mount into a medium without fluorescence. In the original method, mounting is done into liquid paraffin.
- Results—DNA is green fluorescent, cytoplasm RNA is copper fluorescent, and nucleus RNA is orange fluorescent.

2.7.8.5 Propidium Iodide Staining

2.7.8.5.1 Principle

Propidium iodide is an intercalating fluorescent dye. Its excitation wavelength is 370 and 560 nm. It emits a red fluorescence at 623 nm. Propidium iodide can be used with immunofluorescent reactions or with *in situ* hybridization using fluorescent probes. Ethidium bromide is also an intercalating fluorescent dye, similar to propidium iodide. It is excited at 370 and 530 nm, and it emits a red fluorescence at 622 nm.

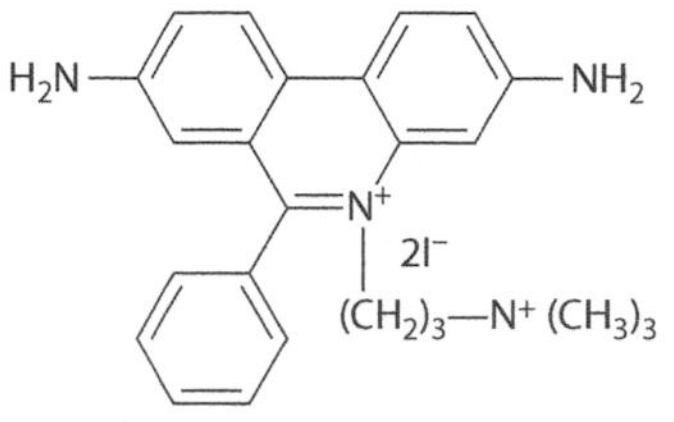

FIGURE 2.33 Propidium iodide.

2.7.8.5.2 Formula
See Figure 2.33.

2.7.8.5.3 Protocol
- Preservation—All the classic fixatives are convenient. The staining can be used on sections, smears, or cell cultures.
- Reagents
  - Propidium iodide, 1 g
  - PBS, 100 mL
  - Phosphate buffer 0.1 M
    - Monopotassium phosphate
      - Monopotassium phosphate, 1.36 g
      - Distilled water, 100 mL
    - Disodium phosphate 0.1 M
      - Disodium phosphate, 1.42 g
      - Distilled water, 100 mL
    - Phosphate buffer 0.1 M, pH 7.4
      - Monopotassium phosphate 0.1 M, 8 mL
      - Disodium phosphate 0.1 M, 42 mL
      - Distilled water, 100 mL
- Method
  - Dewax, hydrate
  - Propidium iodide, 30 min. Operate in darkness.
  - Mount without dehydration with a mounting medium without fluorescence
- Results—Nuclei and cytoplasm RNA are red fluorescent. Preparations can be preserved for several days or weeks by storing at −20°C.

2.7.8.6 Hoechst 33258 Staining
2.7.8.6.1 Principle
Hoechst 33258, a fluorescent dye that reacts by intercalating between adenine and thymine, is DNA specific. It is essentially used in flux cytometry and can be used in photonic microscopy to stain nuclei after a visualization with a fluorescent method. Excitation wavelength is 360 nm. Fluorescent staining is emitted at 470 nm.

2.7.8.6.2 Formula
See Figure 2.34.

2.7.8.6.3 DNA Visualization by Hoechst 33258
- Fixative—All the classic fixatives are convenient. Testing the effects of the fixative is recommended.
- Reagents

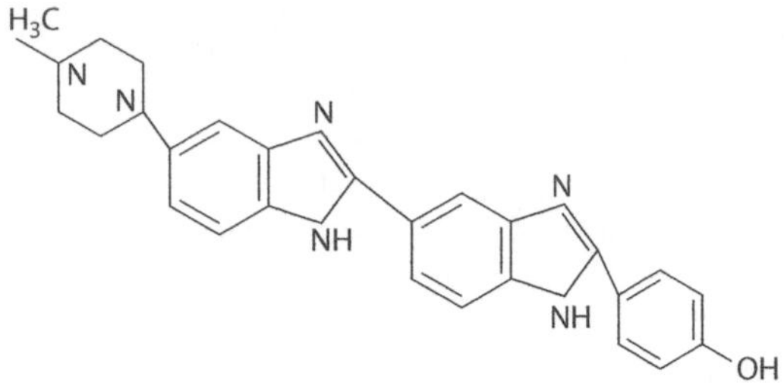

FIGURE 2.34 Hoechst 33258.

- Hoechst 33258. Stocking solution:
  - Sterile distilled water, 50 mL
  - Hoechst 33258, 2.5 mg
  - Thimerosal, 5 mg
- Hoechst 33258 dye. Working solution:
  - Stocking solution, 1 mL
  - Mac Ilvaine buffer pH 5.5, 10 mL
- Mac Ilvaine buffer
  - Citric acid 0.1 M, 42 mL. Preparation:
    - Citric acid, 2.1 g
    - Distilled water, 100 mL
  - Disodium phosphate 0.2 M, 58 mL
    - Disodium phosphate, 3.56 g
    - Distilled water, 100 mL
- Protocol
  - Dewax, hydrate
  - Hoechst 33258, 15 min. Working solution.
  - Distilled water, 2 × 5 min
  - Let air-dry in the dark
  - Mount the slide without dehydration with a glycerin buffer or a mounting medium without fluorescence.
- Results—Nuclei are green fluorescent. Preparations can be saved for several days or weeks at a temperature of –20°C.

2.7.8.6.4 *Visualization of Mycoplasma by Hoechst 33258*

See Figure 2.35 and Figure 2.36.

- Fixative—For cell cultures, Carnoy's fluid is recommended, although other fixatives can be used.
  - Use cell culture on coverslip
  - Carnoy's fluid, 2 min
  - Empty the fixative excess
  - Carnoy's fluid, 5 min
  - Empty the fixative excess
  - Let dry at 37°C
- Reagents
  - Hoechst 33258. Stocking solution:
    - Sterile distilled water, 50 mL
    - Hoechst 33258, 2.5 mg
    - Thimerosal, 5 mg

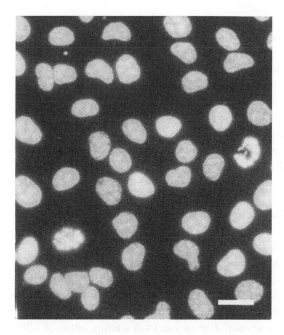

FIGURE 2.35   (See color insert.) Method: Hoechst 33258. Tissue: MRC5 cell strain without mycoplasma. Preparation: Cell culture. Fixative: Buffered formalin. Observations: Nuclei are green fluorescent. Bar = 10 μm.

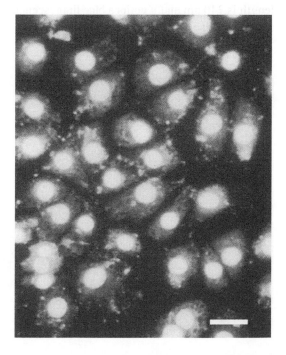

FIGURE 2.36   (See color insert.) Method: Hoechst 33258. Tissue: MRC5 cell strain with mycoplasmae. Preparation: Cell culture. Fixative: Buffered formalin. Observations: Nuclei are green fluorescent. Bar = 10 μm.

- Hoechst 33258 dye
  - Stocking solution, 1 mL
  - Mac Ilvaine buffer pH 5.5, 10 mL
- Mac Ilvaine buffer
  - Citric acid 0.1 M, 42 mL
    - Citric acid, 2.1 g
    - Distilled water, 100 mL
  - Disodium phosphate 0.2 M, 58 mL
    - Disodium phosphate, 3.56 g
    - Distilled water, 100 mL
- Protocol
  - Dewax, hydrate
  - Hoechst (working solution), 15 min
  - Distilled water, 2 × 5 min
  - Let air-dry in the dark.
  - Mount the slide without dehydration with a glycerin buffer or a mounting medium without fluorescence.
- Results—Cell nuclei are green fluorescent. Mycoplasma are visualized as small green points or drops on the cell, the cell membrane, and the space between the cells. Preparations can be conserved for several days or weeks at a temperature of –20°C.

2.7.8.7   Hoechst 33342 Staining

See Figure 2.37.

2.7.8.7.1   Principle

Hoechst 33342 is a fluorescent dye that intercalates between adenine and thymine. It is DNA specific. The excitation wavelength is 340 nm and it emits a blue fluorescence at 450 nm.

2.7.8.7.2   Formula

See Figure 2.38.

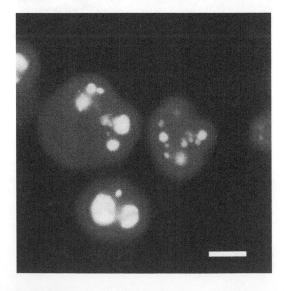

FIGURE 2.37   (See color insert.) Method: Hoechst 33342. Tissue: Cell culture with apoptosis. Preparation: Cell culture. Fixative: Buffered formalin. Observations: Nuclei and apoptotic bodies are blue fluorescent. Bar = 5 μm.

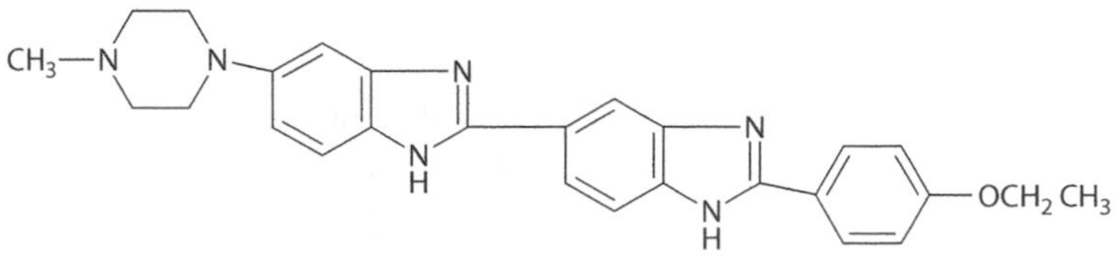

FIGURE 2.38 Hoechst 33342.

2.7.8.7.3 Protocol

- Fixative—All classic fixatives can be used. Staining can be done on sections, smears, and cell cultures. Testing the effects of the fixative is recommended.
- Reagents
  - Hoechst 33342 10^{-3} M, 1 g
  - PBS buffer, 100 mL
- Protocol
  - Dewax, hydrate
  - Hoechst 33342, 30 min. Operate in the dark.
  - Mount the slide without dehydration with a glycerin buffer or a mounting medium without fluorescence.
- Results—Cell nuclei are blue fluorescent. Preparations can be conserved for several days or weeks at a temperature of –20°C.

2.7.8.8 Quinacrine Mustard

2.7.8.8.1 Principle

The use of quinacrine mustard permits the visualization of X chromosome.

2.7.8.8.2 Method

- Fixative—Method is used on smears preserved with ethanol 95% or Carnoy's fluid.
- Reagents
  - Quinacrine
    - Quinacrine, 5 g
    - Distilled water, 100 mL
  - Citric acid
    - Citric acid 0.1 M, 1.92 g
    - Distilled water, 100 mL
  - Disodium phosphate
    - Disodium phosphate 0.2 M, 2.84 g
    - Distilled water, 100 mL
  - Phosphate citric acid buffer 0.01M, pH 5.5
    - Citric acid 0.1 M, 9 mL
    - Disodium phosphate 0.2 M, pH 5.5, 11 mL
  - Phosphate buffer 0.1 M
    - Monopotassium phosphate, 1.36 g
    - Distilled water, 100 mL
  - Disodium phosphate 0.1 M
    - Disodium phosphate, 1.42 g
    - Distilled water, 100 mL
  - Phosphate buffer 0.1 M, pH 7.4
    - Monopotassium phosphate 0.1 M, 8 mL
    - Disodium phosphate 0.1 M, 42 mL
    - Distilled water, 100 mL

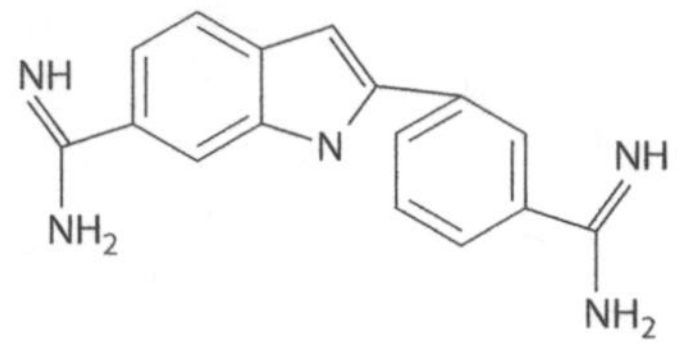

FIGURE 2.39   DAPI.

- Protocol
  - Ethanol 100%, 3 min
  - Ethanol 95%, 3 min
  - Ethanol 80%, 3 min
  - Ethanol 70%, 3 min
  - Ethanol 50%, 3 min
  - Distilled water, 3 min
  - Quinacrine 0.5%, 5 min
  - Distilled water, 2 × 3 min
  - Citric acid phosphate buffer pH 5.5, 3 min
  - Phosphate buffer pH 7.4, 2 × 3 min
  - Mount with phosphate buffer 0.1 M, pH 7.4. Lute with varnish.
- Results—The X chromosome is visualized as a green fluorescent point in the male cells' nuclei.

2.7.8.9   DAPI and DIPI

DAPI (4'6 diaminido-2 imidazolinyl $_4$H-$_5$H) is an intercalating fluorescent dye that intercalates itself between adenine and thymine (Figure 2.39). It is DNA specific. The excitation wavelength is 365 nm. It emits a blue fluorescent staining at 420 nm. These fluorescent dyes have been developed for flux cytometry, and they are rarely used to visualize nucleic acids with a photonic microscope. Only a brief description is given here.

DIPI (4'6 diaminido-2 phenylindole) is a fluorescent dye that intercalates itself between adenine and thymine. It is DNA specific. The excitation wavelength is 340 nm. It emits a green fluorescent staining at 465 nm.

REFERENCES

Adams, C.W.M. 1957. A p-dimethylaminobenzaldehyde-nitrite method for the histochemical demonstration of tryptophane and related compounds. *Journal of Clinical Pathology* 10:56–62.

Adams, C.W.M. 1959. Role of lipids of aortic elastic fibers in atherogenesis. *Lancet* 1:1075.

Adams, C.W.M. 1961. A perchloric acid—Naphthoquinone method for the histochemical localization of cholesterol. *Nature* 192:331–332.

Baker, J.R. 1947. The histochemical recognition of certain guanidine derivatives. *Quarterly Journal of Microscopical Science* 88:115–121.

Barnett, R.J., and A.M. Seligman. 1952. Histo-chemical demonstration of protein-bound sulfhydryl groups. *Science* 116:323–327.

Bayliss, O.B., and C.W.M. Adams. 1972. Bromine-Sudan Black: A general stain for lipids including free cholesterol. *Histochemical Journal* 4:505–515.

Chèvremont, M., and J. Frédéric. 1943. Une nouvelle méthode de mise en évidence des substances a fraction sulphydryle. *Archives de Biologie* 54:589–591.

Cleveland, R., and J.M. Wolfe. 1932. A differential stain for the anterior lobe of the hypophysis. *The Anatomical Record* 51:409–413.

Danielli, J.F. 1950. Studies on the cytochemistry of proteins. *Cold Spring Harbor Symposium of Quantitative Biology* 14:32–39.

De Lamater, E.D., and D. Ulrich. 1948. Basic fuchsin as a nuclear stain. *Stain Technology* 38:161–176.

Dunnigan, M.G. 1968a. Chromatographic separation and photometric analysis of the components of Nile blue sulphate. *Stain Technology* 43:243–248.

Dunnigan, M.G. 1968b. The use of Nile blue sulphate in the histochemical identification of phospholipids. *Stain Technology* 43:249–256.

Exbrayat, J.M. 2001. *Genome visualization by classic methods in light microscopy.* CRC Press, Boca Raton, FL.

Feulgen, R., and H. Rossenbeck. 1924. Mikorskopisch-chemischer nachweis einer nucleinsaure vom typus der thymonucleisaure und die darauf beruhende elektive Farbung vom zellkernen in mikroskopischen praparaten. *Zeitschrift für Physikalische Chemie* 155:203–248.

Feulgen, R., and K. Voit. 1924. Über den Mechanismus der Nuclealfarbung. *Zeitschrift für Physikalische Chemie* 135:249–252.

Gabe, M. 1968. *Techniques histologiques.* Masson et Cie, Paris.

Ganter, P., and G. Jolles. 1969. *Histochimie normale et pathologique*, 2 vol. Gauthier-Villars, Paris.

Himes, M., and L. Moriber. 1956. A triple stain for deoxyribonucleic acid, polysaccharides and proteins. *Stain Technology* 31:67–70.

Holczinger, L. 1959. Histochemischer Nachweis freier. Fettsauren *Acxta Histochemica* 8:167–175.

Hould, R. 1984. *Techniques d'histopathologie et de cytopathology.* Decarie, Montreal, Maloine, Paris.

Korson, R. 1956. A silver stain for deoxyribonucleic acid. *Journal of Histochemistry and Cytochemistry* 4:310–317.

Liebermann, C. 1885. Über das Oxychinoterpen. *Berichte der Deutschen Chemischen Gesellshaft* 18:1803–1885.

Lillie, R.D., and L.L. Ashburn. 1943. Supersaturated solutions of fat stains in dilute isopropanol for demonstration of acute fatty degeneration not shown by Herxheimer's technique. *Archives of Pathology* 36:432-440.

Lison, L. 1960. *Histochimie et cytochimie animals.* Gauthier-Villars, Paris.

Love, R., and R.H. Liles. 1959. Differentiation of nucleoproteins by inactivation of protein-bound amino groups and staining with toluidine blue and ammonium molybdate. *Journal of Histochemistry* 7:164–181.

Love, R., and R.G. Suskind. 1961. Further observations on the ribonucleoproteins of mitotically dividing mammalian cells. *Experimental Cell Research* 22:193–207.

Martoja, R., and M. Martoja-Pierson. 1967. *Initiation aux techniques de l'histologie animale.* Masson, Paris.

Michaelis, L. 1947. The nature of the interaction of nucleic acids and nuclei with basic dye stuffs. *Cold Spring Harbor Symposium of Quantitative Biology* 12:131–142.

Millon, E. 1849. Sur une réactif propre aux composés protéiques. *Comptes rendus des séances de l'Académie des Sciences de Paris* 28:40–42.

Morel, G., and A. Cavalier. 1998. *Hybridation in situ.* Polytechnica, Editions Economica.

Pappenheim, A. 1899. Vergleichende Untersuchungen iiber die elementare Zuzammensetzung des rothen Knochenmarkes einige Saugethiere. *Virchows Archive für Pathogogische Anatomie und für Kilinishche Medizin* 157:19–76.

Perls, M. 1867. Nachweis von Eisenoxyl in gewissen Pigmenten. *Virchows Archive für Pathogogische Anatomie und für Kilinishche Medizin* 39:42–48.

Sakaguchi, S. 1925. A new color reaction of protein and arginine. *Journal of Bichemistry* 5:25–31.

Schultz, A. 1924. Eine Methode des mikrochemichen Cholesterinnachweises am Gewebsschnitt. *Zentralbuch Allgemeine Pathologie und Pathologische Anatomie* 35:314.

Smith, C.A., and E.J. Wood. 1997. *Les biomolécules.* Masson, Paris.

Stoelzner, W. 1905. *Virchows Archive für Pathogogische Anatomie und für Kilinishche Medizin* 180:362.

Trump, B.F., E.A. Smuckler, and E.P. Benditt. 1961. A method or staining epoxy section for light microscopy. *Journal of Ultrastructural Research* 5:343–348.

Turchini, J., P. Castel, and K. Kien. 1944. Contribution à l'étude histochimique de la cellule, une nouvelle réaction nucléale. *Bulletin d'Histologie Appliquée* 21:124.

Unna, P.G. 1913. Die Herkunft der Plasmazellen. *Virchows Archive für Pathogogische Anatomie und für Kilinishche Medizin* 214:320–339.

Hamilton, M.G. (1968) Chromatographic separation and spectrum analysis of the components of Nile blue sulfate. *Stain Technology* **43**:237–239.

Thomson, S.W. (1966) The use of histochemistry in the determination of different stain of lipid droplet. *Stain Technology*, **41**:237–250.

Zaborski, P.M. (2002) Protein interactions by immunochemical techniques. *CRC Press*, New York, NY.

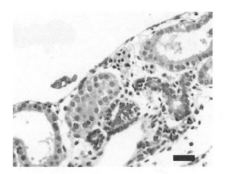

FIGURE 1.11 Method: Hematoxylin–eosine. Tissue: Young amphibian kidney. Preparation: Paraffin section. Fixative: Bouin's fluid. Observations: Nuclei are blue stained, and cytoplasm, connective tissue, and smooth muscle are pink stained. Bar = 90 μm.

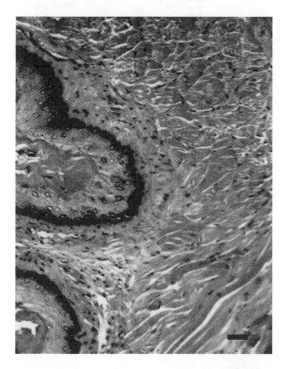

FIGURE 1.12 Method: Hematoxylin–phloxine–saffron. Tissue: Mouse esophagus. Preparation: Paraffin section. Fixative: Bouin's fluid. Observations: Nuclei are dark stained, cytoplasm is pink, connective tissue is yellow, muscle are pink stained. Bar = 100 μm.

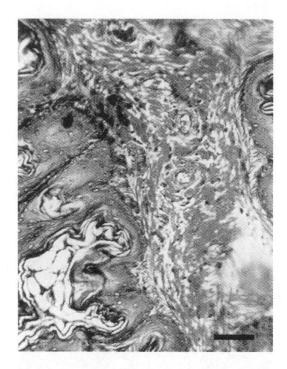

FIGURE 1.13 Method: Masson's trichroma. Tissue: Mouse esophagus. Preparation: Paraffin section. Fixative: Bouin's fluid. Observations: Nuclei are brown stained, cytoplasm is pink, connective tissue is blue, muscles are pink stained. Bar = 200 μm.

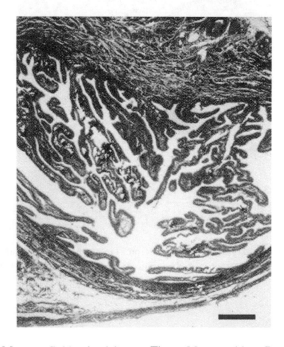

FIGURE 1.14 Method: Masson–Goldner's trichroma. Tissue: Mouse oviduct. Preparation: Paraffin section. Fixative: Bouin's fluid. Observations: Nuclei are dark stained, cytoplasm is pink. Bar = 10 μm.

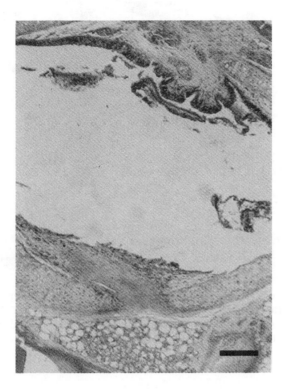

FIGURE 1.15 Method: Prenant's triple staining. Tissue: Mouse trachea. Preparation: Paraffin section. Fixative: Bouin's fluid. Observations: Nuclei are brown stained, cytoplasm is pink, connective tissue is green. Bar = 200 μm.

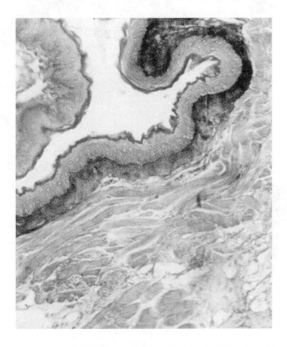

FIGURE 1.16 Method: Ramon y Cajal's trichroma. Tissue: Mouse esophagus. Preparation: Paraffin section. Fixative: Bouin's fluid. Observations: Nuclei are red, cytoplasm are green or yellow, keratin is red stained, connective tissue is purple, muscles are yellow stained. Bar = 200 μm.

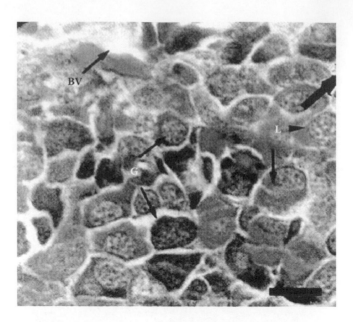

FIGURE 1.17 Method: Cleveland and Wolfe's staining. Tissue: Amphibian adenohyopohysis. Preparation: Paraffin section. Fixative: Bouin's fluid. Observations: Nuclei are red stained, cytoplasm is blue, pink, purple or orange stained depending on cell type, connective tissue is blue, blood cells are orange stained. BV: blood vessel; G: gonadotrophic cell; L: lactotrophic cell. Bar = 10 μm.

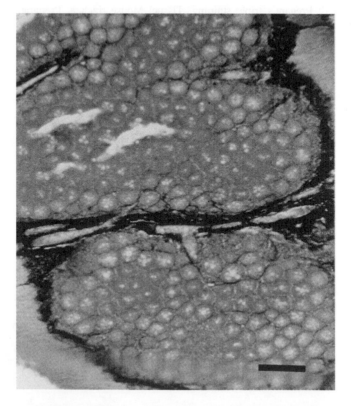

FIGURE 1.18 Method: Herlant's trichroma. Tissue: Mouse intestine (caecum). Preparation: Paraffin section. Fixative: Bouin's fluid. Observations: Nuclei are red, cytoplasm is pink or nonstained, connective tissue is blue. Bar = 30 μm.

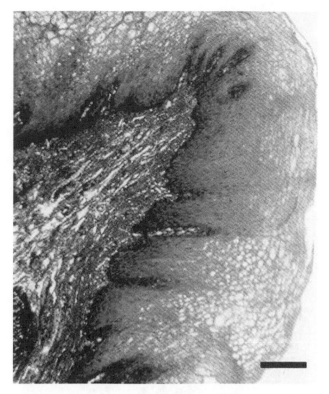

FIGURE 1.19 Method: Romeis's azan. Tissue: Mouse cardia. Preparation: Paraffin section. Fixative: Bouin's fluid. Observations: Nuclei are red stained, cytoplasm is pink, connective tissue is blue. Bar = 200 μm.

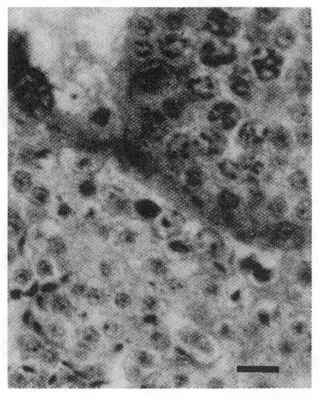

FIGURE 1.20 Method: Modified azan. Tissue: Mammal testis. Preparation: Paraffin section. Fixative: Bouin's fluid. Observations: Nuclei are red stained with details, cytoplasm is pink, connective tissue is blue. Bar = 30 μm.

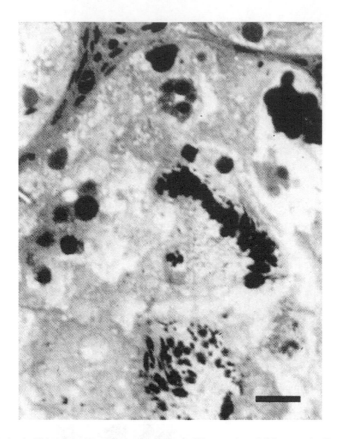

FIGURE 1.21 Method: Toluidine blue (metachromasy). Tissue: Amphibian testis. Preparation: Inclusion in Epon, semithin sections. Fixative: Glutaraldehyde–paraformaldehyde. Observations: Nuclei are purple, cytoplasms are pink. Bar = 30 µm.

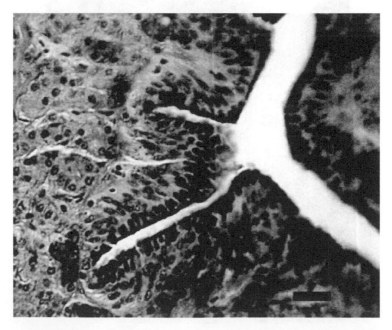

FIGURE 2.3 Method: PAS. Tissue: Amphibian stomach. Preparation: Paraffin section. Fixative: Bouin's fluid. Observations: Nuclei are brown, carbohydrates are red, and collagen is red-purple. Bar = 100 µm.

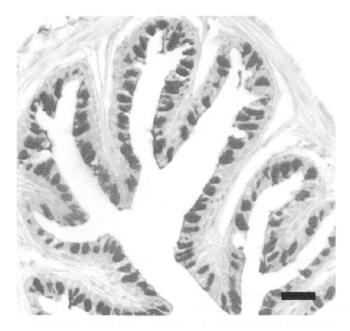

FIGURE 2.6   Method: Alcian blue. Tissue: Amphibian esophagus. Preparation: Paraffin section. Fixative: Bouin's fluid. Observations: Nuclei are brown, acidic carbohydrates arc blue. Bar = 100 μm.

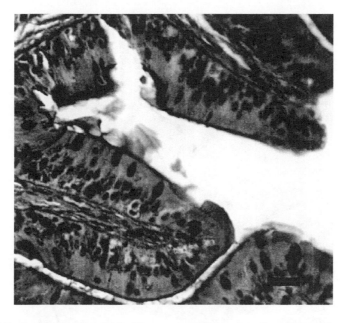

FIGURE 2.7   Method: Alcian blue–PAS. Tissue: Amphibian intestine. Preparation: Paraffin section. Fixative: Bouin's fluid. Observations: Nuclei are brown, carbohydrates are red, and collagen is red-purple. Bar = 100 μm.

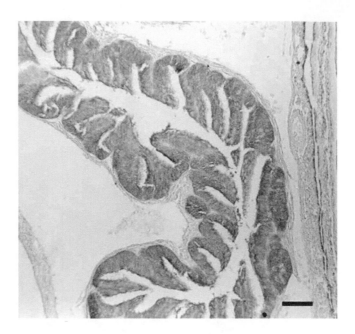

FIGURE 2.8 Method: Ravetto's staining. Tissue: Amphibian oviduct. Preparation: Paraffin section. Fixative: Bouin's fluid. Observations: Nuclei are brown, acidic carboxyled carbohydrates are blue, sulfated carbohydrates are yellow, and mixtures of carboxyled and sulfated carbohydrates are green. Bar = 100 μm.

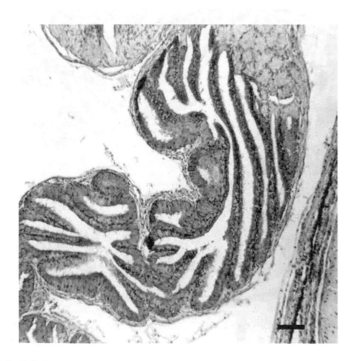

FIGURE 2.9 Method: Ravetto's staining. Tissue: Amphibian oviduct. Preparation: Paraffin section. Fixative: Bouin's fluid. Observations: Nuclei are brown, acidic carboxyled carbohydrates are blue, sulfated carbohydrates are yellow, and mixtures of carboxyled and sulfated carbohydrates are green. Bar = 100 μm.

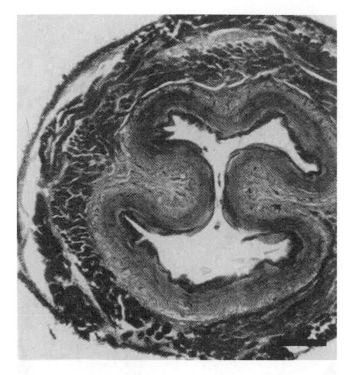

FIGURE 2.13   Method: Danielli's tetrazoreaction. Tissue: Mouse esophagus. Preparation: Paraffin section. Fixative: Bouin's fluid. Observations: Proteins are brown stained. Bar = 100 μm.

FIGURE 2.16   Method: Sudan black and nuclear fast red. Tissue: Amphibian testis. Preparation: Frozen sections. Fixative: Buffered formalin. Observations: Nuclei are red stained and lipids are black. Bar = 30 μm.

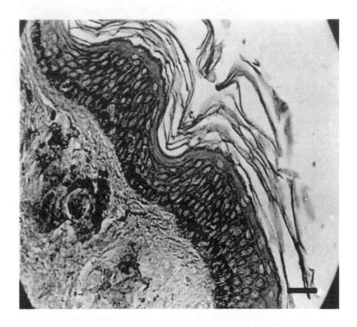

FIGURE 2.17 Method: Argyrophily, hemalum. Tissue: Melanin in mouse skin. Preparation: Paraffin sections. Fixative: Bouin's fluid. Observations: Melanin is dark stained. Bar = 100 μm.

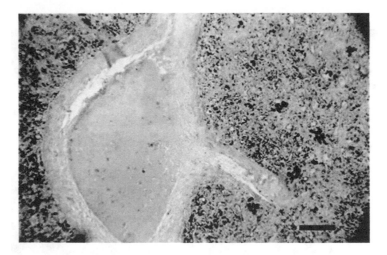

FIGURE 2.19 Method: Perls's staining. Tissue: Spleen. Preparation: Paraffin sections. Fixative: Bouin's fluid. Observations: Iron is blue stained. Bar = 100 μm.

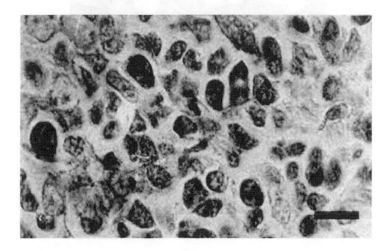

FIGURE 2.27 Method: Pappenheim–Unna's staining or methyl green–pyronine. Tissue: Plasmocytes in connective tissue. Preparation: Paraffin. Fixative: Buffered formalin. Observations: DNA ion nuclei is purple stained and RNAs in cytoplasm are pink. Bar = 30 μm.

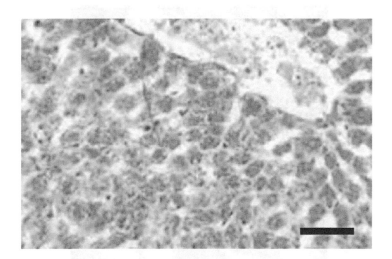

FIGURE 2.30 Method: Feulgen and Rossenbeck's staining. Tissue: Mouse spleen. Preparation: Paraffin section. Fixative: Bouin's fluid. Observations: Nuclei are red stained. Bar = 30 μm.

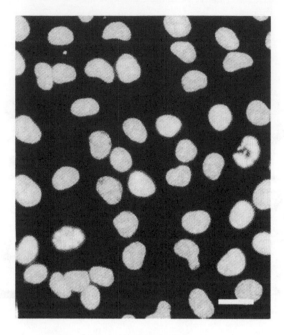

FIGURE 2.35 Method: Hoechst 33258. Tissue: MRC5 cell strain without mycoplasma. Preparation: Cell culture. Fixative: Buffered formalin. Observations: Nuclei are green fluorescent. Bar = 10 μm.

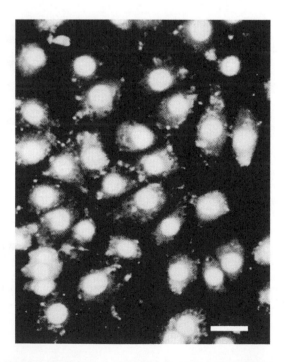

FIGURE 2.36 Method: Hoechst 33258. Tissue: MRC5 cell strain with mycoplasmae. Preparation: Cell culture. Fixative: Buffered formalin. Observations: Nuclei are green fluorescent. Bar = 10 μm.

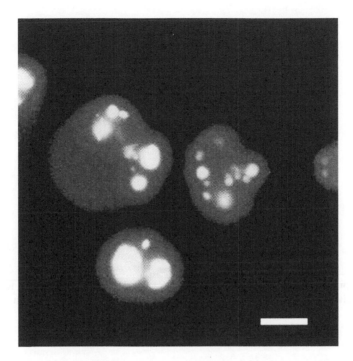

FIGURE 2.37   Method: Hoechst 33342. Tissue: Cell culture with apoptosis. Preparation: Cell culture. Fixative: Buffered formalin. Observations: Nuclei and apoptotic bodies are blue fluorescent. Bar = 5 μm.

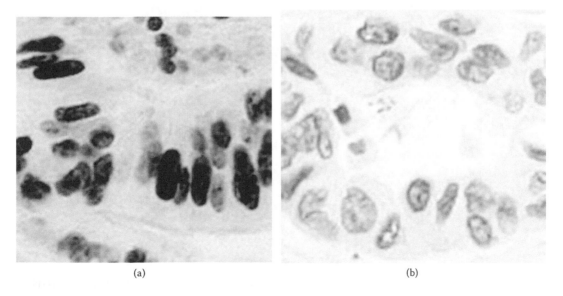

(a)   (b)

FIGURE 6.28   References color RGB images: DAB immunostained reference (a) and hematoxyline counterstained reference (b).

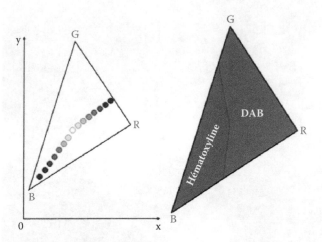

FIGURE 6.29 Representation of the color space used for pixel classification according to their RGB values. The color space is divided in two domains: DAB domain (red) and hematoxyline domain (blue). (From Heus, 2009.)

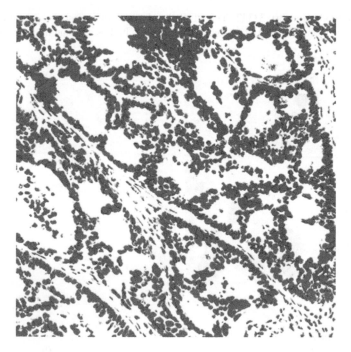

FIGURE 6.30 Segregation of positive immunostained pixels (red) and negative immunostained pixels (blue) in the whole structure of interest presented in Figure 6.27.

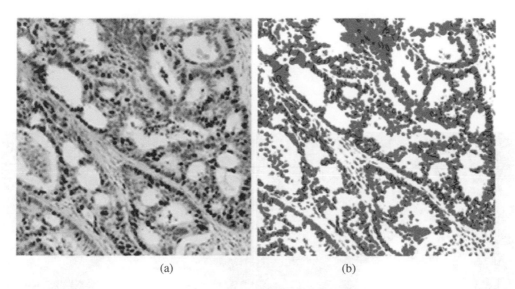

(a)   (b)

FIGURE 6.31 When DAB immunostaining protocol is used, the G component (a) is selected for densito-metric quantification in relation with the corresponding image (Figure 6.26) summarizing the segregation between negative and positive immunostained pixels in the structure of interest (b).

FIGURE 13.3 Representation of the colored filters laid upon the sensor to produce a color image. From left to right, image of the red, blue, and green filters, and of their assembly to create the Bayer matrix.

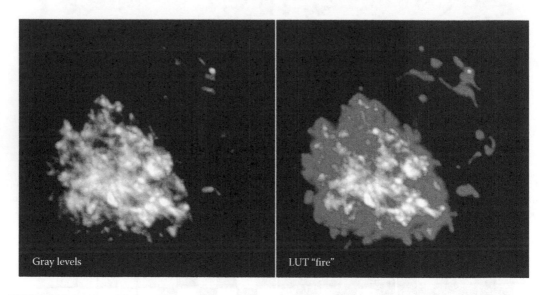

FIGURE 13.8   A LUT applied in a gray-level image transforms low-level intensity differences in differences of colors.

3 Enzyme Histochemistry Methods

Jeanne Estabel

CONTENTS

3.1 INTRODUCTION

Enzyme histochemistry (or histoenzymology) is a morphological technique applied to functional questions in histology and pathology. Enzyme histochemistry constitutes a bridge between morphology and biochemistry. It also provides additional information to conventional histology and immunohistochemistry. This information is used to demonstrate enzyme presence or absence in different animal tissues. Detection of changes in histochemical reactivity is of diagnostic value. These techniques are simple, rapid, and inexpensive and have a vast range of application in pathology and disease diagnosis.

Enzymes are proteins that catalyze chemical reactions. Enzymes are fragile proteins that can be destroyed or inactivated by fixation and long tissue processing. Different parameters (pH, temperature, substrate concentration, etc.) could influence the techniques. The success of the technique also results in using the right pH and the right substrate for the enzyme.

The principle is based on the localization of the enzyme by the metabolization of a specific substrate brought to the enzyme. Following the reaction between the enzyme and the appropriate substrate, an insoluble colored product should develop to provide the visualization. One of the downsides of these techniques is the difficulty of quantification.

The most frequently used reactions allowing the visualization of the final product are metal precipitation techniques like Gomori, coupling azo dye methods, and tetrazolium salt methods.

These techniques also have limitations like diffusion or false localization of the end products and nonspecificity of the reaction.

3.2   GENERAL PRINCIPLES

Enzyme histochemistry combines the biochemical activity of the enzyme with information about its localization. The basics of enzyme histochemistry was described extensively in the 1970s before immunohistochemistry arose as a new technique.

3.2.1   GENERAL REACTION

Reaction between enzyme and substrate produces a primary reaction product (PRP). Using a histochemical reaction involving, for example, chemicals like diazonium salts, produces a stained precipitate product. The method can be represented as follows:

$$\text{Substrate} + \text{Enzyme} = \text{Primary reaction product}$$

$$\text{Primary reaction product} + \text{Diazonium salt} = \text{Final reaction product}$$

A substrate is a molecule upon which an enzyme acts.

3.2.2   FACTORS INFLUENCING THE REACTION

Several factors could influence the success of the enzymatic reaction:

- Fixation—Use the right fixative and control the time of exposure.
- pH—The buffer medium has to be at an ideal pH allowing the enzyme to maximize the reaction.
- Substrate solubility—The substrate should be soluble in water or in the buffer solution to allow maximum hydrolysis by the enzyme.
- Substrate concentration—If the concentration is too high, an inhibition of the reaction can occur.

3.2.3   NOMENCLATURE OF THE ENZYMES

Different enzyme types are described and different nomenclatures can be used.

- Hydrolases including phosphatases, lipases, esterases, and glycosidases.

$$AB + H_2O \rightarrow AOH + BH$$

- Oxidoreductases including oxidases, peroxidases, and dehydrogenases. They catalyze oxidation/reduction reactions by transferring of H and O atoms or electrons from one substance to another.

$$AH + B \rightarrow A + BH \text{ (reduced)}$$

$$A + O \rightarrow AO \text{ (oxidized)}$$

The type of reaction commonly used is the simultaneous capture or coupling. When the final product is not detectable under microscope, the capture reaction has to be followed by a stained reaction resulting in a crystalline precipitate for visualization. Different chemicals can be used like nitro blue tetrazolium (NBT) or potassium ferro- and ferricyanide.

3.2.4 CONTROLS

Controls are necessary to monitor the success and the specificity of the reaction. Some mammal tissues can be used as positive controls such as kidneys or intestines for alkaline phosphatase.

3.3 TISSUE PREPARATION

3.3.1 FIXATION

The consequence of the fixative on an enzyme is the loss of activity. A decrease of 60% activity is not unusual.

Physical or chemical fixation can be used for enzyme histochemistry.

The fixation of the tissue can be done by liquid nitrogen or isopentane or also by a chemical fixative like ethanol or formalin-based fixative. The choice of the fixation should be decided in relation with the enzyme involved. Tougher enzymes such as phosphatases and hydrolytic enzymes can be still stained after formaldehyde fixation.

The fixation and tissue processing should be controlled. Overfixation and high temperature need to be avoided. Almost all the enzyme histochemical reactions are performed on frozen tissues to ensure the integrity of the protein and also to avoid inactivation by fixatives like the formalin.

Each fixative has its own effect on the activity of the enzyme.

3.3.2 SECTIONS

The best choice is the cryosections. This technique avoids the transfer in ethanol and paraffin, which can denaturate the enzyme. The temperature used to melt paraffin can inhibit enzymatic activity. Nevertheless, for some enzymes, the paraffin sections can be used; the only restriction to this is to use paraffin with low melting point around 46°C. For paraffin characteristics, see Chapter 1, Section 1.3.3.1.

3.4 PRINCIPAL ENZYMATIC ACTIVITIES TECHNIQUES

3.4.1 ALKALINE PHOSPHATASE

- Location—Intestine, kidney
- Fixative—10% neutral buffered formalin at 4°C. Time fixation is less than 24 h.

3.4.1.1 Gomori Calcium Method

- Sections—Paraffin sections (paraffin with low melting point)
- Reagents
  - Nuclear fast red: See Chapter 7, Section 7.2.1.12.
  - Incubating medium
    - 2% Sodium β-glycerophosphate, 25 mL

- – 2% Sodium veronal, 25 mL
- – 2% Calcium nitrate, 10 mL
- – 1% Magnesium chloride (activation), 2 mL
- – Distilled water, 50 mL
  - • Solution A
    - – Distilled water, 100 mL
    - – Nitrate cobalt, 2 g
    - – Sodium veronal, 0.5 mL
  - • Solution B
    - – 2% Ammonium sulfide (freshly prepared)
- • Method
  - • Dewax the slides
  - • Incubating medium, pH 9–9.4, 24 h at 37°C
  - • Rinse with distilled water, twice
  - • Incubate in solution A, 5 min
  - • Rinse with distilled water, twice
  - • Incubate in solution B, 2 min
  - • Rinse with distilled water, twice
  - • Nuclear fast red, 1 min
  - • Rinse with distilled water
  - • Dehydrate, mount
- • Control—Replace the incubation medium by water.
- • Results—Alkaline phosphatase activity sites are brownish-black.

3.4.1.2 Azo Dye Coupling Method

- • Sections—Cryosections or paraffin sections (paraffin with low melting point)
- • Reagents
  - • Methyl green 2% (chloroform extracted): See Chapter 7, Section 7.2.3.17.
  - • Incubating medium (freshly made)
    - – Sodium α naphtyl phosphate, 10 mg
    - – 0.2M Tris buffer pH 10, 10 mL
    - – Diazonium salt (fast red TR), 10 mg
    - – The sodium α naphtyl phosphate is dissolved in the buffer, the diazonium is added, and the solution well mixed. The solution is used immediately.
- • Method
  - • Dewax the sections if paraffin embedded
  - • Bring section to water
  - • Incubating medium, 10 to 60 min at room temperature
  - • Wash in distilled water
  - • Counterstain with 2% methyl green, 1 to 2 min
  - • Wash in running tap water
  - • Mount using an aqueous solution
- • Results—Alkaline phosphatase activity sites are reddish-black and nuclei are green.

3.4.2 Acid Phosphatase

- • Location—Liver, kidney
- • Fixative—Cryosections preferred
- • Reagents
  - • Nuclear fast red: See Chapter 7, Section 7.2.1.12.

- Incubating medium
  - 0.05 M acetate buffer pH 5, 10 mL
  - Lead nitrate, 20 mg. The lead nitrate must be dissolved in the buffer before the sodium β glycerophosphate is added.
  - Sodium β-glycerophosphate, 32 mg
- Method
  - Slides in incubating medium, 2 h at 37°C
  - Wash in distilled water
  - Ammonium sulfide (freshly made), 2 min
  - Wash well in distilled water
  - Nuclear fast red, 10 min
  - Wash in distilled water
  - Mount using an aqueous solution
- Results—Acid phosphatase activity sites are black and nuclei are red.

3.4.3 ATPase (Adenosine Triphosphatase)

- Location—Skeletal muscle
- Fixation—Unfixed cryosections or cold formalin followed by sucrose
- Reagents
  - Harris's hematoxylin, optional: See Chapter 7, Section 7.2.1.9.
  - Glycine buffer—Solution 0.1 M glycine buffer with 0.75 M $CaCl_2$
    - 0.1 M glycine buffer (0.75 g glycine + 0.585 g NaCl up to 100 mL with distilled water), 50 mL
    - 0.75 M $CaCl_2$ (11.03 g $CaCl_2.H_2O$/100 mL): 10 mL
    - Add 0.1 M NaOH until pH 9.6–9.8, approximately 22 mL
  - ATP solution
    - ATP, 5 mg
    - Solution glycine/$CaCl_2$, 10mL
    - 2% $CaCl_2$
- Method—Methods at pH 9.6, 4.2, and 4.6 used in combination to distinguish between type 1 and type 2 fibers in muscles.

3.4.3.1 pH 9.4

- Incubate in the ATP solution, 11 min at 37°C; use freshly cut sections
- Rinse well in distilled water
- Immerse in 2% $CoCl_2$, 5 min
- Rinse well in distilled water
- Rinse in tap water, 3 or 4 times
- Ammonium sulfide solution, 30 s. Use diluted ammonium sulfide solution (1:10).
- Rinse well in running tap water
- (Optional) Stain with Harris's hematoxylin, blue in tap water
- Mount in aqueous medium

3.4.3.2 pH 4.2 and pH 4.6

- Preincubate in 0.1 M sodium acetate buffer with 10 mM EDTA added (0.372 g/100 ml buffer) at pH 4.2 or 4.6, 10 min at 4°C; use freshly cut frozen sections
- Wash in distilled water
- Proceed as for the pH 9.4 method

3.4.3.3 Results

ATPase pH 9.4 staining is present mostly in type 1 fibers. The type 1 fibers appeared with the darker staining.

Note: Only the pH 4.2 needs counterstaining in hematoxylin. Sections must be well washed after the cobalt chloride step.

3.4.4 GLUCOSE-6-PHOSPHATASE

- Location—Liver, small intestine, kidney
- Fixation—Unfixed cryosections (15 µm thickness), fixation is done after staining
- Reagents
  - Incubating medium
    - 0.125% glucose-6-phosphate, 4 mL
    - Tris maleate buffer pH 6.7, 4 mL
    - 2% lead nitrate: 0.6 mL
    - Distilled water, 1.4 mL
  - 1% ammonium sulfite
- Method
  - Fresh cryosections into incubating medium, 5 to 20 min at 37°C
  - Rinse in distilled water, twice
  - Ammonium sulfide, 2 min
  - Wash in distilled water
  - Fix the sections with 10% formaldehyde, 15 to 30 min
  - Rinse in distilled water
  - Mount in aqueous medium
- Results—Glucose-6-phosphatase activity sites appear brownish-black.

3.4.5 CHOLINESTERASE AND NONSPECIFIC ESTERASE

3.4.5.1 Nonspecific Esterase

Several methods can be used to demonstrate the activity of nonspecific esterases.

- Location—Kidney, liver
- Fixation—Cold formalin, frozen or wax sections
- Reagents
  - Mayer's hemalum: See Chapter 7, Section 7.2.1.11.
  - Fast blue B solution
    - β-naphtyl acetate, 10 mg
    - Acetone, 0.25 mL
    - Add 0.1 M phosphate buffer pH 7.4, 20 mL
    - Shake until all cloudiness disappears
    - Add fast blue B, 100 mg
    - Shake
    - Filter directly onto slides
- Method fast blue B
  - Cut 10 to 15 µm thick frozen sections
  - Fast blue B solution, 1 to 15 min
  - Wash in running water, 2 min
  - Counterstain in Mayer's hemalum, 4 to 6 min; counterstaining is optional
  - Wash in running water
  - Mount in aqueous medium

- Results—Esterase are black, nuclei are dark blue. Lipase, acetylcholinesterase, and cholinesterase could also hydrolyze the β-naphthol and appear black.

3.4.5.2   Acetylcholinesterase

- Location—Nervous system and muscle
- Fixation—Unfixed cryosections or cold formalin sucrose preserved sections
- Reagents
  - Hematoxylin: See Chapter 7, Section 7.2.1.9.
  - Incubating solution for acetylcholinesterase (add in order, mixing well at each stage)
    - Acetyl thiocholine iodide, 5 mg
    - 0.1 M acetate buffer (pH 6.0), 6.5 mL
    - 0.1 M sodium citrate, 0.5 mL
    - 30 M copper sulfate, 1.0 mL
    - Distilled water, 1.0 mL
    - 5 mM potassium ferricyanide, 1.0 mL
- Method
  - Incubate in solution, 15 to 20 min at 37°C
  - Rinse in distilled water
  - Counterstain with hematoxylin, 1 to 2 min
  - Dehydrate, clear, and mount in aqueous medium
- Results—Enzyme activity appears red/brown.

3.4.6   β-GALACTOSIDASE

The LacZ reporter gene that encodes for bacterial β-galactosidase is one of the most used enzymes to locate genetically modified alleles in cells or organisms.

Using 4% paraformaldehyde as a fixative is a common way. The pH and the length of fixation should be monitored and not exceed several hours.

- Fixation—1 h in 4% cold paraformaldehyde pH 8 at 4°C under agitation.
- Sections—Cryosections
- Reagents
  - LacZ staining solution (incubating medium)
    - 2 mM $MgCl_2.6H_2O$, 1 mL
    - 0.01% deoxycholic acid, 150 μL
    - 0.02% IGEPAL CA-630, 1 mL
    - 5 mM potassium ferrocyanide, 5 mL
    - 5 mM potassium ferricyanide, 5 mL
    - 0.1% X-Gal in DMF, 2.5 mL
    - Cold, freshly made PBS pH 8.0, 85.35 mL
- Method
  - Rinse in PBS pH 8, 2 times
  - For calcified tissues, place them in EDTA solution on rocker for at least 3 days. After one day, change the solution for a fresh one.
  - PBS pH 8, 3 × 20 min
  - Equilibrate in 15% sucrose, until tissues sink to bottom at 4°C
  - Equilibrate in 30% sucrose, overnight at 4°C
  - Remove excess sucrose carefully with a lint-free tissue
  - Embed the tissue in OCT
  - Cut 20 μm sections to stain for β-galactosidase and in a parallel set of sections at 10 μm for H&E (hemalum-eosine) on Superfrost® slides: See Chapter 1, Section 1.3.5.1.

- Allow them to air-dry at room temperature before doing the staining
- LacZ staining solution, overnight at 37°C. Note: The reaction is light sensitive.
- Rinse in PBS, several times
- Post-fix in 4% PFA: 1 h at 4°C
- Mild eosin counterstain, 30 s
- Dehydrate in graded ethanols and xylene less than 5 min
- Coverslip the slides
- Results—The structures expressing the enzyme are blue.

3.4.7 PEROXIDASE

- Location—This enzyme has a wide tissue distribution. The most common locations are blood cells and thyroid follicular cells.
- Fixation—10% neutral buffered formalin, wax sections
- Reagents
  - Hematoxylin: See Chapter 7, Section 7.2.1.9. Hematoxylin can be replaced with methyl green.
  - Incubating medium
    - DAB (diaminobenzidine), 5 mg
    - Hydrogen peroxide, 0.1 mL
    - 0.05 M Tris-HCl buffer pH 7.6, 10 mL
- Control—Control is done by inhibition of the activity of the peroxidase incubation of the sections in 0.3% hydrogen peroxide in methanol for 30 min before doing the staining.
- Method
  - 5 μm sections incubated in the incubating medium, 10 min at 37°C
  - Rinse in distilled water
  - Counterstain with hematoxylin, 5 min; or with methyl green, 10 min
  - Dehydrate and mount
- Results—Peroxidases activity sites are brown stained.

REFERENCES

Adams, N.C., and N.W. Gale. 2006. High resolution gene expression analysis in mice using genetically inserted reporter genes. In *Principle and Practice Mammalian and Avian Transgenesis—New Approaches*, S. Pease and C. Lois, eds., pp. 132–172, Springer Verlag, Germany.

Bancroft, J.D., and M. Gamble. 2002. *Theory and Practice of Histological Techniques*. 5th edition. Churchill Livingstone, Edinburgh.

Meier-Ruge, W., and E. Bruder. 2008. Current concept of enzyme histochemistry in modern pathology. *Pathobiology* 75:233–243.

4 Visualization of Cell Proliferation

Françoise Giroud and Marie-Paule Montmasson

CONTENTS

4.1 INDICATORS FOR CELL PROLIFERATION

Cell proliferation is the increase in cell number as a result of cell division. The accurate assessment of cell proliferation is useful in many biological assays and is a key readout in a wide range of pharmacological and regulatory studies.

The observations of Howard and Pelc led to the introduction of the concept of the cell cycle and its subdivision into several phases. DNA synthesis (replication) and doubling of the genome take place during the S phase. This is preceded by a period of variable duration known as the first gap phase (G1), which separates the S phase from the previous mitosis (M phase). The S phase is followed by a period of apparent inactivity known as the second gap phase (G2), which comes before the next mitosis. Interphase comprises successive G1, S, and G2 phases, and forms the largest part of the cell cycle. For a typical DNA histogram one peak represents the G1 and another (with twice the value) represents the G2/M phase of the cell cycle. S phase cells are spread between the two peaks.

Quiescence is the state of a cell when it is not dividing but that might be stimulated to divide. Such a cell is in G0 phase.

In any tissue there are also cells that, for whatever reason, can no longer divide, thus any cell population can be divided into a cycling and a noncycling compartment. This naturally leads to the definition of the proliferative or growth fraction of any cellular population as the ratio of cycling to the total number of cells (cycling plus noncycling cells). Cell growth fraction stands for percentage of proliferating cells.

Against this background a variety of cellular changes can be identified that can be used to pinpoint cell proliferation in biological samples. In simple terms, mitoses can be counted, the incorporation of nucleotides (or their analogues) into newly synthesized DNA during the S phase can be identified, and the varying levels of structural or functional proteins associated with different aspects of the cell cycle can be assayed.

The Mitotic Index is defined as the ratio between the number of cells in mitose (M phase: prophase, metaphase, anaphase, and telophase) and the total number of cells. The Mitotic Index can be worked out from a slide with light microscopy. The Mitotic Index is related to the relative duration of M phase.

DNA content (Section 4.2), DNA synthesis (DNA content combined with BrdU incorporation; Section 4.3), cycling cells (DNA content combined with Ki-67 protein detection; Section 4.4), or relative cell cycle durations (DNA content combined with AgNORs detection; Section 4.5) are current approaches that provide useful information on cell-cycle-related studies.

4.2 DNA CONTENT

The capacity of some dyes to bind to DNA in a stoichiometric manner means that the amount of DNA present in a nucleus can be determined, at least in comparison with some reference standard.

The reliability of the Feulgen reaction for localizing DNA within the cell was universally accepted many years ago. Today, the Feulgen reaction, when properly controlled, is beyond any doubt specific and stoichiometric for DNA.

The Feulgen reaction is widely used in absorption image analysis for the quantitative determination of nuclear DNA content, especially in tumor pathology. The methods of fluorescence image cytometry are being developed as an alternative for the determination of DNA content on a single-cell basis when using DNA-specific fluorochromes.

The Feulgen reaction is a cytochemical procedure necessarily done in two separate steps:

1. Acid hydrolysis to remove the purine bases from DNA molecules (depurination) and unmask the aldehyde groups of deoxyribose.
2. Staining the aldehyde groups by a chemical reaction between the exposed aldehyde groups and the Schiff's reagent leading to a color complex. Two dyes can be used as Schiff's reagent: pararosalinine or thionine, leading both to colored compounds, respectively, magenta and blue.

4.2.1 THE EUROPEAN CONSENSUS

A task force of invited experts in the field of diagnostic DNA image cytometry, open to any other scientist or physician revealing experience in that diagnostic procedure, agreed upon an updated consensus report during the 5th International Congress of the ESACP 1997 in Oslo (Haroske et al., 1998). The report deals with the following items:

- Biological background and aims of DNA image cytometry
- Principles of the method
- Basic performance standards
- Diagnostic interpretation of DNA measurements
- Recommendations for practical use

4.2.2 METHODOLOGICAL RECOMMENDATIONS

4.2.2.1 Reagents and Solutions
- PBS, Phosphate-buffered saline (Invitrogen) pH 7.2
- PARA4, 4% paraformaldehyde (weight/vol)
  - Polyoxymethylene, 4 g
  - PBS, 100 mL
  - Heat to 70°C
  - Cool to room temperature
  - Filter
  - Alternative: Böhm-Sprenger fixative, methanol 80%, formalin 15%, acetic acid 5%
- HCl 5M
- Either pararosalinine (BDH)
  - Chemical name: Basic red 9
  - Color index 42500
  - Absorption maximum, 560 nm
  - CAS number 569-61-9
- Or thionine
  - Chemical name: No Cl name
  - Color index 52000

- Absorption maximum, 590 nm
- CAS number 78338-22-4

Use of appropriate commercial solutions and so-called Feulgen kits is possible.

Schiff's reagents must be kept refrigerated. The limit of shelf life of pararosaniline and thionine solutions is about 1 year and 2 weeks, respectively.

- Sulfurous solution, 5 g Na or K metabisulfite, 50 mL HCl 1M, make up to 1000 mL with distilled water. Use sulfurous solution within 2 h after preparation.
- XAM mounting medium. Use a neutral mounting medium.

4.2.2.2 Samples

Use every kind of method for collecting cells for cytopathologic analysis. At present, measurements of the DNA content in tissue sections do not meet the performance standards.

- Cell cultures on slides
- Fine needle aspiration biopsy (FNAB)
- Smears from exfoliated cells
- Cytocentrifuged preparations from body fluids
- Cell separation specimens from FNABs, core or other biopsies, or from formaldehyde-fixed, paraffin embedded, after mechanic and/or enzymatic dispersion. Enzymatic procedures have to be carefully worked out to achieve sufficient yields of structurally intact cells. A 1% pepsin digestion for 30 min at 37°C should be adequate.

4.2.2.3 Fixation

Fixation with formaldehyde is necessary before staining by the Feulgen reaction. Prestained smears may also be used. Postfixation after uncovering is necessary. Postfixation is especially important for monolayer preparations prepared from paraffin sections.

- Air-dry the samples for at least 1 h at room temperature.
- Fix in 4% paraformaldehyde for 30 min.
- All other types of fixatives, which should be applied after air-drying and which should contain formaldehyde anyway, should be tested by the appropriate quality control procedure. See Section 4.2.5. Böhm-Sprenger fixative (methanol 80%, formalin 15%, acetic acid 5%) has been tested and found to be suitable.
- Rinse in distilled water for 5 min.
- The samples should be strongly washed after fixation. This step is crucial to remove residual formaldehyde to avoid unspecific staining.

4.2.2.4 Staining

Fixation and staining steps are both in aqueous conditions.

- Hydration: Ethanol 100% twice, 95% twice, 70%, 5 min each bath and distilled water, 10 min
- Step 1: Hydrolysis
  - Hydrolysis has to be performed under controlled temperature and time conditions. Conditions suitable for many routine applications:
  - Hydrolysis: 5 M HCl, 60 min at 25°C
  - Stop hydrolysis by rinsing in distilled water, four baths, 1 min each at room temperature

- Optimal hydrolysis conditions are dependent on tissue type, fixation conditions (type of fixative, time, and concentration), and mode of sample preparation. Optimal hydrolysis conditions have to be worked out, based on hydrolysis curves (time versus integrated optical density). For each new type of material hydrolysis curves should be obtained.
- Avoid treating big batches of slides in the same bath, working out a maximum of 20 slides for 200 mL of solution, including one or more reference slides and a negative control is recommended. Check that surfaces of slides are correctly exposed to reagents. Use gentle agitation.
- Step 2: Staining
  - Schiff's reagent, 60 min at 25°C in the dark. Take out of refrigerator the Schiff's reagent before use to allow equilibrium to 25°C. Change regularly Schiff's reagent, 300 mL for a maximum of 120 slides. Discard Schiff's reagent if it becomes slightly colored.
  - Sulfurous baths, four baths, 1 min each at room temperature. Use sulfite rinse to remove surplus dye from the cell nuclei and cytoplasm.
  - Rinse with tap water, 10 min at room temperature
- Dehydrate, ethanol 70%, 95% twice, 100% twice, 5 min each bath
- Toluene, two baths, 5 min each. Alternative: xylene. Toluene and xylene toxicity causes irritation of skin, eyes, and respiratory system and brain effects.
- Mount in XAM medium. Alternative: Any neutral medium.

4.2.2.5 Negative Control

Specificity control of the Feulgen reaction may be performed by staining an unhydrolyzed specimen, which must remain unstained.

4.2.3 INSTRUMENT REQUIREMENTS

4.2.3.1 Setting Up the System

Before each batch of analyses, the image cytometer system used for quantitation must be correctly set up.

1. Use adequate interference filters for blue (thionine, e.g., 590/10 nm) or red (pararosaniline, e.g., 560/10 nm) stain
2. Perform Köhler illumination
3. Perform analog/digital adjustments (the so-called offset and gain setting) for correct light intensity
4. Adjust offset and gain before each measurement

4.2.3.2 System Quality Assurance

Perform regularly quality assurance tests on the system at least once a year. Corresponding quality assurance protocols are given in Giroud et al. (1998).

- Check stability over time
- Check densitometric linearity
- Check for shading phenomena
- Check for glare phenomena

4.2.3.3 The PRESS Slide

The PRESS slide is a microscopic slide used for image cytometer quality control. It has the advantage to present

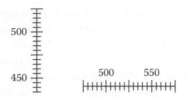

FIGURE 4.1 Vertical and horizontal scales.

- A full game (5% to 100% transmission) of density filters
- Two horizontal and vertical calibration scales (Figure 4.1)
- Regular fiducial marks
- Circles (10 μm diameter) and squares (10 μm side) in a large range of densitometric values (5% to 100% transmission), on black or transparent background

Use the PRESS slide to check for

- Illumination stability—Electronic noise, drift during time
- Shading phenomenon
- Densitometric linearity
- Pixel calibration
- Stage motion characterization
- Densitometric object measurements
- Object segmentation
- Geometric distortion
- Optical transfer function (OTF)

4.2.4 DENSITOMETRIC MEASUREMENTS

Evaluation of the performance of densitometric measurement procedures should be done running the appropriate quality control procedure (Giroud et al., 1998.)

4.2.4.1 Principle

- Nuclei to be measured must be in focus
- No change of instrumentation adjustment during measurements: Köhler-illumination, light intensity, field diaphragm, analogue/digital adjustment
- Correct for shading by software procedures
- Correct for local background per nucleus by software procedures
- Correct for glare (stray light) by software procedures
- Use visual control during and/or after measurements for artifact rejection, correct focus, and appropriate segmentation

For algorithms of correction, see guidelines in Giroud et al. (1998).

4.2.4.2 Reference Cells

The coefficient of variation (CV) of IOD (Integrated Optical Density) measurements on the G0/1 peak of reference cells should not exceed 5% (CV = [SD/Mean] × 100; SD, standard deviation). For recognizing a unimodal DNA histogram usually at least 50 cells have to be measured.

- In clinical samples, internal reference cells should be preferred. Lymphocytes, granulocytes, normal epithelial cells, or stroma cells can be analyzed as internal standards. Diploïd rat liver hepatocytes may be used as external standards.

- External reference cells should be prepared and fixed identically as cells under analysis. Deposition of the external reference cells on the same slide as the clinical sample is recommended. External reference cells should be stained in the same staining baths as the clinical sample.
- All reference cells should be analyzed during the same run as the clinical sample and under the same conditions.

4.2.4.3 Scaling Procedure

The accuracy of each diagnostic DNA evaluation depends decisively on the standard deviation (SD) of the corrective factor used during the scaling procedure.

- Use reference cells to transform the arbitrary unit (a.u.) scale IOD measurements in a reference unit scale (2c, 4c, 8c, for example).
- Make estimation of the ratio between the means of IOD values obtained for both reference cells used and normal cells of the tissue under study; this will define a corrective factor.
- Apply the corrective factor to DNA measurements from the clinical sample before DNA histogram interpretation.

4.2.4.4 Sampling

- Use a systematic random sampling strategy.
- A selective sampling for rare nuclei characterized by a high DNA content is allowed.
- A stochastic sampling should not be combined with a selective one, unless the rare nuclei are clearly flagged as such and are not included in a population-based interpretation.
- The number of cells measured defines the validity of the histogram properties, that is, the number of peaks that can be identified. For recognizing a unimodal DNA histogram usually at least 50 cells have to be measured. The greater the expected peaks, the greater should be the number of cells measured.

4.2.5 The Quality Assurance Protocol

Quality assurance in medical services has become a more important issue. A quality assurance protocol has been agreed by the task force on DNA image cytometry during the 5th International Congress of the ESACP 1997 in Oslo (Giroud et al., 1998).

Six tests are recommended for quality control:

- QC for ICM-DNA protocol
- QC for preparation stability
- QC for corrective factor
- QC for diagnostic DNA interpretation
- QC for IOD measurements
- QC for ICM instrumentation

A flow diagram in Giroud et al. (1998) presents the feedback mechanisms between the six tests.

4.2.6 Diagnostic DNA Image Cytometry

Quantitation of nuclear DNA content by image cytometry has come into practice for assistance in the diagnosis and grading of malignant tumors. The basic aim of diagnostic DNA image cytometry is to identify DNA stemlines outside the euploid regions as abnormal (or aneuploid) at a defined statistic level of significance. The usual precision of DNA image cytometric measurements should at least allow DNA stemlines to be identified as abnormal (or aneuploid), if they deviate more than 10% from the diploid (2c) or tetraploid region (4c), that is, if they are outside 2±0:2c or 4±0:4c.

DNA content measurements represent both DNA ploidy and cell proliferation events. DNA ploidy is related to the number of chromosomes. Euploidy is the state of a cell having an integral multiple of the monoploid number of chromosomes. Aneuploidy is the state of a cell that is not euploid. A euploid diploid cell has two sets of chromosomes; a tetraploid has four sets of chromosomes.

A proliferating diploid cell doubles its DNA content during the S phase. G2 cells have twice the DNA content compared to G1 phase cells, and cells in the S phase will have intermediate amounts of DNA. Thus, a proliferating diploid cell in the G2 phase will have the same DNA content as a tetraploid cell in the G1 phase.

Furthermore, DNA image cytometry should give information about:

- Number of abnormal (aneuploid) DNA stemlines
- Polyploidization of euploid or aneuploid DNA stemlines
- Cell cycle fractions
- Occurrence of rare cells with an abnormally high DNA content

4.2.7 DYNAMIC OF PROLIFERATION

In the case of cell cultures, DNA content is a powerful tool for cell cycle and cell cycle phase duration evaluations, when combined to cell cycle synchronization and sampling over time procedures.

- Plate the cells on the culture Lab-Tek glass chamber slides with appropriate culture medium. Prepare as many slides as needed for sampling.
- Culture the cells until exponential phase is reached.
- Use a synchronization protocol.
- Sample each hour after synchronization.
- Analyze the different samples.
- Appreciate the variations of the proportion of cells in the different cell cycle phases.

The variation along time of the proportions of cells in the different cell cycle phases makes it possible to evaluate cell cycle phase durations.

4.2.8 DNA CONTENT AND HISTOMETRY

At present, measurements of the DNA content in tissue sections do not meet the performance standards. Extraction of cells from formaldehyde-fixed, paraffin-embedded samples is possible.

4.2.9 DNA CONTENT AND FLUORIMETRY

Potentially, most fluorochromes specific for DNA can stain this nucleic acid after a cell fixation step. However, stoichiometry and accuracy, reproducibility, and stability of image cytometry measurements remain to be proved. More than that, neither quality insurance protocol was developed nor metrology tools exist to address this problem.

Most of the fluorescent dyes used to stain DNA are carcinogenic: skin contact and inhalation must be avoided. They must be handled with care, and gloves and fluid-resistant masks must be worn.

4.3 BrdU INCORPORATION

Synthesis of DNA can be assessed by measuring the incorporation of a thymidine analog such as the 5-bromo-2′-deoxyuridine (BrdU). This molecule can then be detected by specific antibodies and used for detailed cell cycle analysis. This technique is actually used as the "gold standard" for cell kinetic studies.

The disadvantage of this procedure is the need for *in vivo* administration or *in vitro* incubation. This restrains BrdU labeling as a practical routine procedure for histopathologists. However, *in vitro* incubation of biopsy or other surgical specimens is feasible. Ethical permission for some *in vivo* studies has been obtained.

When cells are cultured with a labeling medium that contains BrdU, this pyrimidine analog is incorporated in place of thymidine into the newly synthesized DNA of proliferating cells. After removing labeling medium, cells are fixed and the DNA is denatured. Then a BrdU mouse mAb is added to detect the incorporated BrdU. The denaturation of DNA is necessary to improve the accessibility of the incorporated BrdU to the antibody.

The detection and measurement of BrdU-labeled cells permit the identification of cells that were newly synthesizing DNA (cells in S phase) during the incorporation time. This measurement is generally made in association with a measurement of the global DNA content, showing the position of each cell in its phase along the cell cycle.

After a short period of growing cells in the presence of BrdU (BrdU pulse), it is possible to measure the proportion of cells in the S phase. A "pulse and chase" protocol combined to a sampling over time strategy makes it possible to follow the cohort of cells that have newly synthesized DNA during the pulse duration. Such a procedure allows one to measure cell cycle and cell cycle phase durations.

The "pulse and chase" BrdU protocol is 30 min of incorporation followed by sampling each hour during a 24 h period (make adaptation according to cell type). A 30 min BrdU pulse incorporation should be enough for major human cell types.

After continuous BrdU incorporation, it is possible to measure the proportion of cycling cells. In continuous BrdU incorporation protocol, the incubation time should be longer than the cell cycle duration (35 h should be enough for human cells).

4.3.1 MATERIALS AND EQUIPMENT

- Lab-Tek glass chamber slides
- Appropriate culture medium (supplemented with fetal bovine serum [FBS] or serum substitute)
- PBS, phosphate-buffered saline (Invitrogen). Alternative to PBS: Saline water, NaCl 8.0 g + H Na_2PO_4, $12H_2O$ 2.7 g or H Na_2PO_4, $2H_2O$ 1.15 g + H_2 $NaPO_4$, $2H_2O$ 0.4 g, qs 1L demineralized H_2O.
- BrdU, 5-bromo-2′-deoxyuridine, stock solution 1 mM in PBS. BrdU is a toxic agent (mutagen).
- Ethanol 100% or 95% in distilled water
- PBS-Tw0.1, 0.1% Tween-20 (Merck) in PBS
- PBS-Tw0.5, 0.5% Tween-20 (Merck) in PBS
- HCl, HCl 4N
- Borax, Borax 0.1 M, pH 8.5
- Anti-BrdU, mouse monoclonal IgG1 anti-human BrdU (Dako)
- GAM-Cy5, cyanine5-conjugated goat anti-mouse anti-IgG (Jackson); Alternative: GAM-FITC, Fluorescein isothiocyanate (FITC)-conjugated goat anti-mouse anti-IgG (Jackson)
- Ho, Hoechst 1 μM in PBS
- Fluorescence mounting medium (Dako). Specific mounting medium is necessary to prevent photobleaching during image acquisition.
- Alternative: antifading solution (Glycerol 9V, DABCO 1V), 0.233g DABCO in 10 mL tris-HCL 0.2M (1V), add 0.216 g NaN_3 and 90 mL Glycérol (9V). qs 100 mL. Stir with a magnetic stirrer to homogenize the solution. Store at –20°C protected from light. [Tris-HCL 0.2M, i.e., Tris-HCl 1M stock solution 2 mL + deionized water 8 mL]. [Tris-HCL 1 M stock

solution, i.e., 12.1 g trizma base in 80 mL deionized water, adjust to pH = 8 with HCl 12N, qsp 100 mL]. Stock antifading solutions at –20°C in the dark.
- Epifluorescence microscope equipped with filter blocks
  - Cy5 filter set, EX BP 640/30, BS FT 660, EM BP 690/50
  - Ho filter set, excitation filter, EX G 365, BS FT 395, EM BP 445/50
- Alternative: FITC filter set, EX BP 475/40, BS FT 500, EM BP 530/50 (EX = excitation, BS = beam splitter, EM = emission)

4.3.2   BrdU INCORPORATION

- Plate the cells on the culture Lab-Tek glass chamber slides with appropriate culture medium.
- Culture the cells until exponential phase is reached.
- Replace culture medium with fresh culture medium containing BrdU, 20 µM final.
  - Incubate the cultures in a humidified CO_2 incubator at 37°C for 30 min.
  - Make BrdU incorporation during exponential growth stage. Depending on the cells, this may be between the second and the third day after seeding the chamber slides.
  - Increase the pulse incubation time (45 min) if cells develop slowly (cell cycle duration > 30 h).
  - Continuous BrdU incorporation should be longer time than cell cycle duration.
- At the end of incubation time, rinse the cells in PBS, two times, 3 min at room temperature.
- In a pulse and chase strategy
  - Replace by fresh culture medium for the necessary time, then rinse the cells with PBS, two times, 3 min at room temperature before next step.
- Using a pulse and chase strategy will need to prepare as many slides as sampling times investigated. Suggestion: sampling each hour postincubation over a period estimated equivalent to cell cycle duration (15 to 30 h depending on cell type).

4.3.3   BrdU DETECTION

4.3.3.1   Fixation
- Fix the cells in ethanol 70%, 30 min at room temperature. Slides can be stored 5 days in EtOH70 at 4°C.
- Rinse the cells in PBS two times, 5 min

4.3.3.2   Denaturation
- Incubate the cells in HCl 4N, 10 min at room temperature.

Double-stranded DNA denaturation is necessary to improve the accessibility of antibody to incorporated BrdU.

4.3.3.3   Neutralization
- Immerse slides in borax, 5 min at room temperature. Borax is necessary to neutralize HCl.
- Rinse the cells twice in PBS-Tw 0.5, 5 min each.

4.3.3.4   Immunolabeling
- Primary IgG antibody anti-BrdU. Incubate the cells with 300 µL anti-BrdU (diluted 1/50 in PBS-Tw 0.1) for 60 min at room temperature in wet chamber.
- Rinse in PBS two times, 5 min each.
- Secondary antibody GAM-Cy5. Incubate the cells with GAM-Cy5 (diluted 1/30 in PBS-Tw 0.1), 30 min at room temperature. Alternative: GAM-FITC in the same conditions.

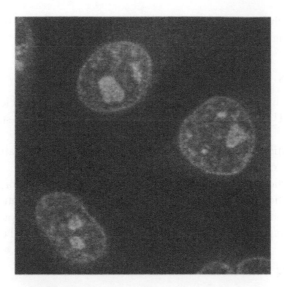

FIGURE 4.2   DNA staining with Hoechst.

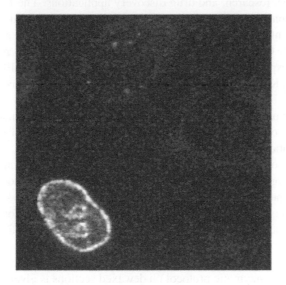

FIGURE 4.3   Immunofluorescence detection of BrdU molecules.

- Rinse in PBS-Tw 0.1 three times, 5 min each.
- Rinse in PBS two times, 10 min each in wet chamber.

4.3.4   DNA COUNTERSTAINING

If no DNA counterstaining is applied, go to step "Dry by slight absorption with filter paper."

- Stain the cells with the Hoechst solution for 30 min at 37°C in the dark.
- Rinse the cells twice in PBS, 5 min each.
- If no DNA counterstaining is applied, go directly to this step.
- Dry by slight absorption with filter paper.
- Add the fluorescence mounting medium. If no DNA counterstaining is applied, go directly to this step.

- Alternative: Use the antifading solution. In case of the use of a fluorescent mounting medium containing glycerol, seal the coverslip with nail polish.

4.3.5 INTERPRETATION

Observe the cells through an epifluorescence microscope. Two types of nuclei can be distinguished: labeled nuclei after BrdU incorporation (Figure 4.3) and nuclei only DNA stained (Figure 4.3).

- Count the proportion of labeled nuclei after BrdU pulse incorporation. Parameter estimated: proportion cell in S phase.
- Count the proportion of labeled nuclei after continuous BrdU incorporation. Parameter estimated: proportion of cycling cells, that is, coefficient of proliferation.
- Count the proportion of labeled nuclei at each sampling time after pulse and chase BrdU incorporation. Parameters estimated: cell cycle phases durations and cell cycle duration.

4.4 Ki-67 PROTEIN

The Ki-67 antigen (Ki-67) is a classic marker of cellular proliferation and has found widespread application in diagnostic, research, and drug discovery applications. The Ki-67 antigen was originally defined by the monoclonal antibody Ki-67, the name being derived from the city of origin (Kiel) and the number of the original clone in the 96-well plate. Ki-67 antigen is preferentially expressed during the late G1, S, G2, and M phase of the cell cycle; whereas resting, noncycling cells (G0 phase) lack Ki-67 expression. Thus, Ki-67 is commonly used as a proliferation marker.

What to expect from Ki-67 assay compared to BrdU incorporation assay?

- Immunodetection of Ki-67 protein needs no denaturation or neutralization.
- Immunodetection of pulse BrdU incorporation leads to labeling of cells in the S phase during pulse duration.
- Immunodetection of Ki-67 protein leads to labeling of all proliferating cells; unlabeled cells are not cycling cells. The proportion of labeled cells gives the coefficient of proliferation of the cell population observed.
- Immunodetection of continuous BrdU incorporation (if incorporation time is longer, i.e., superior to cell cycle duration) leads also to the estimation of the coefficient of proliferation.

The advantage of Ki-67 labeling is that it can be used as a practical routine procedure for histopathologists. The immunoenzymatic protocol on dewaxed sections is given in Section 4.5.2.

4.4.1 MATERIALS AND EQUIPMENT

- Lab-Tek glass chamber slides
- Appropriate culture medium (supplemented with fetal bovine serum [FBS] or serum substitute)
- PBS, phosphate-buffered saline (Invitrogen)
- PFA4, paraformaldehyde (Sigma) 4% in PBS, pH 7.2
- MeOH (Merck)
- PBS-Tw 0.1, 0.1% Tween-20 (Merck) in PBS
- Normal goat serum (Dako)
- Anti-Ki67, mouse monoclonal IgG1 anti-human MIB1 (Dako)
- GAM-FITC, fluorescein isothiocyanate (FITC)-conjugated goat anti-mouse anti-IgG (Jackson).
- Ho, Hoechst 1 µM in PBS
- Fluorescence mounting medium (Dako). Alternative: antifading solution (Glycerol 9V, DABCO 1V). See Section 4.3.1. Stock antifading solutions at –20°C in the dark.

- Epifluorescence microscope equipped with filter blocks
  - FITC filter set, EX BP 475/40, BS FT 500, EM BP 530/50
  - Ho filter set, excitation filter, EX G 365, BS FT 395, EM BP 445/50

4.4.2   KI-67 DETECTION

- Plate the cells on the culture Lab-Tek glass chamber slides with appropriate culture medium
- Culture the cells until exponential phase is reached
- Rinse the cells in PBS, two times, 3 min at room temperature

4.4.2.1   Fixation

- Fix the cells in PFA4, 2 min at room temperature
- Postfix in methanol, 10 min at 20°C
- At this stage, slides can be stored for a maximum of 3 weeks in storage solution. (Storage solution: 250 mL glycerol + 250 mL PBS added with 42.8 g saccharose and 0.33 g magnesium chloride anhydrous. Stock at –20°C a maximum of 3 months.) After storage, rinse twice in PBS, 5 min each.
- Rinse the cells in PBS two times, 5 min.

4.4.2.2   Immunolabeling

- Incubate normal goat serum 10% in PBS, 15 min at room temperature to block nonspecific protein–protein interactions.
- Primary IgG antibody anti-human MIB1. Incubate the cells with 300 μL anti-human MIB1 (diluted 1/50 in PBS), 60 min at room temperature in wet chamber.
- Rinse in PBS two times, 5 min each
- Secondary antibody GAM-FITC. Incubate the cells with GAM-FITC (diluted 1/30 in PBS-Tw 0.1), 30 min at room temperature.
- Rinse in PBS-Tw 0.1 three times, 5 min each
- Rinse twice in PBS, 5 min each

If no DNA counterstaining is applied, go to step "Dry by slight absorption with filter paper." See Section 4.4.3.

4.4.3   DNA COUNTERSTAINING

- Stain the cells in the Ho solution for 30 min at 37°C in the dark.
- Rinse the cells twice in PBS, 5 min each.
- Dry by slight absorption with filter paper. If no DNA counterstaining is applied, go directly to this step.
- Add the fluorescence mounting medium. In case of the use of a fluorescent medium containing glycerol, seal the coverslip with nail polish.

4.4.4   INTERPRETATION

Observe the cells through an epifluorescence microscope. Counting the proportion of Ki-67 positive cells (i.e., cells with nuclei labeled for both DNA content [Figure 4.4] and Ki-67 content [Figure 4.5]) will give the coefficient of proliferation of the cell population.

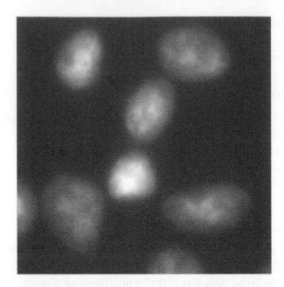

FIGURE 4.4   DNA staining with Hoechst.

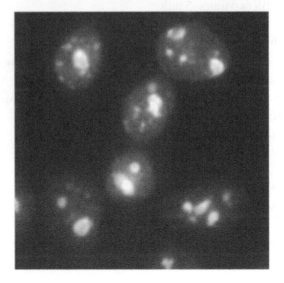

FIGURE 4.5   Immunofluorescence detection of Ki-67 protein.

4.5   AgNORs

Nucleolar Organizer Regions (NORs) are defined as nucleolar components containing a set of argyrophilic proteins, which are selectively stained by silver methods. After silver-staining, the NORs can be easily identified as black dots exclusively localized throughout the nucleolar area, and are called "AgNORs." The NORs argyrophilia is due to a group of nucleolar proteins, which have a high affinity for silver (AgNOR proteins; see Chapter 2, Section 2.1.3.3).

Experimental evidence supports the correlation between the cell cycle time of proliferating cells and the size of the ribogenesis factory assessed by the quantity of its specific proteins revealed by silver affinity. The size of silver precipitates measured by image analysis in cycling cells proved to be inversely proportional to the cell cycle time and provided a significant correlation with prognosis for a large spectrum of cancers.

AgNORs evaluation appears as a static marker of cell cycle time, which in combination with the Ki-67/MIB1 antibody, would provide pathologists with a direct measurement of the two parameters responsible for cell proliferation in normal and malignant tissues: coefficient of proliferation and relative cell cycle duration.

4.5.1   MATERIALS AND EQUIPMENT

- PBS, phosphate-buffered saline (Invitrogen)
- PFA4, paraformaldehyde (Sigma) 4% in PBS, pH 7.2
- Methanol (Merck)
- PBS-Tw 0.1, 0.1% Tween-20 (Merck) in PBS
- Normal goat serum (Dako)
- Anti-Ki67, mouse monoclonal IgG1 anti-human MIB1 (Dako)
- GAM-PO, alkaline phosphatase AP-conjugated goat anti-mouse anti-IgG (Jackson)
- Fast red, visualization kit, fast red TR/naphthol AS-MX tablets (Sigma)
- $AgNO_3$, silver nitrate solution, solve 50 g AgNO3 (Sigma) in 100 mL distilled water. Silver nitrate stains clothes lab benches and fingers black, normally after about 10 to 16 h. It is corrosive, but not really dangerous to human beings and is used occasionally as a disinfectant. The skin sheds and the black spots disappear after a few days. It is advisable to wear gloves, as the black spots are not very attractive. Cloths and lab benches are normally soiled for life.
- Preparation for two slides
  - AgNO, 1.2 g AgNO3 in 2400 µL distilled water
  - GelFA, 1200 µL
- GelFA, gelatin-formic acid solution, solve 2 g of gelatin (Sigma) and 1 g formic acid HCOOH (Merck) in 100 mL of distilled water in a water bath at 58°C to 60°C
- XAM mounting medium

4.5.2   KI-67 DETECTION

4.5.2.1   Fixation and Postfixation

The protocol could be applied to cytological samples as well as histological samples.

- Cytological sample
  - Fix the cells in PFA4, 2 min at room temperature. At this stage, slides can be stored for a maximum of 3 weeks in storage solution.
  - After storage, rinse twice in PBS, 5 min each (storage solution, see Section 4.2.1).
  - Postfix in MeOH, 10 min at −20°C
  - Rinse the cells in PBS, two times, 5 min each
- Histological sample
  - Dewax the sections in toluene (twice, 10 min each). Alternative: Xylene. Toluene and xylene toxicity causes irritation of skin, eyes, and respiratory system and brain effects.
  - Hydrate
    - Ethanol 100%, 10 min
    - Ethanol 95%, 5 min
    - Distilled water, 5 min
  - 10 mmol/L citrate buffer pH 6.0, 1 min at room temperature
  - Autoclave 1 bar 120°C, 20 min. Autoclave for antigen retrieval.
  - Cool to room temperature, 1 min
  - Take out to room temperature, 20 min

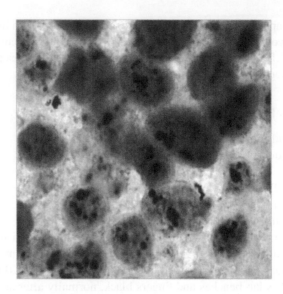

FIGURE 4.6   AgNORs labeling (black spots) with anti-Ki-67/alkaline phosphatase/fast red.

4.5.2.2   Immunolabeling

- Primary IgG antibody anti-human MIB.
- Rinse in PBS two times, 5 min each
- Secondary antibody GAM-AP
- Incubate the cells with GAM-AP (diluted 1/30 in PBS-Tw 0.1) for 30 min at room temperature in wet chamber. Alternative: GAM-PO, peroxydase. In this case desactive endogeneous peroxydase with H_2O_2. (H_2O_2/MeOH 0.3% (v/v) in distilled water, 30 min at room temperature.)
- Rinse in PBS-Tw 0.1, 5 min each
- Rinse in PBS, 5 min
- Alkaline phosphate revelation. Alkaline phosphatase revelation is based on an enzyme–substrate reaction. A fast red alkaline phosphatase kit is used. Fast red alkaline phosphate substrate, 10 min in the dark. Alternative: Peroxidase detection using a DAB peroxidase substrate. DAB is a carcinogen.
- Rinse twice in PBS, 5 min each
- Rinse twice in distilled water, 5 min each

4.5.3   AgNORs Detection

- Apply 1800 μL freshly made mixture, 2 vols AgNO/1 vol GelFA on the slide, 12 min at 37°C in wet chamber.
- Rinse twice in deonized water, 5 min each, with gentle agitation in the dark.
- Immerge in thiosulfate 5%, 30 s at room temperature in the dark. Use thiosulfate to clear the unfixed molecules of AgNORs.
- Rinse twice in deonized water, 5 min each, with gentle agitation in the dark.
- DNA counterstaining is not necessary because nuclei appear yellow to light brown and nucleoli brown.
- Mount with XAM medium. Mount in glycerol and aqueous medium if alkaline phosphatase detection was performed.

4.5.4 INTERPRETATION

Observe the cells through a light microscope (Figure 4.6). Counting the proportion of Ki-67 positive cells will give the coefficient of proliferation of the cell population. Measuring the relative area of AgNORs on Ki-67 positive nuclei will give an approximation of relative cell cycle durations.

REFERENCES

Giroud, F., G. Haroske, A. Reith, and A. Böcking. 1998. 1997 ESACP consensus report on diagnostic DNA image cytometry. Part II: Specific recommendations for quality assurance. *Analytical Cellular Pathology* 17:201–208.

Haroske, G., F. Giroud, A. Reith, and A. Böcking. 1998. ESACP consensus report on diagnostic DNA image cytometry. Part I: Basic considerations and recommendations for preparation, measurement and interpretation. *Analytical Cellular Pathology* 17:189–200.

5 Visualization of Apoptosis

Claire Brun and Elara Moudilou

CONTENTS

5.1 PRINCIPLE

Apoptosis, the programmed cell death, is used to eliminate cells in excess, damaged, or infected and potentially dangerous for the organism. The cells govern their own death, as a consequence of internal molecular signals, leading to a series of events requiring from a few hours to several days to overcome. The apoptotic cells detach from their neighboring cells and are readily phagocyted by macrophages or adjacent cells without generating an inflammatory response.

Generally, apoptosis is opposed to necrosis, as though there were only two modes of cell death. Today, other modes of cell death have been described that can be intermediary between apoptosis and necrosis. The distinction between apoptosis and necrosis is based on morphological, biochemical, and molecular changes. The most common morphological modifications of apoptosis are the loss of cell surface structures, condensation of cytoplasm and nuclei, margination of condensed chromatin at the nuclear membrane, and apoptotic body formation. At the biochemical and molecular level, externalization of phosphatidylserine, alteration of mitochondrial integrity and function, caspases activation, and loss of DNA integrity have been characterized. Some of these characteristics could be highlighted on tissue sections or on cell cultures.

Consequently, histological techniques used for detection of apoptotic cell death are based on:

- The nuclear morphology observed by routine histological staining and fluorescent DNA-binding dyes. The altered morphology of nuclei can also be identified by electron microscopy.
- The DNA fragmentation into characteristic internucleosomal (180–200 pb) and large (50–150 kb) fragments pointed out by an enzymatic labeling with modified nucleotides (TUNEL, or terminal deoxynucleotidyl transferase mediated dUTP nick end labeling) or with an immunohistochemical labeling after a formamide denaturation step (Apostain).
- The disruption of the mitochondrial transmembrane potential highlighted with fluorescent probes that accumulate in mitochondrion.

- The alteration in phospholipids distribution detected with Annexin V. Annexin V labeling cannot be done on tissue sections, because the outer and the inner membrane leaflets are exposed.
- The detection of apoptosis-related proteins by immunochemistry. Antisera against activated caspase-3 or caspase-cleavage products are actually available and allow the detection of biochemical alterations specific for apoptotic cells.

The techniques are very different depending on models used by scientists. The work on cell cultures allows a real-time study just after the apoptotic stimulus, whereas the use of tissue sections allows one to work on a whole organism and in a more physiologic system.

5.2 CELL CULTURES

Staining can be done on suspension cells or adherent cells. We will focus on adherent cells grown in chamber slides or on sterile coverslips placed in a Petri dish or well plate. For suspension cells, the staining method is equivalent but the work is done by successive centrifugations.

5.2.1 NUCLEAR MORPHOLOGY

The fragmentation of genomic DNA is a hallmark of apoptosis, and cellular DNA can be stained with routine histological staining or DNA fluorochromes capable of intercalating into DNA strands. Like routine histological staining has been described in Chapter 1 (see Section 1.4.3; see also Chapter 2, Section 2.7.8.), only specific DNA fluorochromes used to detect apoptosis will be detailed in this part. All the fluorescent-dyes mentioned in Chapter 2 are commonly used to study the nucleus morphology of fixed cells by fluorescent microscopy. All of them are supposed to be mutagens and should be handled with the appropriate precautions. These fluorochromes become highly fluorescent upon DNA binding and allow the study of DNA morphology. All the fluorescent staining described here can be realized on cell suspension and analyzed by flow cytometry, which permits one to determine the DNA contents of the cells.

An apoptotic nucleus is characterized by a chromatin condensation with aggregation along the nuclear envelope during early stages and the appearance of apoptotic bodies in the later stages.

5.2.1.1 Basic Staining

Basic staining is realized after a fixation of the cells and it can be done with all the fluorescent dyes mentioned in earlier chapters (see Chapter 2, Section 2.7.8). Only the two most frequently used—DAPI (4′,6-diamidino-2-phenylindole dihydrochloride) and propidium iodide—will be mentioned here.

5.2.1.1.1 DAPI Staining

- Fixatives—Paraformaldehyde 3.7% prepared in PBS
- Reagents
  - Paraformaldehyde 3.7%
  - Phosphate buffered saline (PBS): See Chapter 12, Section 12.3.7.
  - Absolute methanol
  - DAPI stock solution
    - DAPI, 10 mg
    - Distilled water, 10 mL. Prepare aliquots and stock at –20°C.
  - Pretreatment solution
    - Citric acid, 2.1 g
    - Tween 20, 0.5 mL
    - Distilled water, 99.5 mL
  - Staining solution
    - Citric acid, 11.8 g
    - DAPI stock solution, 200 μL
    - Distilled water, 99.8 mL

- – The staining solution can be centrifuged at 13,000 g for 1 min to remove particles. The staining takes about 30 min.
- Protocol
  - Wash cells twice with PBS
  - 3.7% paraformaldehyde, 15 min at room temperature (fixation of cells)
  - Wash twice with PBS
  - Absolute methanol, 5 min at room temperature (permeabilization of cells)
  - Wash in PBS
  - Cover cells with 1 volume of pretreatment solution
  - Incubate 5 min at room temperature
  - Add 6 volumes of staining solution
  - Incubate 5 min at room temperature
  - Remove the liquid and the culture chamber. If cells have grown on sterile coverslips, invert coverslips in the fluorescent mounting medium.
  - Mount in fluorescence mounting medium. Mounting medium helps to reduce fading of fluorescence during the microscopic observation and to preserve fluorescence for longer periods if the slide is stored in the dark at 4°C. There are a lot of commercially available preparations.
- Results—DAPI is a blue-emitting fluorochrome. Excitation, 358 nm; emission, 461 nm. The blue fluorescence is brighter in apoptotic cells than in healthy cells. Nonapoptotic cells have round and a uniformly stained nucleus with a clear margin. At the beginning of apoptosis, chromatin is condensed on a nucleus margin, and later, apoptotic bodies can be observed.

5.2.1.1.2   Propidium Iodide Staining

- Fixatives—Paraformaldehyde 3.7% prepared in PBS
- Reagent
  - Paraformaldehyde 3.7%
  - PBS: See Chapter 12, Section 12.3.7.
  - Absolute methanol
  - Propidium iodide
  - Propidium iodide stock solution
    - – Propidium iodide, 10 mg. Protect from light and store at 4°C (stable for 6 months).
    - – PBS, 10 mL
  - Staining solution
    - – Stock solution, 10 μL
    - – PBS, 990 μL
- Protocol—Staining takes about 30 min
  - Wash cells twice with PBS
  - 3.7% paraformaldehyde, 15 min at room temperature (fixation of cells)
  - Wash twice with PBS
  - Absolute methanol, 5 min at room temperature (permeabilization of cells)
  - Wash in PBS
  - Cover cells with the staining solution
  - Incubate, 5 min at room temperature
  - Wash cells in PBS
  - Remove the liquid and the culture chamber
  - Mount in fluorescence mounting medium. If cells have grown on sterile coverslips, invert coverslips in the fluorescent mounting medium.
- Results—Propidium iodide is a red-emitting fluorochrome (excitation, 536 nm; emission, 617 nm). All nuclei are red, but nonapoptotic cells have round and uniformly stained nuclei with a clear margin. At the beginning of apoptosis, chromatin is condensed on nucleus margin, and later, apoptotic bodies can be observed.

5.2.1.2 Dye Exclusion Method

The dye exclusion method uses the differential staining of cells according to their viability and their membrane permeability. The permeability of the plasma membrane is a key difference between necrosis and apoptosis. Exclusion dyes, such as propidium iodide, do not allow one to label apoptotic cells until the final lysis stage, because viable cells cannot be entered because of their large size. On the other hand, smaller dyes, like DAPI, Hoechst 33258 or 33342, or acridine orange, are permeable DNA stains that enter in cells and label both apoptotic and normal cells. Consequently, a double staining using one of each category of dyes is a rapid method to detect apoptotic cells.

The most used association is Hoechst 33342 and propidium iodide. Hoechst reagent stains all nuclei (apoptotic ones being brighter than healthy ones), and propidium iodide stains only dead cells (late apoptosis or necrosis).

- Reagent
  - Hoechst stock solution
    - Hoechst 33342, 10 mg
    - Distilled water, 2 mL. Protect from light and store at 4°C (stable for at least 1 month).
  - Propidium iodide stock solution
    - Propidium iodide, 5 mg
    - Distilled water: 5 mL. Protect from light and store at 4°C (stable for 6 months).
  - Staining solution—Dilute 1 µl of each dye in 1 ml of PBS (see Chapter 12, Section 12.3.7).
- Protocol—Staining takes about 30 min
  - Induce apoptosis and set negative controls. Different controls should be performed without apoptosis induction, without propidium iodide, without Hoechst 33342, and finally without any of the two dyes.
  - Wash cells twice with PBS
  - Stain cells, 30 min at room temperature
  - Wash cells with PBS
  - Remove the liquid and the culture chamber
  - Mount in fluorescence mounting medium. If cells have grown on sterile coverslips, invert coverslips in the fluorescent mounting media.
- Results—Hoechst 33342 is a blue fluorescent dye (excitation, 350 nm; emission, 461 nm) and propidium iodide is a red fluorescent dye (excitation, 535 nm; emission, 617 nm). Viable cells show only a low level of fluorescence. Dead cells show a high red fluorescence and a low blue fluorescence, whereas apoptotic cells show a condensed bright blue chromatin.

5.2.2 Disruption of the Mitochondrial Transmembrane Potential

The loss of the mitochondrial transmembrane potential is one of the earliest events of apoptosis, and it can be detected by the use of mitochondrial potential sensors (lipophilic cationic dyes as JC-1 (5,5′,6,6′-tetrachloro-1,1′,3,3′-tetraethylbenzamidazolocarbocyanin iodide), JC-9 (3,3′-dimethyl-α-naphthoxacarbocyanine iodide), TMRE (tetramethylrhodamine ethyl ester), and TMRM (tetramethylrhodamine methyl ester). These dyes exhibit a potential dependent accumulation in mitochondria.

- Principle—In healthy cells, JC-1 concentrates in mitochondrial matrix and aggregates, giving a red fluorescence. When the mitochondrial membrane potential collapses, JC-1 disperses in the cytoplasm under its monomeric form (monomeric form is green fluorescent).
- Reagent—Prepare a 1000X stock solution of JC-1 in DMSO (5 mg/ml). Divide the stock solution in aliquots and store at –20°C.

- Protocol—Staining takes less than 30 min
  - Wash cells twice with PBS
  - Induce cell apoptosis and set a negative control
  - Dilute JC-1 in prewarmed culture medium to a final concentration of 5 μg/mL (vortex for better solubilization)
  - Staining solution, 15 min at 37°C
  - Wash twice with PBS
  - Remove the liquid and the culture chamber
  - Mount with medium for fluorescence staining. Mount with medium for fluorescence. (See Chapter 1, Section 1.5.6.) If cells have grown on sterile coverslips, invert coverslips in the fluorescent mounting media.
- Results—In healthy cells, polarized mitochondria are marked with a spotted orange-red fluorescent staining. According to their stages of apoptosis, apoptotic cells show a diffuse green with less and less red fluorescence. For fluorescence microscopy use a band-pass filter, detecting FITC (excitation, 488 nm; emission, 530 nm) and rhodamine (excitation, 535 nm; emission, 617 nm). For confocal laser scanning microscopy, the monomer and J-aggregate forms can be excited simultaneously by 488 nm argon-ion laser sources.

5.2.3 CASPASES ACTIVATION

There are principally two systems available:

- Use of permeate and nontoxic fluorescent inhibitors of caspases in order to detect exclusively active caspases by a specific binding to their active sites
- Use of caspase substrates that only fluoresce after cleavage

These products are distributed by different companies in kits and are usable on cell culture and cell suspension.

5.2.4 PLASMA MEMBRANE CHANGES

Cell undergoing apoptosis shows a phospholipid asymmetry within a particular phosphatidylserine translocation from the cytoplasmic side to the external side of the plasma membrane. Annexin V preferentially binds to negatively charged phospholipids in the presence of Ca^{++}. When Annexin V is conjugated to FITC (fluorescein isothiocyanate), it allows identification of early apoptotic cells. Furthermore, an association with propidium iodide staining permits distinguishing between early (only Annexin V positive cells) and late apoptosis and necrosis (Annexin V and propidium iodide positive cells). Cells become Annexin V positive after the apoptotic nuclear condensation has started. Annexin V apoptosis detection kits are commercially available.

- Reagent
  - Annexin V-FITC (150 μg/mL)
  - Annexin V binding buffer
  - Propidium iodide
  - Annexin V binding buffer (200 mL)
    - HEPES, 476 mg
    - NaCl, 1.636 g
    - $CaCl_2$, 110 mg
    - Dissolve in 150 mL
    - Adjust pH to 7.4 with NaOH
    - Complete volume to 200 mL

- Propidium iodide stock solution
  - Propidium iodide, 5 mg. Protect from light and store at 4°C (stable for 6 months).
  - Distilled water, 5 mL
- Protocol—The procedure takes about 30 min
  - Wash cells twice with PBS: See Chapter 12, Section 12.3.7.
  - Induce cell apoptosis and set negative controls. Different controls should be performed without apoptosis induction, without propidium iodide, without Annexin V, without any of the two molecules, and finally with unlabeled Annexin V.
  - Place cells in 500 μL of binding buffer
  - Add 1 μL Annexin V-FITC and 1 μL propidium iodide in the binding buffer
  - Incubate in the dark, 5 to 15 min
  - Wash twice with PBS
  - Remove the liquid and the culture chamber
  - Mount in fluorescent mounting medium. If cells have grown on sterile coverslips, invert coverslips in the fluorescent mounting medium.
- Results—FITC is a green fluorescent dye (excitation, 488 nm; emission, 530 nm) and propidium iodide is an orange-red fluorescent dye (excitation, 535 nm; emission, 617 nm). Annexin V can only be employed on adherent cells or on cell suspension, because the enzymatic treatment used to detach cells can affect Annexin V binding. Furthermore, this labeling is not useful for use on tissue sections because the two leaflets are exposed. Early apoptotic cells show a green staining on the plasma membrane. Cells of the membrane integrity that have been lost (late apoptosis or necrosis) have a red nucleus and a green plasmic membrane.

5.3   TISSUE SECTIONS

The detection of apoptosis on tissue sections principally utilizes the DNA fragmentation or the immunolabeling of apoptosis-related proteins.

5.3.1   THE TUNEL ASSAY

The TUNEL assay is the historical method, which is still widely used today. TUNEL apoptosis detection kits are commercially available. All the procedures take about 4 h.

5.3.1.1   Original Method

- Fixatives—Paraformaldehyde 4% diluted in PBS is recommended for all types of samples (frozen sections, cell cultures, cell smears, or paraffin-embedded tissues). Sections must be affixed on silane-coated or superfrost+® slides.
- Principle—The TUNEL reaction permits the detection of the characteristic internucleosomal double-strand breaks by TdT (terminal deoxynucleotidyl transferase). The TdT enzyme work is template independent. This technique is based on the incorporation of labeled nucleotides (digoxigenin, fluorescein nucleotides) at free 3′-OH ends of cleaved DNA.
- Reagents
  - Proteinase K nucleases free
  - PBS: See Chapter 12, Section 12.3.7.
  - H_2O_2
  - TUNEL kit. TUNEL apoptosis detection kits are commercially available.
  - Diaminobenzidine (DAB)
  - Hematoxylin
- Protocol—Procedure takes about 4 h

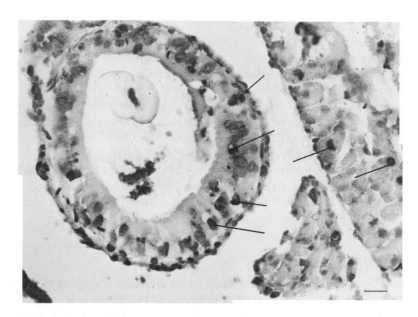

FIGURE 5.1 Visualization of apoptosis with TUNEL method. Amphibian intestine. Labeled nuclei appear as intense black (lines). Bar = 25 μm.

- Deparaffinize and rehydrate. Add additional tissue sections for the negative and positive labeling controls in each experiment to validate the technique.
- Nuclease-free proteinase K (20 μg/mL in 10 mM Tris/HCl buffer, pH 7.4) treatment, 30 min at 37°C. Proteinase K allows the high condensed chromatin structure accessible to reagents and facilitates TdT enzymatic activity. An alternative microwave pretreatment (350 W) in 0.1 M citrate buffer, pH 6.0 for 5 min, could be useful on material with a long postfixation time.
- Rinse twice in PBS
- Quench endogenous peroxidase activity in 3% H_2O_2 diluted in PBS, 10 min at room temperature
- Rinse
- Dry area around sample with a tissue paper or apply hydrophobic resin.
- Prepare the TUNEL reaction mixture (TdT and nucleotides labeled with fluorescein in appropriate concentration) and store on ice. For negative control, TUNEL mixture is prepared without TdT. For positive control, the sections are incubated with DNase I (3000 U/mL in 50 mM Tris-HCl, 10 mM MgCl2, pH 7.5, 1 mg/mL BSA) for 10 min at 15°C to 25°C to induce DNA strand breaks. The DNase I reaction is stopped by washing with PBS three times for 5 min.
- Cover sections (samples and positive control) with 50 μL TUNEL mixture and the negative control with 50 μL TUNEL mixture without TdT and incubate in a humidified chamber for 40 to 60 min at 37°C.
- Rinse in PBS. Labeled cells can be directly detected with fluorescence microscopy or flow cytometry for suspension cells (excitation wavelength in the range of 450 to 500 nm and detection under a wavelength in the range of 515 to 565 nm). But on tissue section, the use of immunohistochemistry is particularly sensitive to visualized incorporated nucleotide (Figure 5.1).

5.3.1.2 Chromogenic Method
- Protocol
  - Peroxydase-conjugated antifluorescein antibody, 50 μL

- Incubate, 30 min at 37°C
- Wash in PBS, 3 × 5 min
- Apply DAB and monitor staining closely with a microscope
- When desired staining is achieved, rinse to stop development
- Counterstain with hematoxylin
- Dehydrate and mount. To quantify a number of apoptotic cells per square micrometer (μm^2), be aware to dilute hematoxylin to obtain a maximal contrast between the dark brown positive nuclei and the pale blue negative ones. Use Mayer's or Lily's hematoxylin, which give blue nuclei. See Chapter 7, Section 7.2.1.11.
- Results—Apototic nuclei are brown and the rest of the tissue is blue. The labeled nucleotides can also be detected by alkaline phosphatase-labeled secondary antibody with NBT/BCIP (nitro blue tetrazolium/5-bromo-4-chloro-3-indolyl phosphate) as chromogen. In this case, use levamisole to block endogenous enzymes.

5.3.2   THE APOSTAIN METHOD

- Principle—The Apostain technique is based on the increased sensibility of DNA in apoptotic cells to thermal denaturation. DNA is denatured by heating at a relatively low temperature (56°C) in the presence of formamide and stained with an anti single-stranded DNA (ssDNA) monoclonal antibody. Formamide is suspected to be carcinogenic (gloves and protection clothes should be worn).
- Fixatives—Neutral buffered formalin 4% to 10% or paraformaldehyde 4% diluted in PBS are recommended (for 24 hours at 4°C). Different types of samples can be used (frozen sections, cell suspensions, or paraffin-embedded tissues). Frozen sections must be fixed in ice-cold paraformaldehyde diluted in PBS for 10 min and dehydrated in 80% methanol diluted in PBS for 30 min. After air-drying, the permeabilization step can only be done without proteinase K (only in 0.2 mg/mL saponin for 20 min). Cell suspension can be used for flow cytometry or fluorescent microscopy. For archival samples fixed for a long time, an additional enzymatic treatment with pronase E (20 min at 20 μg/mL) can be done after the permeabilization step with saponin and proteinase K.
- Reagents
  - Nonfat dry milk
  - Deionized formamide
  - PBS: See Chapter 12, Section 12.3.7.
  - Saponin
  - Proteinase K
  - H_2O_2
  - Monoclonal antibody F7-26
  - Peroxydase antimouse IgM antibody. Anti-ssDNA monoclonal antibody belonging to Apostain kit.
  - DAB
  - Hematoxylin
- Protocol
  - Dewax and rehydrate
  - During deparaffinization and rehydration, prepare:
    - 3% nonfat milk solution in distilled water and prewarm at 37°C
    - 50% formamide solution in distilled water and prewarm at 56°C to 58°C
    - PBS tank and cool down at 4°C
  - Incubate slides in 0.2 mg/mL saponin and 20 μg/mL proteinase K diluted in PBS, 20 min at room temperature
  - Rinse in distilled water

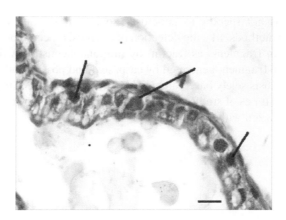

FIGURE 5.2 Visualization of apoptosis with Apostain method. Larval amphibian gill. Labeled nuclei appear as intense black (lines). Bar = 25 μm.

- Wipe around the sections with a tissue paper or apply hydrophobic resin
- Apply preheated 50% formamide on the slides and incubate in a humidity chamber, 20 min at 56°C to 58°C. Formamide is a gentle DNA denaturing agent. Hydrogen bonds are formed between formamide and DNA bases, preventing the formation of inter-base bonds. Consequently, DNA melting temperature is linearly lowed (approximately 0.6°C per% formamide).
- Transfer slides in ice-cold PBS, 5 min
- H_2O_2 3% diluted in PBS, 15 min. This operation is used to block endogenous peroxydases.
- Rinse in PBS, 3 min
- Preheated 3% nonfat dry milk, 20 min at 37°C
- Wash in PBS
- Add mouse monoclonal anti-ssDNA antibody and incubate in humidity chamber, 30 min at room temperature. Normal serum can be added on negative controls.
- Wash in PBS, 3 × 3 min
- Apply peroxydase anti-mouse IgM and incubate in a humidity chamber, 30 min at room temperature
- Wash in PBS, 3 × 3 min
- Apply DAB and monitor staining closely with a microscope
- When desired staining is achieved, rinse to stop development
- Counterstain with hematoxylin. To quantify number of apoptotic cells per square micrometer, be aware to dilute hematoxylin to obtain a maximal contrast between the dark brown positive nuclei and the pale blue negative ones. (Use Mayer's or Lily's hematoxylin: See Chapter 4, Section 7.2.1.11.)
- Dehydrate and mount.
- Results—Apoptotic nuclei are brown and the rest of the tissue is blue. If no signal is detected with an anti-IgM antibody, a polymer-based detection reagent or an avidin-biotin amplification (but cytoplasmic background can appear) can be planned (Figure 5.2).

5.3.3 Conclusions

From these two methods, it has been proven that TUNEL positive cells can be observed during replication, DNA reparation, or necrosis, which is not the case with Apostain. So, it seems that Apostain is safer and can detect apoptosis to an earlier stage.

DNA fragmentation is not specific to apoptosis, so it is important to associate TUNEL or Apostain with additional methods, like the detection of apoptosis specific proteins (active caspase-3 or caspase-cleaved protein fragments as fractin) by immunochemistry. A double staining can be envisaged. Fractin is actin fragment generated by caspase 3 and possibly other caspases.

Quantification of apoptosis solely based on histological staining is sensitive, but observer dependent and tedious, so these types of studies must be carried out on a large numbers of samples (6 to 12 per condition) and analyzed with statistics.

6 *In Situ* High-Speed Tissue Proteomic

Françoise Giroud and Marie-Paule Montmasson

CONTENTS

6.1   TISSUE MICROARRAY (TMA) DESIGN AND SAMPLING

Tissue microarray (TMA) technology provides an efficient way to analyze many different histological samples in a single experiment. Preparatory work must be done by the pathologist according to the following steps:

- Selecting cases for TMA construction
- Selecting donor blocks
- Making annotation of representative areas
- Designing the array

6.1.1   TISSUE COLLECTION

The initial identification and collection of tumor samples represents the greatest portion of the work associated with TMA construction (Figure 6.1). Most of the work (approximately 95%) devoted to TMA manufacturing is traditional pathology work that cannot be accelerated by machinery. A typical project involving 200 cases may take 2 months for case identification and collection but only 2 days for array building.

The preparatory work done by the pathologist includes:

- Identifying appropriate patient cohorts to address the proposed scientific or clinical question and with accessible archival tissue materials.

FIGURE 6.1   Collection of selected donor paraffin-embedded tissue blocks.

- Reviewing all HES (hematoxylin–eosin–saffron) sections from all candidate blocks to select the appropriate ones.
- Identifying the clinical material from which is issued the HES section, which is most often formalin-fixed paraffin-embedded tissue (the gold standard of diagnostic histopathology), usually paraffin blocks from a surgical pathology repository.
- Ensuring that the slide is representative of the block; if some blocks have been cut into for other clinical or research purposes, prepare fresh surface HES slides.
- Selecting the appropriate donor blocks. The number of blocks involved can be an order of magnitude higher in TMA studies compared to traditional studies involving "large" tissue sections.
- Checking each block by visual inspection. Each block will be visually controlled to ensure the quality and the quantity of the embedded tissue expected for the TMA preparation.
- Arranging the blocks in a comfortable manner consistent with the TMA map (see Section 6.1.3.5).

6.1.2 DONOR BLOCKS ANNOTATION

With the use of TMAs, most of the work time is now focused on the preparation of the TMA.

- Make a first section at each donor block surface
- HES stain these first sections
- Review by a pathologist
- Determine areas of interest
- Make a second serial section at each donor block surface
- Ki-67 label these second sections. Ki-67 protein expression is a marker for cell proliferation activity (Ki-67 protein; see Chapter 4, Section 4.4). Alternative: Any biological marker of major interest with the subject under study.
- Evaluate proliferative activity of areas of interest. Alternative: Evaluate activity of the marker under study.

6.1.2.1 HES Slide Annotation

- Reagents and solutions
  - Hematoxylin, Mayer's hemalum solution (Merck): See Chapter 7, Section 7.2.1.11.
  - Acid alcohol, 1% HCl in 70% ETOH
  - Eosine, Eosine 1% (Merck) in distilled water: See Chapter 7, Section 7.2.1.11.
  - Saffron, alcoholic solution
    - Saffron of Gâtinais (filaments), 10g
    - Dry in oven, 24 h at 56°C
    - Pulverize in a mortar
    - Infuse saffron powder in ethanol 100%, 500 mL
    - The solution is kept in an oven where it gets better
  - Alternative
    - Saffron of Gâtinais (filaments), 5 g
    - Ethanol 100%, 200 mL
    - Let boil 1 h
    - Keep in the dark at room temperature in a bottle tightly closed
- HES protocol
  - Dewax the sections in toluene (3 min, then two times, 2 min each). Alternative: Xylene. Toluene and xylene toxicity causes irritation of skin, eyes, and respiratory system, and brain effects.

- Hydrate
  - Ethanol 100%, 10 min
  - Ethanol 95%, 5 min
  - Distilled water, 5 min
- Hematoxylin, 2 min
- Rinse in running tap water, 2 min. Stop the rinsing, the water is no longer colored.
- Rinse in acid alcohol, 30 s. The sections turn pink.
- Rinse in running tap water, 1 min
- Eosine, 2 min
- Rinse in running tap water 2 min, 100% ethanol (2 baths, 1 min each)
- Saffron, 1/3 diluted, 2 min
- Dehydrate
  - Distilled water, 5 min
  - Ethanol 70%, 2 min
  - Ethanol 95%, 2 min
  - Ethanol 100%, 3 min
- Toluene, two baths, 1 min each. Alternative: Xylene. Toluene and xylene toxicity causes irritation of skin, eyes, and respiratory system and brain effects.
- Mount in XAM medium
- Pathologist annotations
  - A pathologist reviews each HES slide. HES slides can be scanned with a high-resolution microscope and the image stored.
  - Annotate representative areas to be punched for TMA construction. Annotations can be superimposed to the scanned HES slide (Figure 6.2).
  - Give a codification for areas of interest
- Register in the TMA datasheet the different areas of interest with codification and annotations

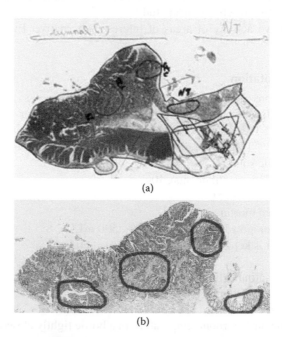

(a)

(b)

FIGURE 6.2 Example of a scanned HES slide with annotations superimposed. Example of correspondence between an HES slide and its Ki-67 partner (slides with two serial sections). (a) Scanned HES slide with annotations superimposed. (b) Scanned Ki-67 slide with report of regions of interest as annotated in the corresponding HES slide.

6.1.2.2 Proliferative Activity

- Alternative: Activity of the specific marker under study
- Reagents and solutions
  - PBS, Phosphate-buffered saline (Invitrogen)
  - MeOH (methanol)
  - PBS-Tw 0.1, 0.1% Tween-20 in PBS
  - Normal goat serum
  - Anti-Ki67 (Merck), Mouse monoclonal IgG1 antihuman MIB1 (Merck)
  - GAM-PO, Peroxydase PO-conjugated goat anti-mouse anti-IgG (Dako)
  - DAB, visualization kit FAST DAB enhancer tablets (Dako)
  - Hematoxyline (Jackson)
  - Ammoniacal water, 1/250 in distilled water (Sigma)
  - XAM mounting medium, Mayer's hematoxylin (Dako)
- Ki-67 labeling
  - Dewax the sections in toluene (twice, 10 min each). Alternative: Xylene. Toluene and xylene toxicity causes irritation of skin, eyes, and respiratory system and brain effects.
  - Hydrate
    - Ethanol 100%, 10 min
    - Ethanol 95%, 5 min
    - Ethanol 70%, 5 min
    - Distilled water, 5 min
  - H_2O_2/methanol 0.3% (v/v) in distilled water, 30 min at room temperature. H_2O_2 desactive endogeneous peroxydase.
  - Autoclave: 20 min at 1 bar 120°C. Autoclave for antigen retrieval.
  - Drop off the temperature, 15 min
  - Take out to room temperature, 20 min
  - Rinse the cells in PBS, two times, 5 min each
  - Primary IgG antibody anti-human MIB1. Incubate the cells with 300 μL anti-MIB1 (diluted 1/50 in PBS) for 60 min at room temperature in wet chamber.
  - Rinse in PBS two times, 5 min each
  - Secondary antibody GAM-PO. Incubate the cells with GAM-PO (diluted 1/30 in PBS-Tw 0.1), 30 min at room temperature.
  - Rinse in PBS-Tw 0.1, 5 min each. Operate in wet chamber.
  - Rinse in PBS, 5 min
  - Peroxidase revelation. DAB peroxidase substrate: 10 min in the dark. Peroxidase revelation is based on an enzyme–substrate reaction, a DAB peroxidase substrate kit is used.
  - Rinse twice in PBS, 5 min each
  - Rinse twice in distilled water, 5 min each
  - Hematoxyline, 2 min
  - Rinse in tap water, 1 min
  - Rinse in ammoniacal water, 30 s
  - Rinse in tap water three times, first time 1 min and twice for 5 min
  - Dehydrate
    - Ethanol 70%, 2 min
    - Ethanol 95%, 5 min
    - Ethanol 100%, 5 min
  - Toluene, two baths, 1 min each. Alternative: Xylene. Toluene and xylene toxicity causes irritation of skin, eyes, and respiratory system and brain effects.
  - Mount in XAM medium
  - Make observation under a light microscope (Figure 6.3)
  - Make correspondence with the area of interest identified on HES slides (Figure 6.2)

FIGURE 6.3   Example of a scanned Ki-67/DAB labeled (hematoxylin counterstained) slide.

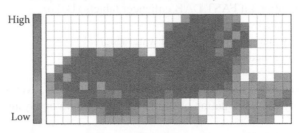

FIGURE 6.4   Density map of Ki-67 protein expression, visualization of levels of proliferative activity (from low to high).

6.1.3 SAMPLING STRATEGY AND TMA DESIGN

- Make final decision about donor blocks
- Make final decision about region of interest (ROI)
- Make decision about the core size and number of cores for each ROI
- Make decision about references
- Design geometry of TMA blocks
- Produce the corresponding datasheet

6.1.3.1 Density Map

- Make correspondence between the HES annotated first section and the Ki-67 labeled second serial section (Figure 6.2).
- Appreciate the repartition of proliferative activity in the whole section. Sampling strategy is exemplified here with respect to interest in proliferative activity of tissues under study.
- Appreciate heterogeneity in proliferative activity for the different regions of interest.
- If possible, visualize the density map for help in appreciation of marker heterogeneity in the full section. The density map, that is, visualization of marker levels, can be built from the scanned images using image analysis tools (Figure 6.4).
- Adjust the choice in the area of interest with respect to the information on density map.

6.1.3.2 Donor Block Selection

- Following:
  - The review of tissue collection (Section 6.1.1)
  - The annotation of selected donor blocks (Section 6.1.2.1)
  - The control of regions of interest with respect to the question under study (Section 6.1.2.2)
- Select the final appropriate donor blocks
  - That are of good quality
  - That are correctly annotated with respect to the objective of the study

The selected donor blocks were supposed to be controlled to ensure the quality and the quantity of the embedded tissue expected for the TMA preparation (Section 6.1.1). Among the selected donor blocks, retain those that are correctly documented after the annotation step.

- Arrange and keep the relevant tissue donor blocks in the order that is represented in the TMA map (Section 6.1.3.5).

6.1.3.3 Mini-TMA References Inclusion

Include mini-TMA references as internal references to reduce intra- and interlaboratories variabilities (Section 6.2.4.2).

- Use mini-TMA tissue references, or
- Use mini-TMA cell lines references, or
- Both mini-TMA references

Mini-TMA references sections should be placed on each slide with a TMA section.

- Alternative: Control cores (tissue or/and cell lines) should be placed in each TMA block during its construction.

6.1.3.4 TMA Geometry and Design

All these parameters can be user adjusted upon needs.

- Define core size and TMA replicas
- Define spacing between cores
- Define density core
- Define the dimension of TMA block
- Include internal references
- Core size and TMA replicas. Small 0.6 mm cores are made as a standard practice.

The TMA approach has been criticized for its use of small punches of usually only 0.6 mm diameter from tumors with an original size of up to several centimeters in diameter.

When compared with the original size of a tumor with a diameter of up to several centimeters, a 2 mm punch is hardly more "representative" than a 0.6 mm punch (0.6 mm diameter punch corresponds to an area of 0.28 mm^2; 2 mm diameter punch corresponds to an area of 3 mm^2). Larger punches (2 mm) cause considerable damage to the donor and acceptor blocks. They also lead to a greater chance of the blocks to break or crack during TMA processing. Smaller punches (0.6 mm) cause a lower incidence of lost cores during sectioning of the TMA. Use of smaller core diameters allows reduction in tissue material extracted from the donor block, a greater number of cores to be extracted, and construction of replicas of TMA blocks.

- Replicate at least three times the same TMA block. The greater the degree of intratumoral heterogeneity for any given marker, the greater the number of cores will be required (i.e., the greater the number of TMA replicas).
- Spacing between cores should not exceed the core diameter. It is easier for the microscopist to follow the rows and columns if he or she can "lead" from one core to another. It avoids false recording of data when performing manual interpretation.
- Density core and grid size
  - Put between 100 and 300 0.6 mm cores in a TMA block. Depending on the procedure used in the laboratory to immunostain slides, avoid placing so many cores on a TMA that the surface section of cores becomes larger than the antibody coverage area on the slide.

- Put the core at 3 mm (at least) away from the block edges. Diminishing too much the amount of paraffin at the edge of the block creates difficulties in sectioning.
- Internal references
  - Integrate negative and positive controls in your TMA, or
  - Add mini-TMA reference sections on each TMA slide (Section 6.2.4.2)
  - Control reference are chosen with respect to the study under investigation

6.1.3.5 TMA Map and Datasheet

Generate a TMA map and corresponding datasheet.

- TMA map is a virtual "row–column" representation of the TMA grid (Figure 6.5a)
- TMA datasheet should contain
  - TMA donor blocks identification number (Figure 6.5b).
  - TMA donor block associated basic patient data (or patient identification number) and clinical information
  - Hyperlink to corresponding full slide HES image and annotations (Figure 6.2a)
  - Hyperlink to eventual full reference marker images (i.e., Ki-67, Figure 6.2b) and corresponding density map
  - TMA core location (and real coordinates) on donor block
  - ROI histological and reference marker (i.e., Ki-67 marker) characterization
  - And all additional data available

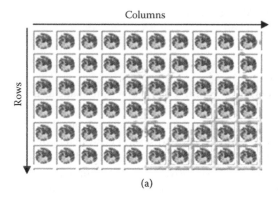

FIGURE 6.5   TMA map. (a) Virtual "rows/columns" TMA grid representation. (b) Visualization of donor blocks identification numbers to be used in the corresponding TMA grid.

6.2 TMA CONSTRUCTION AND PROCESSING

- Building of TMA block using an arrayer
- Sectioning of TMA block
- HES staining of TMA section to confirm morphology
- Labeling of serial TMA sections for biomarkers evaluation
- Producing datasheet with TMA grid, morphological details, and quantitative data

6.2.1 BUILDING A TMA BLOCK

6.2.1.1 Manual Tissue Arrayer

For the physical construction of the TMA, a tissue microarrayer is required and commercially available (Figure 6.6). Carousels with a larger number of positions are proposed to allow performance and productivity. A typical tissue microarrayer includes two hollow needles, one with the outer diameter equal to the inner diameter of the other, and a lock holder that operates on a manual basis. A manual punch extractor offers a good alternative for casual application. Automated arrayers are available for laboratories requiring higher throughput in array production.

6.2.1.2 Selection of Coring Coordinates

Selection of the coring positions in the donor blocks is easily and quickly achieved.

- Take the donor block and its corresponding annotated HES slide (Figure 6.7).
- Align the HES annotated slide on the block and locate the areas for puncture.

Depending on the system used, automated procedures are proposed to help for selection coring. For example, an image of the donor block displayed at the screen, superimposed with its matching HES control slide image is shown in Figure 6.8.

6.2.1.3 Transfer to Recipient Block

The TMA manufacturing is a simple three-step procedure that is repeated for each sample to be incorporated into the TMA:

- Prepare a blank paraffin block (Figure 6.9).
- Calibrate the punches for correct positioning at the surfaces of donor and recipient blocks. Exact positioning of the tip of the tissue cylinder at the level of the recipient block surface is crucial for the quality and the yield of the TMA block (Figure 6.10). The punches are coaxial, that is, the donor punch, recipient punch, and stylus are concentric. Procedure for calibration of the needles for exact positioning is dependent on the system used.

FIGURE 6.6 Punching equipment from Beecher instruments with a four-position recipient block.

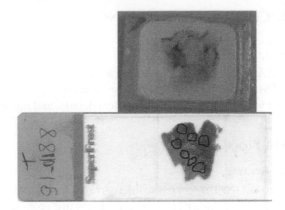

FIGURE 6.7   Donor block image and HES slide. The marks made by the user on the HES slide are shown.

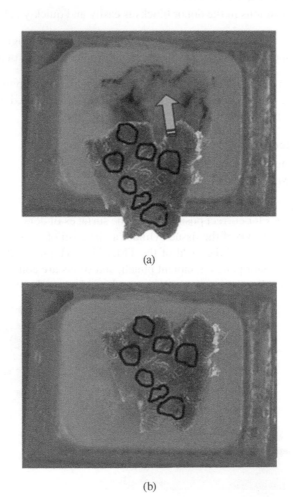

(a)

(b)

FIGURE 6.8   Sliding the annotated image of HES slide to be aligned and superimposed over the donor block.

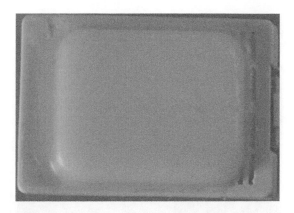

FIGURE 6.9   Blank paraffin block.

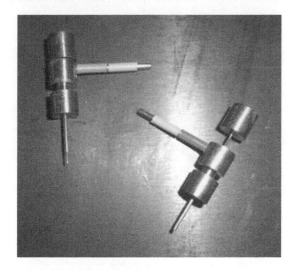

FIGURE 6.10   Punching needles with stylet and depth stopper.

- Remove a core from the blank paraffin "recipient" block (Figure 6.11).
- Use the second needle to remove a core of representative tissue from the donor block with respect to the planned TMA previously established. Remember the location read on the HES annotated slide (Figure 6.12).
- Insert the tissue core into the previously created hole in the recipient block (Figure 6.13). Placing the tissue too deep into the recipient block results in empty spots in the first sections taken from the TMA block. Not positioning the tissue cylinder deep enough causes empty spots in the last sections taken from this TMA. The success of this procedure depends on how the operator has calibrated and maneuvered the needles and the blocks.

6.2.1.4   TMA Matrix
- Prewarm the TMA block, 15 min at 37°C
- If tissue cylinders protrude, gently press deeper into the TMA block using a glass plates on the top of the TMA block (Figure 6.14). To create and improve the adherence between tissue cores and the recipient block, the TMA block has to be lightly heated to melt both. During this phase cores may misalign or the fusion may not be complete, resulting in core losses when sectioning.

FIGURE 6.11   Making a hole in an empty "recipient" paraffin block with the recipient punch.

FIGURE 6.12   Taking a cylindrical sample from the tissue sample (donor) paraffin block with the donor punch.

FIGURE 6.13   Placing the cylindrical tissue sample in the premade hole in the recipient block by pushing the stylus inside the donor punch.

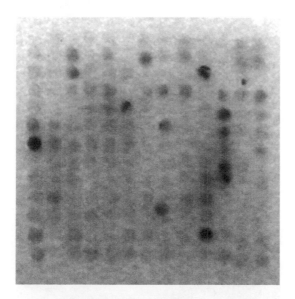

FIGURE 6.14 TMA matrix block, 169 cores, 0.6 mm punch.

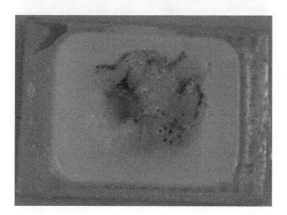

FIGURE 6.15 Donor block after coring.

6.2.1.5 Quality and Traceability

It is necessary to ensure complete traceability from tissue array design to tissue array interpretation through construction. Donor block handling should ensure quality and traceability:

- Annotate in the model datasheet with the different cores punched.
- Take a picture of the donor block after coring (Figure 6.15). Also note any comment made by the operator.
- Label the different cores with respect to the model TMA.

6.2.2 SECTIONING THE TMA BLOCK

The TMA block is subsequently sectioned on a microtome, resulting in TMA slides. Certified and skilled technologists should conduct this step.

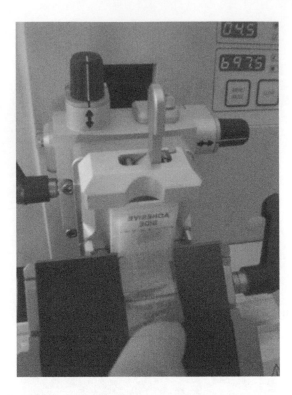

FIGURE 6.16   Cutting the section, the tissue slice is adhering to the tape.

6.2.2.1   Classical Method

TMAs can be sectioned like standard paraffin blocks using regular microtomes, but TMAs do not always section suitably, particularly if multiple tissue types are included in the array. Some cores could be preferentially lost depending on the hardness of the tissue. As a rule, it can be said that smaller cores are related to greater losses.

6.2.2.2   The Tape Technique

The tape sectioning technique facilitates cutting and leads to highly regular nondistorted sections.

- Place the TMA block on a microtome.
- Place an adhesive tape on the top of the TMA block immediately before cutting (Figure 6.16). Always use a hand roller in order to avoid bubbles.
- Cut section, usually 5 μm thick. The tissue slice is now adhering to the tape.
- Transfer the tissue slice on an adhesive-coated PAS slide using the hand roller (Figure 6.17). An adhesive-coated tape sectioning system reduces the tissue losses, especially when working with a 0.6 mm core. Store the adhesive-coated PSA slide in the dark before use. Adhering slide properties deteriorate rapidly under light exposure.
- Expose the slide, tissue on the bottom, to UV light for 30 s in order to polymerase the PSA between the tape and the slide.
- Dip the slide into TPC solvent at room temperature for a few seconds.
- Gently remove the tape from the slide. The tissue remains on the slide.
- Air-dry at room temperature.

The tape technique:

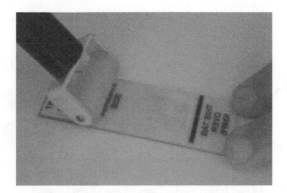

FIGURE 6.17 Transfer to adhesive coated PSA slice using hand roller to prevent bubbles.

- May prevent the organization of the core samples in the section (no stretch or distortion) assuring precise alignment of spots on the slide, and reduce folding of the section, tears, and sample losses.
- May prevent arrayed samples from floating off the slide if very harsh pretreatment methods are used.

The tape technique:

- Can produce up to 3 to 5 times more cuts per TMA block compared to the conventional method.
- Can produce a number of sections, which is dependent on the depth of the donor blocks, histotechnologist skill, and the thickness of individual sections. The typical number of sections obtainable in practice ranges from 50 to 150. TMA sections can be maximized by cutting the block as infrequently as possible, producing many sections in each cutting batch. If we cut the block each time as required, there will inevitably be some trimming of the block, leading to loss of tissue.

6.2.2.3 Preservation and Storage of TMA

- Air-dry overnight in a vertical position.
- Place in a closed container protected from dust. The slides are precious and need to be stored in a fashion that will protect the quality of all the tissue characteristics. Stored slides should not be stacked directly on top of each other or in direct contact. Some labs choose to store their slides in ziplock bags with desiccant in the refrigerator (4°C) or freezer (–20°C).
- Store at 4°C in the dark to prevent slide oxidation. Antigenicity is greatly affected by ambient air over time.

6.2.3 TMA Staining and Labeling

- HES stain the first TMA section for each TMA block (Section 6.1.2).
- HES stain one TMA slide at regular intervals (on every 15th slide) when sectioning each TMA block. HES stain for histopathological examination and diagnostic confirmation of tumor tissue compared to the initial full HES slide annotations (Section 6.1.2.1).
- Review HES-stained TMA slide by a pathologist. Morphological heterogeneity might occur in the surface of each donor block as well as in the depth. Histopathological examination and diagnostic confirmation is necessary when progressing in the depth of the TMA block (Figure 6.18). Immunohistochemistry (IHC) or other applications can be performed on TMAs with minimal changes to standard protocols. Functionally, the slides are nearly identical to regular slides.

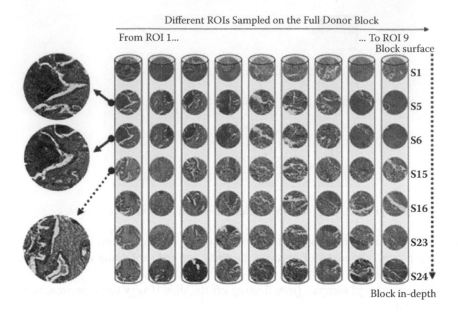

FIGURE 6.18 Illustration of morphological variations in the depth of a TMA block by examination of consecutive numbered sections (S1 to S24) through a TMA block in different regions of interest (ROI1 to ROI9).

FIGURE 6.19 Overview of an immunostained TMA tape section. Note that only a few cores are deteriorated; none are lost.

- Immunostain the serial TMA block sections with a large range of immunostaining.
- Perform FISH (fluorescent *in situ* hybridization) as well.
- Ensure that the labeling procedures that work in the laboratory are suitable for TMA slide labeling. Harsher antigen retrieval techniques or enzymatic digestions (e.g., deproteination for FISH) may cause some tissue to detach from the TMA slide and be lost to analysis. We do not recommend using a microwave for antigen retrieval with the array slides (Figure 6.19). The tape sections are quite hardy and withstand antigen retrieval well. But the sticky slides may increase background signal and slightly deteriorate morphology of

some tissue samples because of the small visible glue dots within the samples. Automated immunostainers can be used with care. Autostainers that use a liquid coverslip technology or a capillary method may not work as effectively with TMAs.

6.2.4 QUALITY CONTROL

6.2.4.1 Histological Characterization

An experienced anatomic pathologist examines the donor and TMA blocks through HES-stained sections (Section 6.1) for tissue content and agreement with associated clinicopathological data for proper histopathological characterization. Use HES staining and histological control is fundamental to the successful contribution of the TMA technology.

6.2.4.2 Mini-TMA Reference

Even though IHC has been in use now for decades, there is still a high variability of intralaboratory and interlaboratory results, mainly because of interlaboratory differences in antigen retrieval, staining protocols, antibodies used, and in the interpretation of staining results. TMAs can facilitate the standardization of immunohistochemical staining procedures and interpretation, and reduce intralaboratory variability.

Introduction of internal references are necessary for quality assurance. The use of internal references reduces the interlaboratory variability. These internal references are prepared as mini-TMA reference blocks. They must include positive and negative controls.

Sections from mini-TMA reference blocks and TMA blocks under investigation are stuck on the same slide. They are simultaneously processed and analyzed, providing a convenient measure for quality control and quantitation.

Because each staining procedure needs to include appropriate negative and positive controls, references should be selected with a wide antigen profile (i.e., involving a large number of different antigens).

6.2.4.2.1 Tissue Control

As internal negative and positive controls in IHC, a mini-TMA tissue reference block with well-characterized tissue cores is recommended. Archived blocks are used as a source of control tissues. Control tissue blocks should be chosen with a wide antigen profile (negative and positive ones). Control tissue cores should be eventually chosen to represent different fixation and embedding protocols. Mini-TMA tissue references will be tested for antigenic output before inclusion in a specific study.

6.2.4.2.2 Control Cell Lines

- Reagents and solutions
  - PBS, phosphate-buffered saline (Invitrogen) pH 7.2
  - Formalin, 10% paraformaldehyde (weight/vol)
    - Polyoxymethylen, 10 g
    - PBS, 100 mL
    - Heat to 70°C
    - Cool to room temperature
    - Filter
  - Agarose
- Protocol
  - Resuspend cells in 10% formalin and fix overnight. Use 0.8% agarose. Cell line arrays consist of normal or cancer cell lines that are grown in culture.
  - Pellet the cells and then wash in PBS.
  - Pellet and resuspend cells in 0.8% agarose at 42°C.
  - Prefill tapered end of a 0.6 ml microfuge tube with agarose and let solidify.
  - Transfer the agars cell mixture to the microfuge tube with the solidified agarose.

- Add a wooden toothpick to the tube; this facilitates removal of the plug form in the microfuge tube.
- Remove the plug after it solidifies.
- Cut in half to yield two blocks of cells.
- Process the plugs using standard TMA processing schedules to build a mini-TMA cell line reference block with well characterized cell line cores. Cell lines should be chosen with a wide antigen profile. The major function of these mini-TMA cell line references is to represent the major proteins that are involved in the study. Mini-TMA cell line references will be tested for antigenic output before inclusion in a specific study.
- Monitoring ICH variables—With the use of serial sections of an internal mini-TMA reference block, variations in staining results will also be discovered immediately, regardless of whether they are caused by differences in reagents or variations in the staining procedure. Monitoring the ICH procedural variables is one of the most important requirements for quantitative IHC.

6.2.4.3 Immunohistochemistry Scoring

Variation of manual IHC scoring is very common. The recommendation is to evaluate the score at least twice by two pathologists. The subjective inconsistence can be controlled by switching to automated quantitative image analysis scoring techniques (Section 6.3).

6.2.4.4 TMA Datasheet

The TMA datasheet includes all the workflow from immunostained core to the original location in the tissue donor block. It includes morphological details and quantitative data. It makes it possible to track each TMA core to the original material and the associated clinical information.

6.3 HIGH-PERFORMANCE IMAGE ANALYSIS

High-throughput molecular profiling of tissues using TMA technology gives rise to an increasing volume of data that is image based:

- Full HES block section images
- Full annotated images
- Full labeled block images
- Density map images
- Multiple TMA section images

Full slides and TMA slides can be manually analyzed; alternatively, automated image analysis programs that can assist the users are available. This leads to the following protocols:

- Acquiring images
- Making annotations by a pathologist
- Making marker quantification
- Handling and processing data
- Making data interpretation

The challenge is to acquire, store, manage, analyze, and associate images to other data in a meaningful way.

6.3.1 Virtual Microscopy

Virtual microscopy is now widely applied in pathology.

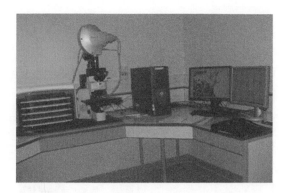

FIGURE 6.20   A virtual microscopy station includes a microscope, a camera, a motorized stage (alternatively: a scanner), a slide loader, and a dedicated software for whole slide management.

6.3.1.1   Instrument Characteristics

The main characteristics to compare between systems concerns:

- Hardware—Microscope, objective, illumination, camera, motorized stage and slide loader (Figure 6.20)
- Scanning speed—Scanning time lies between 1 and 5 min for a biopsy, and between 5 and 20 min for a surgical specimen
- Image acquisition and software functions
- Image resolution
- Densitometric homogeneity and linearity
- High capacity and automatic throughput—All system components are designed to interact seamlessly, producing a fully automated, high-speed scanning system with excellent flexibility and simple operation
- Speed of virtual microscope
- Software functions

6.3.1.2   Virtual Slides

- Acquiring a whole full section. The virtual slide image is stored with all the relevant information in such a way that it enables instant, controlled access from anywhere. Virtual slides are digital facsimiles that, when viewed with a pan and zoom viewer, can emulate viewing a glass slide with a traditional microscope (Figure 6.21).
- Acquiring a TMA section (Figure 6.22).

6.3.1.3   High-Resolution Images

Selecting a region of interest (ROI) to be acquired by a higher resolution:

- In case of whole full sections, select ROI manually
- In case of TMA section, automatic TMA spot localization can be run

The images can be digitized directly at low resolution or be obtained from a subsampled high-resolution.

6.3.1.3.1   Regions of Interest (ROIs)
See Figure 6.23.

6.3.1.3.2   TMA Spot Localization
See Figure 6.24 and Figure 6.25.

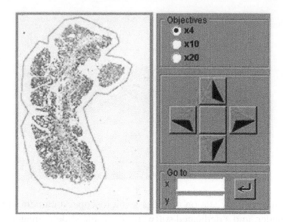

FIGURE 6.21 Pan and zoom viewer of a virtual microscope to focus in different regions of interest on the virtual full section shown on the left at low magnification.

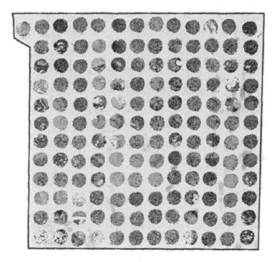

FIGURE 6.22 A virtual TMA section at low magnification.

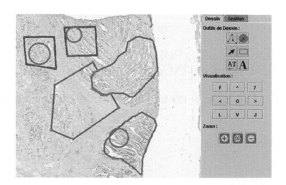

FIGURE 6.23 Dedicated software interface for manual selection of regions of interest (ROI) before running high-resolution acquisition.

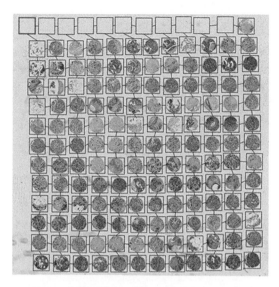

FIGURE 6.24 Dedicated software interface for automatic selection (with manual interaction) of spots on a TMA section.

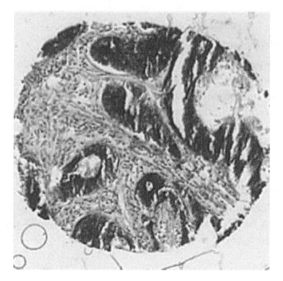

FIGURE 6.25 TMA spot viewed at high resolution.

6.3.2 EXPERTISE OF THE PATHOLOGIST

Annotations can be made on the basis of an expert-based ontology. An ontology is a formal representation of knowledge as a set of concepts within a domain and the relationships between those concepts.

6.3.3 MARKER QUANTIFICATION

Image analysis protocol for marker quantification are usually run according to three steps: segmentation, segregation, and quantification. These three steps are necessary whatever the label, leading either to a color or a fluorescent signal. Running image analysis procedures for marker quantification:

- Running structure segmentation
- Running marker segregation
- Running marker quantification

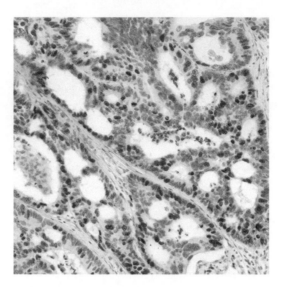

FIGURE 6.26   Image of a DAB immunostained sample with an hematoxyline nuclear counterstained.

6.3.3.1   Color Image Processing

Color image processing is performed when immunoenzymatic protocols are used. The red, green, and blue (RGB) color model is used in most of the color-image-producing technology. It had a solid theory behind it, based on human perception of colors. The RGB color model is an additive color model in which red, green, and blue light are added together in various ways to reproduce a broad array of colors. The name of the model comes from the initials of the three additive primary colors: red, green, and blue. (See Figure 6.26.)

6.3.3.1.1   Structure Segmentation

Image segmentation is the process of assigning a label to every pixel in an image such that pixels with the same label share certain visual characteristics. The result of a simple image segmentation method (thresholding) is a binary image resulting in the detection of the pixel representing the structure of interest (Figure 6.27).

6.3.3.1.2   Marker Segregation

Marker segregation (separation of positively immunostained versus negatively immunostained in the whole structure of interest) is based on color image analysis. Depending on the system used, the image analysis protocols (if some are proposed) may be based on concepts different than those proposed in the following section.

Learning the system for how to recognize immunostained versus counterstained pixels:

- Prepare control reference slides
    - Control reference DAB immunostained slide, only immunostained without counter-staining (Figure 6.28)
    - Control reference counterstain, only counterstained (Figure 6.28b)
- Make acquisition of reference fields
- Visualize segregation of immunostained and counterstained pixels in the color space (Figure 6.29)
- Use the color classification space for segregation of pixels (Figure 6.30)

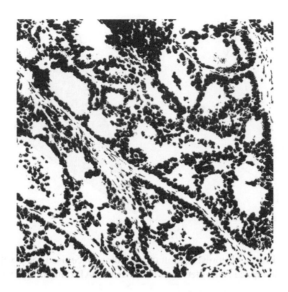

FIGURE 6.27 Binary image obtained by thresholding the RGB image presented in Figure 6.26 (336 × 336 px). The whole nuclear area (structure of interest) is represented in black.

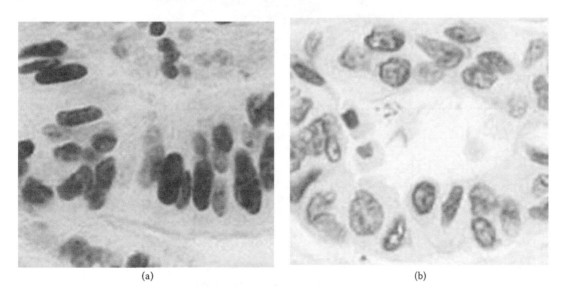

(a)   (b)

FIGURE 6.28 **(See color insert.)** References color RGB images: DAB immunostained reference (a) and hematoxyline counterstained reference (b).

6.3.3.1.3 Marker Quantification
- On the whole structure of interest
  - Use the binary image (Figure 6.27) that delineates the whole structure of interest.
  - Select the R, G, or B component (densitometric image) of the RGB color image in which the immunostaining is present at a high contrast (Figure 6.31). The image acquisition procedure should be controlled for its linear response at least in the channel corresponding to the selected component.
  - Make integration of intensity values from the densitometric image for positive immunostained pixels detected in the whole structure of interest (Figure 6.30).

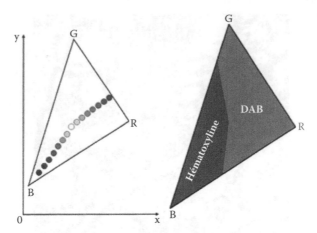

FIGURE 6.29   (See color insert.) Representation of the color space used for pixel classification according to their RGB values. The color space is divided in two domains: DAB domain (red) and hematoxyline domain (blue). (From Heus, 2009.)

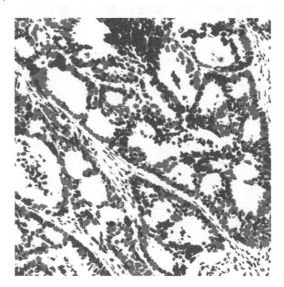

FIGURE 6.30   (See color insert.) Segregation of positive immunostained pixels (red) and negative immuno-stained pixels (blue) in the whole structure of interest presented in Figure 6.27.

- On cell basis—Someone may be interested in densitometric measurements at a cellular level rather than at a structural level.
  - Use an algorithm for cell segmentation. For example, the Voronoï algorithm is a good algorithm for cell segmentation (Figure 6.32).
  - Apply the same protocol described in the previous section considering, instead of the whole structure of interest, each individual cell extracted from the Voronoï decomposition of the original image.

6.3.3.2   Fluorescence Image Processing

Fluorescence image processing is performed when immunofluorescence protocols are used. Immunofluorescence protocols combine a specific immunostaining and a counterstaining. Both stainings are visualized using different fluorochromes. Each fluorochrome is characterized by its absorption and emission spectra.

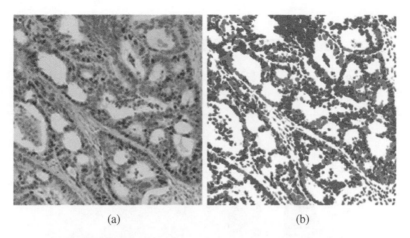

(a)   (b)

FIGURE 6.31   **(See color insert.)** When DAB immunostaining protocol is used, the G component (a) is selected for densitometric quantification in relation with the corresponding image (Figure 6.26) summarizing the segregation between negative and positive immunostained pixels in the structure of interest (b).

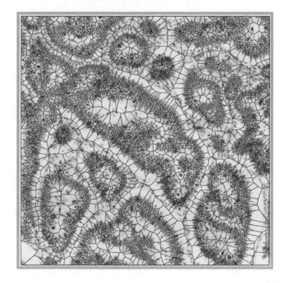

FIGURE 6.32   Cell segmentation, Voronoï decomposition of the image presented in Figure 6.26.

6.3.3.2.1   Structure Segmentation

Image segmentation of the structure of interest is performed on the fluorescent image obtained for the counterstaining.

- Select adequate condition for image acquisition of counterstaining (Figure 6.33).
- Obtain binary image by thresholding the image from counterstaining.

6.3.3.2.2   Marker Visualization

Marker visualization is performed on the fluorescent image obtained for the specific immunostaining. Select a fluorochrome compatible with counterstaining fluorescence (i.e., without overlap in spectral characteristics) (Figure 6.33).

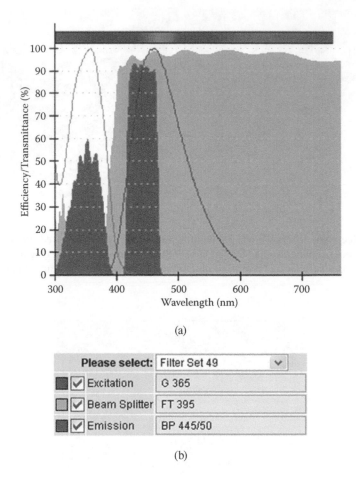

(a)

		Please select:	Filter Set 49	⌄
■	☑	Excitation	G 365	
□	☑	Beam Splitter	FT 395	
■	☑	Emission	BP 445/50	

(b)

FIGURE 6.33 Example of spectral conditions for visualization of the DAPI nuclear counterstaining. (From Carl Zeiss S.A.S., Germany.).

6.3.3.2.3   Marker Quantification

The quantitative evaluation of the marker is the result of integrating the fluorescent signal from segregated areas included in the structure of interest.

- Use the binary image that delineates the whole structure of interest on the basis of counterstaining.
- Select the fluorescent image corresponding to immunostaining.
- Make integration of fluorescent values from the immunofluorescent image in the whole structure of interest.
- Use a Voronoï algorithm (or equivalent) if you are interested at the cellular level.

6.3.4   QUALITY CONTROL

Quality control procedures are necessary for intra- and interlaboratory comparisons.

- Making instrumentation calibration
- Making marker calibration

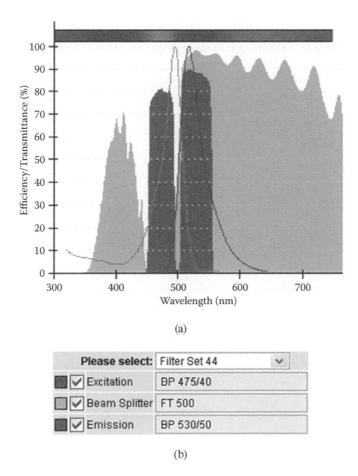

(a)

(b)

FIGURE 6.34 Example of spectral conditions for visualization of the FITC fluorochrome coupled to antibodies. FITC is compatible with DAPI. (From Carl Zeiss S.A.S., Germany.).

6.3.4.1 Colorimetry and Densitometry

6.3.4.1.1 Instrument Calibration

Apply all quality control protocols given in Chapter 4, Section 4.2.3 and Section 4.2.4.

6.3.4.1.2 Measurements Calibration

For quantitative or semiquantitative evaluation, develop appropriate procedures using internal and external references, and negative as well as positive references.

6.3.4.2 Fluorescence

6.3.4.2.1 Instrument Calibration

Quality and control procedures are not yet proposed for fluorimetric calibration. At least, linearity of response of the sensor should be tested.

6.3.4.2.2 Measurements Calibration

Same recommendations as for densitometric evaluation.

REFERENCES

Heus, R. 2009. Approches virtuelles dédiées à la technologie des puces à tissus "Tissue MicroArrays" TMA: Application à l'étude de la transformation tumorale du tissu colorectal. Ph.D., University of J. Fourier, Grenoble (France).

7 Preparation of Products for Histology and Histochemistry

Jean-Marie Exbrayat

CONTENTS

7.1 FIXATIVES FOR LIGHT MICROSCOPY

7.1.1 PREPARATION OF FIXATIVES

7.1.1.1 Alcohol–Formalin
- Ethanol 100%, 90 mL
- Formalin, 10 mL

This fixative has long been used for museum collections.

7.1.1.2 Baker's Fluid
See also Section 7.1.1.8.

- Neutral formalin, 10 mL
- Anhydrous calcium chloride 10% in distilled water, 10 mL
- Distilled water, 80 mL

This fixative is also called formalin–calcium fixative. It is used for the nervous system, lipids, proteins, and amines.

7.1.1.3 Bouin's Fluid
- Water saturated solution of picric acid, 300 mL
- Formalin, 100 mL
- Acetic acid, 20 mL

Bouin's fluid is a classic fixative that is acidic and cannot be used for visualization of certain components (such as mineral components).

7.1.1.4 Bouin–Hollande
- Distilled water, 100 mL
- Copper acetate, 2.5 mL
- Formalin, 10 mL
- Acetic acid, 1 mL
- Picric acid, 4 g

Bouin–Hollande's fluid is a classic fixative that is acidic and cannot be used for visualization of certain components (such as mineral components).

7.1.1.5 Carnoy's Fluid
- Ethanol 100%, 60 mL
- Chloroform, 30 mL
- Acetic acid, 10 mL

Ethanol 100% can be replaced with methanol. This fixative must be prepared at time of use. It is recommended for nucleic acid visualization and for study of the nervous system (Nissl's bodies).

7.1.1.6 Flemming's Fluid
- Chromic oxide 1%, 75 mL
- Osmium tetroxide 2%, 20 mL
- Acetic acid, 5 mL

This fixative must be prepared at time of use. It is used for cytological studies and can be used for lipid visualization.

7.1.1.7   Formalin
- 36% to 40% formalin, 25 mL
- Distilled water, 100 mL

Formalin purchased is 36% to 40%.

7.1.1.8   Formalin–Calcium
See also Section 7.1.1.2.

- 36% to 40% formalin, 10 mL
- Calcium chloride, 1 g. Calcium acetate can also be used.
- Distilled water, 90 mL

7.1.1.9   Neutral Formalin
- Dilute formalin at 39% in distilled water to obtain formalin at 10%. Then neutralize with saturated calcium carbonate.
- Verify with pH paper (pH 7).

Neutral formalin has multiple uses. It can be advantageously replaced with buffered formalin.

7.1.1.10   Salt Formalin
- Neutral formalin, 10 mL
- Physiological serum, 10 mL

Salt formalin has multiple uses. It can be advantageously replaced with buffered formalin.

7.1.1.11   Buffered Formalin
- Fixative (0.1 M, pH 7)
  - Formalin 35%, 200 mL
  - Phosphate buffer, 1000 mL
- Phosphate buffer
  - Solution A, 400 mL
  - Solution B, 700 mL
- Solution A, disodium phosphate 0.1 M (Na_2HPO_4, 12 H_2O, molecular weight: 358.17 g)
  - Disodium phosphate, 35.817 g
  - Distilled water, 1000 mL
- Solution B, monopotassium phosphate, 0.1 M (KH_2PO_4, molecular weight: 136.09 g)
  - Monopotassium phosphate, 13.609 g
  - Distilled water, 1000 mL

Buffered formalin has multiple uses.

7.1.1.12   Halmi's Fluid
- Water saturated picric acid, 10 mL
- Heidenhain's Susa, 90 mL

7.1.1.13 Helly's Fluid

This fixative is also called Zenker's formalin.

- Zenker's fluid, 100 mL. Use stock solution.
- Neutral formalin, 5 mL

This fixative can be used for cytological studies.

7.1.1.14 Heidenhain's Susa

- Mercuric chloride, 4.5 g
- Sodium chloride, 0.5 g
- Distilled water, 80 mL
- Trichloracetic acid, 2 g
- Acetic acid, 4 mL
- Formalin, 20 mL

7.1.1.15 Zenker's Fluid

- Stock solution
  - Potassium dichromate, 2.5 g
  - Mercuric chloride, 5 g
  - Sodium sulfate, 1 g
  - Distilled water, 100 mL
- For a working solution: Add acetic acid, 5 mL

This fixative can be used for cytological studies.

7.1.2 FIXATION DURATION

Determination of the optimal fixation duration is recommended for each fixative.

- Bouin's fluid, 24 to 48 h
- Carnoy's fluid, 4 h
- Formalin, indefinite

7.2 DYES

7.2.1 NUCLEAR DYES

7.2.1.1 Acetocarmine

This staining is used on a block before sectioning.

- Stock solution
  - Carmine, 1 g
  - Acetic solution 45%, 200 mL
  - Let boil 5 min
- Working solution
  - Stock solution, 50 mL
  - Acetic solution 45%, 50 mL
  - Ferric chloride

With the addition of ferric chloride, chromosomes are violet stained.

7.2.1.2 Acid Hematein (First Formula)

- Hematoxylin, 0.05 g
- Distilled water, 48 mL. Distilled water contains 1% sodium iodide.
  - Distilled water, 100 mL
  - Sodium iodide, 1 g
- Let boil
- Let cool
- Acetic acid, 1 mL

7.2.1.3 Acid Hematein (Second Formula)

- Hematoxylin 0.1%, 0.1 g
- Distilled water, 100 mL
- Sodium periodate 1%, 2 mL
- Let boil
- Let cool
- Acetic acid, 2 mL

This formula is used to visualize lipids with dichromate–acid hematein or sodium hydroxide–dichromate–acid hematein.

7.2.1.4 Azocarmine

- Azocarmine (G or B), 0.1 g.

Staining with G azocarmine is done at 60°C. Staining with B azocarmine is done at room temperature.

- Distilled water, 200 mL
- Let boil
- Let cool
- Acetic acid, 2 mL

7.2.1.5 Borated Carmine

- Carmine, 2 g
- Sodium tetraborate, 4 g
- Distilled water, 100 mL at 100°C
- Ethanol 70%, 100 mL
- Let rest 3 weeks
- Filter

7.2.1.6 Carmalum

- Potassium alum, 10 g
- Distilled water, 100 mL
- Carminic acid, 1 g
- Dissolve at high temperature
- Let cool
- Filter
- Salicylic acid, 0.2 g. Salicylic acid is used to prevent development of microorganisms. It can be replaced by formalin (1 mL).
- Potassium alum, 1 g

7.2.1.7 Groat's Hematoxylin

- Solution No. 1
  - Iron and ammonium alum, 1 g
  - Distilled water, 50 mL
  - Concentrated sulfuric acid, 0.8 mL
- Solution No. 2
  - Hematoxylin, 0.5 g
  - Ethanol 95%, 50 mL
- After dissolution, mix the two solutions
- Let set 1 h
- Filter: Can be stored for 1 month.

7.2.1.8 Ferric Hematoxylin

- Solution A
  - Hydrochloric acid, 2 mL
  - $FeCl_3$, 6 H_2O, 2.5 g
  - $FeSO_4$, 7 H_2O, 4.5 g
  - Distilled water, 298 mL
  - Acetic acid, 2 mL
- Solution B
  - Hematoxylin, 1 g
  - Distilled water, 100 mL
- Working solution (ferric hematoxylin)
  - Solution A, 30 mL
  - Solution B, 10 mL

This formula is used to visualize sphingomyelin with sodium hydroxide–ferric hematoxylin.

7.2.1.9 Hematoxylin

- Original protocol
  - Hematoxylin, 10 g
  - Ethanol 95%, 100 mL
  - Let the hematoxylin artificially "grow old" by adding:
    - Potassium iodinate, 0.2 g
    - Hematoxylin, 1 g
- Variant: Harris's hematoxylin
  - Hematoxylin, 5 g
  - Ethanol 100%, 50 mL
  - Let dissolve
  - Potassium or ammonium alum, 100 mL
  - Distilled water, 1000 mL
  - Mix hematoxylin and alum
  - Mercuric acid red, 2.5 g

Other variants are available.

7.2.1.10 Masson's Hemalum

- Hematein, 0.2 g
- Potassium alum, 5 g
- Distilled water, 100 mL

- Let boil, then let cool and filter
- Acetic acid, 2 mL

7.2.1.11   Mayer's Hemalum

- Hematoxylin, 1 g
- Distilled water, 1000 mL
- Warm
- Sodium iodate, 0.2 g
- Potassium alum, 50 g
- Dissolve without warming
- Chloral hydrate, 50 g
- Citric acid, 1g
- Conservation, several weeks at 4°C

7.2.1.12   Nuclear Fast Red

- Nuclear fast red, 0.1 g
- Aluminum sulfate, 5 g
- Distilled water, 100 mL
- Let boil
- Let cool
- Filter: Can be stored indefinitely in a flask.

7.2.1.13   Regaud's Hematoxylin

- Old solution of hematoxylin, 10 mL
- Glycerin, 10 mL
- Distilled water, 80 mL
- Place the solution at 37°C for 12 h
- Let cool
- Filter

7.2.2   BACKGROUND COLORATION

7.2.2.1   Acetic Light Green

- Light green, 0.1 g
- For light green 0.5%
  - Light green, 0.5 g
  - Distilled water, 100 mL
  - Acetic acid, 2 mL
- Distilled water, 100 mL
- Acetic acid, 2 mL

Storage is unlimited.

7.2.2.2   Acidic Fuchsin

- Acid fuchsin, 0.1 g
- Distilled water, 200 mL
- Acetic acid, 1 mL

7.2.2.3 Acidic Fuchsin and Culvert
- Acidic fuchsin, 0.1 g
- Culvert, 0.2 g
- Distilled water, 300 mL
- After dissolution, acetic acid, 0.6 mL

Storage is unlimited.

7.2.2.4 Alizarin Acid Blue
- Alizarin acid blue, 0.5 g
- Distilled water, 100 mL
- Aluminum sulfate, 10 g
- Let boil 5 min
- Let cool 24 h
- Filter

Storage is unlimited.

7.2.2.5 Altmann's Fuchsin
- Acidic fuchsin, 10 g
- Old aniline solution at 1.5% then filtered, 100 mL

Storage is possible for 3 months.

7.2.2.6 Aniline Blue
- Aniline blue, 0.5 g
- Distilled water, 100 mL
- Let boil
- Let cool
- Filter
- Acetic acid, 8 mL

Storage time of the stock solution is unlimited.

- Before use, dilute.
  - Stock solution, 100 mL
  - Distilled water, 200 mL

The diluted solution can be stored for several months.

7.2.2.7 Calleja's Picroindigocarmine
- Indigocarmine, 0.4 g
- Picric acid saturated aqueous solution, 100 mL
- Let rest
- Filter

Can be stored for 6 months.

7.2.2.8 Eosin
- Eosin, 1 g
- Distilled water, 100 mL

Storage time is unlimited.

7.2.2.9   Eosin–Light Green

- Eosin, 1 g
- Light green, 0.2 g
- Phosphotungstic acid, 0.5 g
- Distilled water, 100 mL
- After dissolution, acetic acid, 0.5 mL

Storage time is unlimited.

7.2.2.10   Erythrosin

- Erythrosin, 1 g
- Distilled water, 100 mL
- After dissolution, acetic acid, 1 drop

Storage time is unlimited.

7.2.2.11   Erythrosin–G Orange

- Erythrosin, 0.2 g
- G orange, 0.6 g
- Distilled water, 100 mL
- Acetic acid, 1 drop
- After dissolution, formalin, 1 mL

Storage time is unlimited.

7.2.2.12   Fast Green

- Fast green, 1 g
- Distilled water, 100 mL
- Acetic acid, 0.5 mL

Storage time is unlimited.

7.2.2.13   Heidenhain's Blue

- Aniline blue, 0.2 g
- G orange, 0.5 g
- Distilled water, 100 mL

Storage time of the stock solution is unlimited.

- Before use, dilute
  - Stock solution, 100 mL
  - Distilled water, 200 mL

The diluted solution can be stored for several months.

7.2.2.14   One-Time Trichroma (Gabe's Formula)

- S azorubin, 0.5 g
- Phosphomolybdic acid, 0.5 g
- Solid green FCF, 0.5 g. Solid green can be replaced by fast green.

- Distilled water, 100 mL
- Acetic acid, 1 mL
- Martius's yellow at saturation. Martius's yellow can be replaced by hydrophilic naphthol yellow. In this case, use:
  - Naphthol yellow, 0.01 g
  - Let dissolve 2 h. In all the cases, dissolution time can be prolonged. It will be more and more efficacious.
- Filter

Storage time is unlimited.

7.2.2.15 Methylene Blue
- Methylene blue, 2 g
- Distilled water, 100 mL

7.2.2.16 One-Time Trichroma (Martoja's Formula)
- Azorubin S, 0.5 g
- Phosphomolybdic acid, 0.5 g
- Solid green, 0.1 g. Fast green can also be used.
- Martius's yellow at saturation, 100 mL
- Acetic acid, 1 mL
- Let rest several hours
- Filter

Storage time is unlimited.

7.2.2.17 Paraldehyde Fuchsin (Gabe's Formula)
- Stock solution
  - Basic fuchsin, 1 g
  - Boiling water, 200 mL
  - Let boil 1 min
  - Let cool
  - Filter
  - Hydrochloric acid, 2 mL
  - Paraldehyde, 2 mL
  - Place 1 drop of solution on a paper filter; when the fuchsin red staining disappears, filter.
  - Let the precipitate dry, then dissolve in ethanol 70%, at saturation (about 150 mL ethanol).

Storage time is unlimited.

- Working solution
  - Stock solution, 25 mL
  - Ethanol 70%, 75 mL
  - Acetic acid, 1 mL

Storage time is possible for several months.

7.2.2.18 Phloxin
- Phloxin, 1 g
- Distilled water, 100 mL

Storage time is unlimited.

7.2.2.19   Saffron
- Gatinais's saffron, 10 g
- Ethanol 100%, 250 mL
- Dry the saffron, 12 h at 37°C. It can be useful to let the saffron dry overnight.
- Crush saffron in a mortar
- Ethanol, 100%
- Warm 12 h at 37°C. It is possible to warm the solution overnight in a well-corked flask to avoid solvent evaporation. It is also possible to extract the saffron with ethanol 70% using a Soxhlet engine. In this case, proceed to the extraction after several hours.
- Filter

Storage is possible for 6 months.

7.2.2.20   Van Gieson's Picrofuchsin
- Picric acid saturated aqueous solution, 100 mL
- Acidic fuchsin 1%, 5 mL. Fuchsin quantity can be 5 to 15 mL for 100 mL of water. This depends on the expected results and on the nature of the tissue.

Can be stored for several months.

7.2.2.21   Ziehl's Fuchsin
- Stock solution
  - Basic fuchsin, 1 g
  - Phenol, 5 g
  - Ethanol 95%, 10 mL
  - Add progressively, distilled water, 90 mL
  - Let rest 1 h
  - Filter

Only the stock solution can be stored for a long time.

- Working solution
  - Stock solution, 30 mL
  - Distilled water, 90 mL

7.2.3   Histochemical Reagents

7.2.3.1   Acid H 2% in Veronal Buffer pH 9.2
- Veronal buffer
  - Hydrochloric acid 8.35 g/L, 231 mL
  - Sodium veronal 20.618 g/L, 769 mL
- Dye
  - Acid H, 2 g
  - Veronal buffer, 100 mL

7.2.3.2   Alcian Blue
- Alcian blue, 1 g
- Acetic acid, 1 mL
- Distilled water, 100 mL

This is the preparation method for alcian blue at pH 2. Other preparation modes exist to obtain alcian blue at different pH by adding acetic or hydrochloric acid. This dye is essentially used to visualize acidic mucopolysaccharides. Storage is possible for several months.

- Variant: Alcian yellow
  - Alcian yellow, 0.5 g
  - Buffer pH 0.5, 100 mL

7.2.3.3 Ammoniacal Silver Nitrate
- Ammoniac 28%, 10 mL
- Silver nitrate, 80 mL
- Continue to pour ammoniacal silver nitrate to dissolve the brown precipitate
- Use distilled water to dissolve opalescent solution

7.2.3.4 Azure A–SO$_2$-
- Azure I or A, 1 g
- Hydrochloric acid M, 5 mL
- Sodium metabisulfite 5%, 1 mL
- Distilled water, 90 mL

This solution can be conserved for a few weeks.

7.2.3.5 Black Sudan B
- Black Sudan B at saturation
- Ethanol 70%

This solution is a lysochrome that dissolves into lipids, giving them a dark staining.

7.2.3.6 Chloramine-T
- Chloramine-T, 1 mL
- Phosphate buffer pH 7.5, 100 mL

7.2.3.7 Coomassie's Blue
- Coomassie's blue, 0.2 g
- Methanol, 46.5 mL
- Acetic acid, 7 mL
- Distilled water, 46.5 mL

7.2.3.8 D D D
- D D D, stocking solution
  - D D D, 0.15 g
  - Ethanol 100%, 100 mL

Solution must be stored at 4°C.

- Buffer veronal pH 8.5
  - HCl M/10 (8.35 mL/L), 129 mL
  - Sodic veronal M/10 (20.618 g/L), 871 mL

Phosphoric buffer can also be used: See Chapter 2, Section 2.3.6.

- D D D
  - Stocking solution, 15 mL
  - Veronal buffer, 35 mL
  - Warm up, 60°C

7.2.3.9 Diamine Reagent
- N,N-dimethyl-meta-phenylenediamine dihydrochloride, 120 mg
- N,N-dimethyl-para-phenylenediamine dihydrochloride, 20 mg
- Distilled water, 50 mL
- Ferric chloride, 1.4 mL

Diamine salts are toxic. They must be handled with precaution.

7.2.3.10   DNFB (or Dinitrofluorobenzene)
- 2,4-dinitro-1-fluorobenzene, 0.74 g
- Ethanol 100%, 52 mL
- Distilled water, 40 mL
- Sodium bicarbonate, 1 M, 8 mL

7.2.3.11   Fast Blue B (Also Called Orthodianisidine)
- Orthodianisidine, 0.2 g
- Veronal buffer pH 9.2, 100 mL

7.2.3.12   Fluorone
- 9-methyl, 2, 3, 7 trihydroxy fluorine, 0.5 g
- Ethanol 95%, 100 mL
- Sulfuric acid, 1 mL

This technique is used to visualize nucleic acids with Turchini's method.

7.2.3.13   Fuchsin
Leucofuschin for Ziehl's method.

- Basic fuchsin, 1 g
- Phenol, 0.5 g
- Ethanol 100%, 10 mL
- Distilled water, 100 mL

7.2.3.14   Gallocyanine Chromic Lac
- Gallocyanine, 0.15 g
- Chrome alum 5%, 100 mL
- Let boil 3 min
- Let cool 24 h

7.2.3.15   Lugol
- Iodine, 1 g
- Potassium iodide, 2 g
- Distilled water, 100 mL

Solution must be stored at 4°C.

- Adjust to pH 3.2 with addition of hydrochloric acid 0.01 N

7.2.3.16 Luxol Fast Blue
- Luxol blue, 0.1 g
- Ethanol 95%, 100 mL

7.2.3.17 Methyl Green
Methyl green always contains impurities, which must be eliminated with chloroform extraction. Methyl green contains methyl violet.

- Methyl green, 2 g
- Distilled water, 100 mL
- Chloroform, 50 mL
- Decant until chloroform remains colorless

Purified methyl green can be stored for a long time at 4°C. Before a new use, extract impurities again with chloroform.

7.2.3.18 Methyl Green (Pollister's Formula)
- Purified methyl green 1%, 25 mL
- Ethanol 95%, 20 mL
- Glycerol, 25 mL
- Phenol, 0.5 g
- Distilled water, 100 mL

Storage is possible for several months.

7.2.3.19 Methyl Green–Pyronine (First Formula)
- Methyl green 2%, 7.5 mL
- Pyronine 2%, 12.5 mL
- Distilled water, 30 mL

7.2.3.20 Methyl Green–Pyronine (Second Formula)
- Methyl green 1%, 15 mL
- Pyronine, 0.25 g
- Phenol, 0.5 g
- Ethanol 95%, 2.5 mL
- Glycerol, 20 mL
- Distilled water, 85 mL

7.2.3.21 Naphthol Yellow
Naphthol yellow is also called Martius's yellow or Mars's yellow.

- Saturated solution in ethanol for ethanol-soluble salts, or
- Saturated solution in acetic water (1%) for water-soluble salts

7.2.3.22 Nile's Blue Sulfate
- Nile's blue sulfate, 1 mL. Nile's blue chlorhydrate can also be used.
- Distilled water, 100 mL

7.2.3.23   Perchloric Acid–Naphthoquinone

- 1, 2 naphthoquinone sulfonic acid, 40 mg
- Ethanol, 20 mL
- Perchloric acid 60%, 10 mL
- Formalin 40%, 1 mL
- Distilled water, 9 mL

7.2.3.24   Phosphomolybdic Acid–G Orange

- G orange, 2 g
- Distilled water, 100 mL
- Phosphomolybdic acid, 1 g. Phosphomolybdic acid can be replaced with phosphotungstic acid.

7.2.3.25   Pyronine (First Formula)

- Aniline, 4 mL
- Pyronine, 0.1 g
- Ethanol 40%, 100 mL

7.2.3.26   Pyronine (Second Formula)

- Pyronine, 0.2 g
- Distilled water, 100 mL

7.2.3.27   Red Oil O

- Red oil at saturation
- Ethanol 70%

Other preparative modes are available, for example:

- Red oil, 0.2 g
- Propanol-2 (isopropanol) 60% in water, 100 mL

7.2.3.28   Schiff's Reagent

- Basic fuchsin, 2 g
- Distilled water, 400 mL at 100°C
- Let cool to 50°C
- Filter
- Hydrochloric acid 2 M, 10 mL
- Let cool to 25°C
- Potassium metabisulfite 4 g
- Let rest 12 h at 4°C
- Mortar-pounded charcoal, 1 g
- Strongly agitate, 2 min
- Filter
- Hydrochloric acid 2 M, 12 mL; or hydrochloric acid M, 20 mL

7.2.3.29   Silver Methenamine

- Silver nitrate 5%, 5 mL
- Tetramine hexamethylene, 100 mL
- Borate buffer 0.2 M, pH 8, 5 mL
  - Solution A, sodium borate, 0.2 M
    - Boric acid, 12.404 g

- Sodium hydroxide, 100 mL
- Distilled water, 900 mL
- Solution B, hydrochloric acid 0.1 M
- Buffer
  - Solution A, 59.9 mL
  - Solution B, 44.1 mL
- Distilled water, 90 mL

Can be stored for several weeks. Add several drops of metabisulfite 10% before each use.

7.2.3.30 Thionine

- Thionine, 1 mL. The quantity can be modified depending on the concentration required.
- Tartaric acid, 0.5 g
- Distilled water, 100 mL. Distilled water can be replaced by buffer.

Storage time is unlimited.

7.2.3.31 Toluidine Blue

- Toluidine blue, 0.5 g
- Distilled water, 100 mL

Section 2

Cytological Methods in Electron Microscopy

8 General Techniques in Electron Microscopy

Jean-Marie Exbrayat

CONTENTS

8.1 INTRODUCTION

8.1.1 GENERAL PRINCIPLE OF ELECTRON MICROSCOPY

For a long time, the light microscope was the only means to observe tissues and cells. But the light microscope has its limitations; it was not possible to obtain a magnification higher than 1000 and 2000, for the highest performing microscopes. The principle of the microscope is based upon the resolving power, still called resolution. Resolution is the smallest size of a detectable object, that is, the smallest distance between two points at which the two points can be separately observed. Under this distance, only one point is observed, one of them being hidden by the other.

In 1878, Abbe showed that the limitations of the light microscope was directly linked to the wavelength of the light source used. In 1895, the corpuscular nature of electricity was shown by Jean Perrin (1820–1842); in 1897, the electron was discovered by Joseph John Thomson (1856–1940), and one of the applications of this new discovery was to obtain higher magnifications of tissues and cells.

The formula for resolution is

$$d = 0.61 \, \lambda/n \sin \alpha$$

in which d is the resolution, λ is the wavelength of light source, n is the refractory index of the medium in which photons or electrons are moving, and α is the angle of aperture of beam.

For the light microscope, resolution is 0.2 μm; for the transmission electron microscope (TEM), it is 0.2 nm; for the scanning electron microscope (SEM), it is 20 nm.

In electron microscopes, the light source is substituted with electrons. Electrons go throughout a column in which a high vacuum has been done. Electrons are submitted to several electromagnetic lenses, used to obtain a good contrast, magnification, and focus. In TEM, the image of correctly prepared sections is observed on a fluorescent screen placed at the bottom of a column. In SEM, the electrons scan the surface of the specimen and the surface images are observed with a cathode tube.

8.1.2 TRANSMISSION ELECTRON MICROSCOPE (TEM)

The first electron microscope was built about 1930 by Ernst Ruska. It was a transmission electron microscope (TEM). The image observed on a fluorescent screen was obtained by electrons transmitted across an ultrathin section. The first microscopes were in use at the end of the 1940s.

Preparation of sections is a very delicate operation. Fixation of tissues must be particularly precise to avoid deformations of organelles. For that, the fixative must be isoosmotic to the tissue, with the same pH and with an equivalent ionic composition. Then, organs are embedded in a plastic resin that must be hard enough to obtain sections with a thickness between 60 and 90 nm. If the first sections were made with a metallic knife; today, the ultramicrotomes are equipped with a glass or a diamond knife. Sections are then disposed on metallic grids, the first one being copper grids. After this operation, sections are contrasted. Contrast is the equivalent to staining for light microscopy. It consists of putting down heavy metals to increase the contrast of a tissue that is not electron dense. This operation allows one to obtain several degrees of electron density and, finally, to visualize several organelles of tissues and cells.

After the first staining with potassium permanganate, several techniques used in light microscopy have been adapted to TEM. After generalization of the use of lead and uranium salts, adaptations to cytochemistry, immunocytochemistry, cytoenzymology, and *in situ* hybridization were performed. For that, contrast was done with metallic salts or with immunogold for immunocytochemistry or *in situ* hybridization. The use of these substances of contrast is to localize with both specificity and precision the presence of the element researched. In all these techniques, dyes were replaced with electron-dense substances.

TEM allows magnifications between 2,000 and 60,000, 200,000, and even more than 500,000 according to the performance of the microscope.

8.1.3 SCANNING ELECTRON MICROSCOPE (SEM)

The scanning electron microscope is another type of microscope invented by Knole in 1935. When matter is submitted to electrons, secondary electrons are emitted. They are absorbed with a photomultiplicator coupled to a cathode tube. So these electrons give more or less bright points. If a correctly prepared heterogeneous piece is scanned with incident electrons, a surface image with reliefs is obtained according to the magnification of the tissue surface.

Biological tissue must be prepared to be observed. Organs must be preserved and dehydrated without any deformation. Then they are covered with heavy metallic atoms (gold or/and palladium) to obtain a conductor surface. After this preparation, SEM gives a surface image.

8.1.4 OTHERS

Other applications of electron microscopy are available. For example, use of retrodiffused electrons, or X-ray microanalysis that has been performed by Castaing (1951), which allows obtention of the chemical composition of tissues as well as to visualize a particular element (calcium and phosphorus for bones, for example).

8.2 PREPARATION OF TISSUES

8.2.1 GENERAL CONSIDERATIONS

Fixation constraints are to respect physicochemical characteristics of cell fluids to avoid movements of ions and water. A modification of cell structure is the consequence of water and ions displacements. The physicochemical parameters to be considered are osmolarity, pH, and ionic concentration.

Fixatives for electron microscopy are osmotically and ionically equilibrated mixtures. For light microscopy (see Chapter 1, Section 1.2), they act by blocking enzymatic systems and stabilizing molecular structures. The fixative must penetrate quickly and specimens must be small to avoid a default of fixation in the center of the specimen. If it is not the case, the risk is that only the periphery of the sample will be correctly fixed, but the observation of the center of organs will not be good.

8.2.1.1 Osmolarity

A living cell is semipermeable. It is in osmotic equilibrium with the extracellular medium. Most fixatives do not avoid the semipermeability. Only osmium tetroxide stops semipermeability. So the fixatives and rinsing solution must be prepared with an osmotically equilibrate solution before use of osmium tetroxide, also called osmic acid. After action of osmium tetroxide, tissue can be submitted to all kinds of solvent, even organic.

Calculation of osmolarity. When a molecule is in solution in water, the osmolarity depends on the number of dissolved particles. For example,

$NaCl \rightarrow Na^+ + Cl^-$, so NaCl 1 Mol $\rightarrow$ 2 OsMol.
$C_6H_{12}O_6$ does not dissociate, so $C_6H_{12}O_6$ 1 Mol $\rightarrow$ 1 OsMol.
2 Mol NaCL + 3 Mol $C_6H_{12}O_6 \rightarrow$ 7 OsMol.

Effects of osmolarity. If osmolarity of fixative is weak, the water penetrates into the tissues, ions go out of tissue, so the structures inflate. Inflation of structure can provoke their bursting.

Fixative osmolarity. To respect osmolarity, the fixative is a mixture of several components:

Fixative mixture = Additive + buffer + fixative (Additive + buffer = vector)

The fixative is used to immobilize cell and tissue structures; buffer is used to obtain a pH and a osmolarity equivalent to that of organ studied; additive is used to adjust at the good ionic concentration. For action of pH, see Section 8.2.1.2. For importance of ionic concentration, see Section 8.2.1.3.

Calculation of osmolarity is complex. Fixative and vector are not equivalent for osmolarity calculation, due to their different rapidity of penetration into the tissues. It is usually admitted that the vector fluid represents two-thirds of referent osmolarity, and the fixative, one-third.

Determination of osmolarity. Determination of osmolarity is based upon Raoult's law:

$\Delta t = kN$
N   = number of particles/L of solvent
k   = solvent constant (k_{water} = 1.85)
Δt = cryoscopic decrease

The higher concentrated the solution, the lower is the frozen temperature.

To determine the osmolarity of tissue, an osmometer is used. Osmolarity of an organism is measured on liquid such as blood or lymph. For tissue, osmolarity can be measured on crushing.

8.2.1.2   pH

Importance of pH. A pH variation has several repercussions on proteins. Variations can affect conformation, polymerization, and even provoke dissolution and liberation of amino acids. So, structural variations can be observed.

Measure of pH is done on internal fluids (e.g., blood, hemolymph). General pH of vertebrates is 7.3 or 7.4. But invertebrates or aquatic animals can have a different pH. So, it is indispensable to measure pH of a new material. pH of a lot of marine animals is that of seawater. In a given animal, each tissue possesses its own pH. In mammals, pH of the kidney is about 7.0; stomach and intestine have a pH of 7.6.

Effects of pH. pH is linked to protein dissociation. In a general manner, pH is about 5. Above this pH, proteins are little dissociated; below this pH, proteins are very dissociated. If proteins are dissociated, osmolarity varies.

For pH equilibrium, several buffers can be used. These buffers are mixtures of both basic and acid salts. They must be soluble in water, but not in other solvents. They must not be toxic, with a large capacity. pH of buffers is generally comprised between 5 and 8.

Mineral buffers. Several mineral buffers are available. We give here only the most employed.

Sörensen's buffer—They are the oldest buffers used for electron microscopy. Their pH is comprised between 5.0 and 7.6. Their preparation is easy but dissolution of salt is difficult, and crystallization can occur. Bacteria develops easily in Sörensen's buffer, even at low temperatures.

Sodium cacodylate—Sodium cacodylate ($Na(CH_3)_2AsO_2$, 3 H_2O) is a buffer with a high capacity. Arsenic salts are toxic, so intracellular membranes can be inflated. The inflation of membranes can be useful to emphasize visualization of some cell types. Sodium cacodylate buffer is often chosen to observe lymphocytes and other blood cells, by means of electron microscopy.

Organic buffers. Several organic buffers can be used for preparation of fixative.

Celloidin-hydrochloric acid—Inflates mitochondria. Unstable buffer.

PIPES—Penetration of fixative slackens.

HEPES—Organic acids dissociate this buffer.

8.2.1.3 Ionic Composition

The ionic composition must be equivalent to that of plasma. Its determination is empirical. To obtain the best results, use a soft osmotic pressure with a strong ionic concentration.

8.2.2 FIXATION

8.2.2.1 Fixatives in Electron Microscopy

The action of fixative agent is not very well known. Several fixatives can be used for electron microscopy investigations.

Potassium permanganate—$KMnO_4$ has certainly been one of the first fixatives used for electron microscopy. Contrast was good, but tissues were not very well fixed.

Aldehydes—The two main fixatives used for electron microscopy are formaldehyde and glutaraldehyde. Avoid glutaraldehyde for immunocytochemistry. Fix small fragments (about 3 mm^3). Glutaraldehyde reacts with $-NH_2$ groups. It penetrates slowly. It reacts essentially with globular and fibrillar proteins, histones, and nucleic acids. It reacts very little with carbohydrates and not with lipids. See Chapter 12, Section 12.1.3. Formaldehyde is a good fixative.

Osmium tetroxide—Osmium tetroxide also called osmic acid and is strongly linked with organic molecules. It is a good fixative. It penetrates and reacts slowly. It is generally used such as a postfixative after fixation with aldehydes. OsO_4 reacts with unsaturated lipids and glycogen. It does not react very well with proteins or nucleic acids. Fix small fragments (about 3 mm^3). Osmium tetroxide is used like fixative as well as staining molecule. See Chapter 12, Section 12.1.4.

8.2.2.2 Karnowsky's Fluid

See Chapter 12, Section 12.1.1.

8.2.2.3 Washing Solution

Washing solution is a Sörensen's buffer at pH 7.4 and 0.175 M.

- Preparation
  - Solution 1
    - NaH_2PO_4, 2.73 g
    - Distilled water, 100 mL
  - Solution 2
    - Na_2HPO_4, 6.265 g
    - Distilled water, 100 mL
  - Buffer
    - Solution 1, 19 mL
    - Solution 2, 81 mL

- Protocol
  - Wash at least 2 h; washing can be prolonged overnight at 4°C
  - Use a new washing solution after 1 h

8.2.2.4   Fixation by Immersion

- Immerse fragments of organs into the fixative 1 to 2 h at 4°C
- Washing solution, 2 h or overnight at 4°C.

8.2.2.5   Fixation by Perfusion

See Chapter 1, Section 1.3.2.5.3.

8.2.2.6   Postfixation with Osmium Tetroxide

Postfixation with osmium immobilizes the phospholipid bilayer of cell membranes. Osmium also immobilizes and stains a lot of lipids, especially unsaturated ones.

- Stocking solution
  - Osmium tetroxide, 0.5 g: See Chapter 12, Section 12.1.4.
  - Distilled water, 25 mL. Osmium tetroxide at 2% is obtained.
- Postfixative
  - Osmium tetroxide 2%, 100 mL
  - Sörensen's buffer 0.2 M, pH 7.4, 100 mL
- Protocol
  - Osmium tetroxide, 1 h at 4°C
  - Distilled water

8.2.2.7   Stocking Solution

It is sometimes useful to leave pieces in waiting solution.

- Stocking solution
  - Sörensen's buffer 0.2 M, pH 7.4, 30 mL
  - Distilled water, 70 mL
  - Sucrose, 7 g

Even this technique is not always considered valuable; it can give good results. Pieces can also be stocked in ethanol 70% after fixation and postfixation with osmium tetroxide.

8.3   EMBEDDING

8.3.1   General Considerations

Tissues must be embedded into a sufficiently hard resin to allow 60 nm thick sections. To be available for observation with TEM, thickness of section must not exceed 90 nm. Generally, epoxides are used; the general protocol is to dehydrate the tissue, then to immerse it into an intermediate molecule before embedding into the wax. Inclusion media must allow a good conservation of ultra-structures; they must be stable under the electrons, and they must allow the obtaining of sections. The first medium used was methacrylate, but the first protocols did not give very good results. Today, new protocols are used.

8.3.2 Dehydration

Dehydration is indispensable if waxes are not hydrophilic. Dehydration must be as quick as possible to avoid modifications of tissues.

8.3.2.1 Standard Dehydration
- Reagents
  - Ethanol 30%, 50%, 70%, 95%, 100%
  - Propylene oxide, also called 1,2-epoxypropane
- Protocol
  - Ethanol 30%, 10 min
  - Ethanol 50%, 10 min
  - Ethanol 70%, 10 min
  - Ethanol 95%, 10 min
  - Ethanol 100%, 2 × 10 min
  - Propylene oxide, 3 × 10 min at 4°C

8.3.2.2 Quick Dehydration
- Reagents
  - Ethanol 70%, 95%, 100%
  - Propylene oxide
- Protocol
  - Ethanol 70%, 2 × 10 min
  - Ethanol 95%, 2 × 10 min
  - Ethanol 100%, 2 × 10 min
  - Propylene oxide, 10 min

8.3.3 Inclusion

8.3.3.1 Epon Embedding
See Chapter 1, Section 1.3.3.6.2.

8.3.3.2 Durcupant Embedding
See Chapter 1, Section 1.3.3.6.3.

8.3.3.3 Methyl Methacrylate Embedding
See Chapter 1, Section 1.3.3.7.3.2.

8.3.3.4 Glycol Methacrylate Embedding
- Solution A
  - Glycol methacrylate, 97 mL
  - Distilled water, 3 mL
- Solution B
  - Butyl methacrylate, 98 mL
  - Luperco, 2 mL. Luperco is used for final polymerization of wax.
- Working solution
  - Solution A, 14 mL
  - Solution B, 6 mL

8.4 SECTIONS

8.4.1 KNIVES

8.4.1.1 Glass Knives

Preparation of glass knives is done from glass bars, cleaned with ethanol–ether. To obtain a good visualization of structures, it is necessary to obtain thin sections (50 to 70 nm thick is recommended). For that, glass or diamond knives will be used (Figure 8.2). A glass-knife maker is used to prepare a glass knife. Knives are then equipped with a gutter made with an adhesive ribbon, being watertight with a wax such as paraffin.

8.4.1.2 Diamond Knives

Diamond knives are usually used. They are equipped with a gutter to recuperate sections, and they allow for very thin sections. It is useful to stock a sufficient number of knives protected from dust (Figure 8.2). Diamond knives are fragile and they must be periodically sharpened.

8.4.2 CUTTING BLOCK

The block is cut with a pyramidal shape. The surface of tissue must be cleared from any embedding pellicle (Figure 8.1).

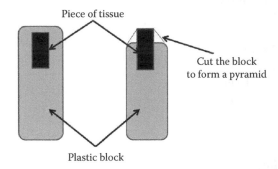

FIGURE 8.1 Cutting the block.

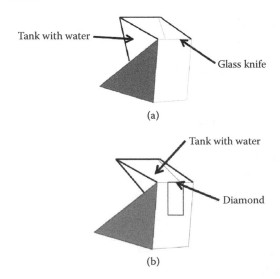

FIGURE 8.2 Knives for transmission electron microscopy. (a) Glass knife. Water tank is prepared with adhesive paper. (b) Diamond knife.

8.4.3 Semithin Sections

(See also Chapter 1, Section 1.3.4.3.) Before executing ultrathin sections, it is useful to obtain semithin sections (0.5 to 1 μm thick) to localize the interesting part of the tissue. Semithin sections are recuperated with grips and disposed on a drop of water, situated on a slide. The observation of semithin sections sometimes allows for cutting the block again in order to obtain the smallest surface for semithin sections.

8.4.4 Ultrathin Sections

The knife is disposed on a special support of the ultramicrotome. The thickness of sections is appreciated according to the Fräunhöfer staining scale. An ultramicrotome allows for ultrathin sections (70 nm thickness).

Ultrathin sections are directly recuperated on a grid (Figure 8.3). Ultrathin sections must be silver or gray. Golden sections are thick, but they also can be used. Grids are made of a metal that will not react with electrons. Generally, copper grids are used, but other metals can also be used. Grids are characterized with the number of mesh/inch square. The larger the number, the smaller the surface of mesh. In certain cases, only one mesh is used to visualize the tissue (Figure 8.4).

8.4.5 Support Films

In certain cases, sections are not disposed on a grid but on a window. In this case, a membrane must cover the window.

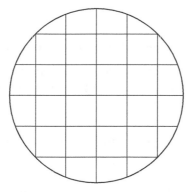

FIGURE 8.3   Grid with meshes.

FIGURE 8.4   Grid with large window.

8.4.5.1 Formvar Film

- Reagents
  - Formvar, 0.5 mL
  - Dichloroethane, 100 mL
- Protocol
  - Dispose formvar on a coverslip. A film of formvar floats on the surface of water.
  - Spread formvar
  - Drive on the coverslip with an angle of 45° in distilled water
  - Dispose the grid on the film of formvar
  - Recuperate the grid with paper filter
  - Let dry
  - Recuperate the grid with grips

8.4.5.2 Collodion Film

- Reagents
  - Collodion, 1 g
  - Isoamyl acetate, 100 mL
- Protocol
  - Dispose the collodion solution on water
  - Dispose the grid on the film of collodion
  - Recuperate the grid with paper filter
  - Let dry. Collodion becomes solid.
  - Recuperate the grid with grips

8.5   STAINING

8.5.1   Standard Contrast with Uranyl Acetate and Lead Citrate

In electron microscopy, contrast or staining is performed with electron dense elements. Some methods of staining are used to visualize all the structures. Cytochemical techniques give a specific staining for precise molecules.

- Reagents
  - Uranyl acetate saturated in ethanol 50%. See Chapter 12, Section 12.2.29.
  - Lead citrate. See Chapter 12, Section 12.2.8.
- Protocol
  - Uranyl acetate, 20 min. Put the grids on a drop of filtrated uranyl acetate, cover and leave in the dark.
  - Distilled water, 20 times. Cut the surface of distilled water with the side of the grid.
  - Let dry.
  - Lead citrate, 10 min. Put the grids on a drop of lead citrate, cover, and leave in the dark.
  - Let dry.
  - Distilled water, 6 times. Cut the surface of distilled water with the side of the grid.
  - Let dry.
- Results—Cell components are stained. This standard method gives good results to appreciate general structure and ultrastructure of tissues (Figure 8.5 and Figure 8.6).

8.5.2   Visualization of Golgi Apparatus

8.5.2.1 Visualization of Golgi Apparatus with Osmium Tetroxide

- Fixative—Tetroxide osmium is both a fixative and dye used to visualize the tissues. Reaction is done on blocks.

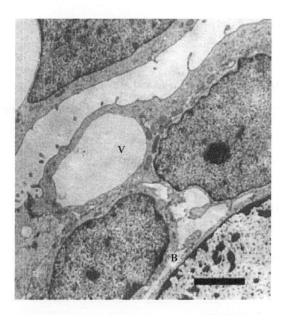

FIGURE 8.5 Classical staining with uranyl acetate and lead citrate. Method: Uranyl acetate–lead citrate. Tissue: Amphibian gill. Preparation: Epon embedding. Fixative: Glutaraldehyde–paraformaldehyde. Observations: V, vacuol; B, basal cell. Bar − 2.5 μm.

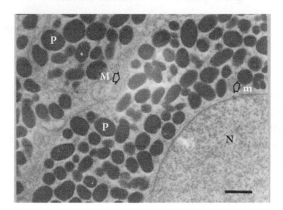

FIGURE 8.6 Classical staining with uranyl acetate and lead citrate. Method: Uranyl acetate–lead citrate. Tissue: Amphibian retina. Preparation: Epon embedding. Fixative: Glutaraldehyde–paraformaldehyde. Observations: M, cell membrane; m, mitochondria: N, nucleus; P, pigment. Bar = 2.5 μm.

- Reagents
  - Osmium tetroxide: See Chapter 12, Section 12.1.4.
  - Uranyl acetate: See Chapter 12, Section 12.2.29.
- Protocol
  - Osmium tetroxide, 24 h at 40°C
  - Osmium tetroxide, 16 h at 40°C
  - Uranyl acetate, 30 min
  - Dehydrate
  - Embed in wax
  - Section
- Results—Golgi complexes are electron dense.

8.5.2.2   Visualization of Golgi Apparatus with Periodic Acid–Silver

See Chapter 9, Section 9.1.4.1. Periodic acid–silver is used to visualize mucopolysaccharides.

8.5.3   VISUALIZATION OF SYNAPSES WITH BISMUTH IODIDE

- Fixative—Fixation with glutaraldehyde 6.5% in buffer pH 7.4. Staining is done on blocks.
- Reagents
  - Bismuth iodide: See Chapter 12, Section 12.2.1.
  - Potassium iodide
  - Formic acid
- Protocol
  - Staining solution, 2 h
  - Formic acid, rinse
  - Uranyl acetate, 30 min
  - Lead acetate: 10 min
- Results—Synapses are visualized with electron density.

8.5.4   VISUALIZATION OF EXOCYTOSIS

8.5.4.1   Visualization of Exocytosis with TAGO

TAGO is short for tannic acid–glutaraldehyde–osmium tetroxide.

- Fixative—Fixation and staining are combined. Tissues are first fixed with a tannic acid–glutaraldehyde mixture, then postfixed with osmium tetroxide.
- Reagents
  - Tannic acid
  - Glutaraldehyde: See Chapter 12, Section 12.1.3.
  - Osmium tetroxide: See Chapter 12, Section 12.1.4.
  - Cacodylate buffer pH 7.3: See Chapter 12, Section 12.3.1.
  - Staining–fixation solution: See Chapter 12, Section 12.2.25.
  - Osmium solution for TAGO
- Protocol
  - Staining-fixation solution, 2 h
  - Cacodylate buffer, 15 min
  - Osmium tetroxide, 1 h at 4°C
- Results—Exocytosed secretory products are electron dense. Lipids are also electron dense.

8.5.4.2   Visualization of Exocytosis with TARI

TARI is short for tannic acid in Ringer's solution.

- Fixative—Tissues are first stained with tannic acid in Ringer's solution, then they are fixed with a classic method.
- Reagents
  - Tannic acid: See Chapter 12, Section 12.2.26.
  - Osmium tetroxide: See Chapter 12, Section 12.1.4.
  - Lead citrate: See Chapter 12, Section 12.2.8.
  - Uranyl acetate: See Chapter 12, Section 12.2.29.
  - Staining solution: See Chapter 12, Section 12.2.26.
  - Ringer's solution: See Chapter 12, Section 12.3.4.

- Protocol
  - Staining solution, 2 h
  - Ringer's solution, 15 min
  - Classic fixation with glutaraldehyde
  - Postfixation with osmium
  - Impregnation
  - Embedding
  - Section
  - Uranyl acetate (classic method). Can be replaced with another uranium salt.
  - Lead citrate (classic method). Can be replaced with another lead salt.
- Results—Exocytosed secretory products are electron dense.

8.5.5 Negative Staining

This technique is used for direct observation of small particles. A layer of contrasting material is disposed around the object. Structures appear light on a dark bottom.

- Reagents
  - Phosphotungstic acid: See Chapter 12, Section 12.2.21 or Section 12.2.22.
  - Osmium tetroxide: See Chapter 12, Section 12.2.19.
- Protocol for unfixed material
  - Dispose a drop of solution to study on a grid
  - Phosphotungstic acid, 3 min
  - Suck up with paper filter
  - Let dry at air temperature
- Protocol for fixed material
  - Dispose a drop of unfixed solution to study on the bottom of a recipient
  - Phosphotungstic acid
  - Osmium tetroxide
  - Let incubate, 3 min
  - Dispose a grid with membrane on the drop
  - Phosphotungstic acid
  - Suck up with paper filter
  - Let dry at air temperature
- Results—Structures appear light on a dark bottom.

REFERENCES

Bjorkman, N., and B. Hellström. 1965. Lead-ammonium acetate; a staining medium for electron microscopy free of contamination by carbonate. *Stain Technology* 40:169–171.

Castaing, R. 1951. *Application des sondes électroniques à une méthode d'analyse ponctuelle chimique et cristallographique. Thèse de doctorat d'état*, Université de Paris, 1952.

Dalton, A.J., and R.F. Zeigel. 1960. A simplified method of staining thin sections of biological material with lead hydroxide for electron microscopy. *Journal Biophysics Biochemistry Cytology* 7:409–410.

Hayat, M.A. 1972. *Basic electron microscopy techniques.* Rheinold Company, New York.

Hayat, M.A. 2000. *Principles and techniques of electron microscopy biological applications.* Cambridge University Press, Cambridge.

Karnovsky, M.J. 1961. Simple methods for staining with lead at high pH in electron microscopy. *Journal of Biophysical and Biochemical Cytology* 11:729–732.

Lever, J.D. 1960. A method of staining sectioned tissues with lead for electron microscopy. *Nature* 186:810-811.

Millonig, G. 1961. A modified procedure for lead staining of thin sections. *Journal of Biophysical and Biochemical Cytology* 11:736-739.

Pottu-Boumendil, J. 1989. *Microscopie électronique. Principes et méthodes de préparation.* INSERM, Paris.

Reynolds, E.S. 1963. The use of lead citrate at high pH as an electron-opaque stain in electron microscopy. *Journal of Cell Biology* 17:208–212.

Sato, T. 1967. A modified method for lead staining of thin sections. *Journal of Electron Microscopy* 16:133–141.

Venable, J.H., and R. Coggeshall. 1965. A simplified lead citrate stain for use in electron microscopy. *Journal of Cell Biology* 25:407–408.

Watson, M.L. 1958. Staining of tissue sections for electron microscopy with heavy metals. *Journal of Biophysical and Biochemical Cytology* 4:475–478.

9 Cytochemical Techniques

Jean-Marie Exbrayat

CONTENTS

9.1   VISUALIZATION OF MUCOPOLYSACCHARIDES

9.1.1   VISUALIZATION OF STARCH WITH IODINE

- Fixative—Classical fixation can be used; glutaraldehyde or glutaraldehyde–paraformalde-hyde followed by osmium tetroxide.
- Reagents
  - Iodine
  - Lead citrate: See Chapter 12, Section 12.2.8.
- Protocol
  - Mount sections on grids
  - Expose to vapors of iodine, 12 h at 55°C
  - Lead citrate, 45 min
- Results—Starch is electron dense.

9.1.2 Acidic Mucopolysaccharides

9.1.2.1 Use of Iron

- Fixative—Blocks must be fixed with glutaraldehyde or glutaraldehyde–paraformaldehyde. Then they are postfixed with osmium tetroxide, embedded in wax, and section will occur after staining.
- Reagents
  - Ferric chloride: See Chapter 12, Section 12.2.5.
  - Buffered osmium tetroxide 2%: See Chapter 12, Section 12.1.5.
- Protocol
  - Stain before embedding, 4 to 12 h
  - Distilled water
  - Osmium tetroxide, 1 h at 0°C
  - Embedding is done according to classical protocol: See Chapter 1, Section 1.3.3.6.
  - Sectioning
- Results—This method can be used to visualize acidic mucopolysaccharide and sialic acid. Acidic mucopolysaccharides and sialic acid are electron dense.

9.1.2.2 Use of Thorium before Embedding

- Fixative—Blocks must be fixed with glutaraldehyde or glutaraldehyde–paraformaldehyde and postfixed with osmium tetroxide. They are embedded in wax, and sectioning will occur after staining.
- Reagents
  - Colloidal thorium in acetic acid: See Chapter 12, Section 12.2.2.
  - Acetic acid 3%, pH 2.5
- Protocol
  - Distilled water
  - Colloidal thorium, 24 h
  - Acetic acid 3%, rinse
  - Dehydration
  - Embedding is done according to classical protocol: See Chapter 1, Section 1.3.3.6.
  - Sectioning
- Results—Acidic mucopolysaccharides are electron dense.

9.1.2.3 Thorium: Staining of Sections

- Fixative—Sections are fixed with glutaraldehyde or glutaraldehyde–paraformaldehyde, then postfixed with osmium tetroxide. Embedding in methacrylate.
- Reagents
  - Colloidal thorium in acetic acid: See Chapter 12, Section 12.2.2.
  - Acetic acid 3%, pH 2.5
  - Uranyl acetate: See Chapter 12, Section 12.2.29.
- Protocol
  - Acetic acid 3%, 2 min. Sections are floated on the reagents. Time of rinsing with acetic acid can be prolonged to 5 min.
  - Colloidal thorium, 5 min
  - Acetic acid 3%, 2 x 3 min
  - Distilled water, 5 min
  - Uranyl acetate, 10 min
  - Mount on grids
- Results—Acidic mucopolysaccharides are electron dense.

9.1.2.4 Sulfomucopolysaccharides: Method with Diaminobenzidine–Osmium Tetroxide

Used at an acidic pH, diaminobenzidine (DAB) forms a specific linkage with sulfated functions that can be oxidized with osmium tetroxide. The result is a thick precipitate at the level of sulfomucopolysaccharides.

- Fixative—Classical preserved sections with glutaraldehyde or glutaraldehyde–paraformaldehyde. Glycol methacrylate embedding. Use gold grids.
- Reagents
  - Boric acid 5%
  - Diaminobenzidine: See Chapter 12, Section 12.2.4.
  - Osmium tetroxide 2%: See Chapter 12, Section 12.2.19.
- Protocol
  - Boric acid, 5 min. Incubation can be prolonged to 30 min.
  - DAB, 20 min
  - Boric acid 5%, 2 × 1 min. Incubation can be prolonged to 20 min.
  - Osmium tetroxide 2%, 10 min
  - Distilled water rinse
- Results—Sulfated mucopolysaccharides are stained as heavy electron-dense precipitates.

9.1.3 EXTERNAL MUCOPOLYSACCHARIDES

9.1.3.1 Staining with Lanthanum

- Fixative—Fixation is done with the staining solution.
- Reagents
  - Glutaraldehyde 25%: See Chapter 12, Section 12.1.3.
  - Lanthanum nitrate: See Chapter 12, Section 12.2.7.
  - Cacodylate buffer pH 7.3
  - Osmium tetroxide: See Chapter 12, Section 12.2.19.
  - Solution A
    - Glutaraldehyde 25%, 10 mL
    - Lanthanum nitrate, 1 g
    - Cacodylate buffer pH 7.3, 100 mL
  - Solution B
    - Lanthanum nitrate, 1 g
    - Cacodylate buffer pH 7.3, 100 mL
  - Solution C
    - Osmium tetroxide, 1 g
    - Lanthanum nitrate, 1 g
    - Cacodylate buffer pH 7.3, 100 mL
- Protocol
  - Solution A, 2 h at 4°C
  - Solution B, 30 min at 4°C
  - Solution C, 30 min at 4°C. Time can be prolonged to 1 h.
  - Dehydrate
  - Embedding
- Results—Extracellular acidic mucopolysaccharides are electron dense (Figure 9.1).

9.1.3.2 Staining with Ruthenium Red

- Fixative—Fixation is done with the staining solution.
- Reagents
  - Glutaraldehyde 25%: See Chapter 12, Section 12.1.3.

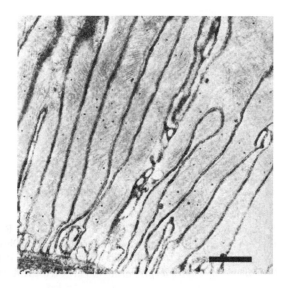

FIGURE 9.1   Staining with lanthanum nitrate. Tissue: Distal tube of rat kidney. Preparation: Lanthanum nitrate. Fixative: Fresh sections. Observations: Lanthanum nitrate diffused between cell membranes. (From Pottu-Boumendil, J., 1989, *Microscopie électronique: Principes et méthodes de préparation*, Paris: INSERM. With permission.)

- Ruthenium red
- Cacodylate buffer pH 7.3: See Chapter 12, Section 12.3.1.
- Osmium tetroxide
- Solution A
  - Glutaraldehyde 25%, 10 mL
  - Ruthenium red, 50 mg
  - Cacodylate buffer pH 7.3, 100 mL
- Solution B
  - Osmium tetroxide, 1 g
  - Ruthenium red, 1 g
  - Cacodylate buffer pH 7.3, 100 mL
- Protocol
  - Solution A, 1 h at 0°C. Temperature can be increased to 4°C.
  - Solution B, 3 h
  - Dehydrate
  - Embedding
- Results—Extracellular acidic mucopolysaccharides are electron dense (Figure 9.2).

9.1.4   PAS-Like Methods for Mucopolysaccharide Visualization

9.1.4.1   Periodic Acid–Silver

- Fixative—Classical fixatives can be used.
- Reagents
  - Silver methenamine solution: See Chapter 12, Section 12.2.24. This solution must be prepared just before use because it is unstable.
  - Periodic acid 1%
  - Sodium thiosulfate 3%

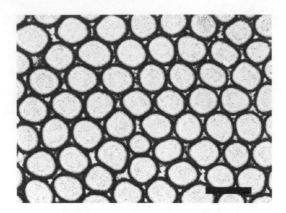

FIGURE 9.2 Staining with lanthanum nitrate. Method: Ruthenium red. Tissue: Rat intestine (enterocytes). Preparation: Epon embedding. Fixative: Glutaraldehyde–paraformaldehyde. Observations: Cross-section of enterocytes. Glycoproteins are visualized in black around microvilli. (From Pottu-Boumendil, J., 1989, *Microscopie électronique: Principes et méthodes de préparation*, Paris: INSERM. With permission.)

- Protocol
    - Periodic acid, 15 min
    - Distilled water, rinse 3 times
    - Silver methenamine solution, 1 h at 60°C. Avoid use of copper grids because silver solution precipitates with copper.
    - Distilled water, several times
    - Sodium thiosulfate, 5 min
    - Distilled water
- Results—Mucopolysaccharides, glycoproteins, and Golgi apparatus are electron dense.

9.1.4.2 Periodic Acid–Chromic Acid Silver

- Fixative—Classical fixatives can be used.
- Reagents
    - Chromic acid 10%. Chromic acid is H_2CrO_4 but in histology the name "chromic acid" generally is used to qualify "chromic oxide" CrO_3, which is the case here.
    - Periodic acid 1%
    - Sodium bisulfate 1%
    - Silver methenamine solution: See Chapter 12, Section 12.2.24. This solution must be prepared just before use because it is unstable.
- Protocol
    - Periodic acid, 20 min
    - Distilled water, 30 min
    - Chromic acid, 5 min
    - Sodium bisulfate, 1 min
    - Distilled water, 30 min
    - Silver methenamine solution, 30 min at 60°C
    - Distilled water, 20 min
- Results—Mucopolysaccharides, glycoproteins, and Golgi apparatus are electron dense.

9.1.4.3 Periodic Acid–Thiosemicarbazide or Thiosemihydrazide–Silver Proteinate

This method decreases the nonspecific staining of nuclei and ribosomes.

- Fixative—Classical fixatives can be used.

- Reagents
  - Thiosemicarbazide: See Chapter 12, Section 12.2.27.
  - Periodic acid 1%
  - Sodium thiosulfate 3%
- Protocol
  - Periodic acid, 30 min
  - Distilled water, rinse 3 times
  - Thiosemicarbazide, 30 min. Thiosemicarbazide can be replaced with thiocarbohydra-zide: See Chapter 12, Section 12.2.28.
  - Acetic acid 10%, 10 min
  - Acetic acid 5%, 10 min
  - Acetic acid 1%, 10 min
  - Distilled water
  - Silver proteinate, 30 min
  - Distilled water
- Results—Mucopolysaccharides, glycoproteins, and glycogen are electron dense.

9.1.5 Visualization of Glycogen

9.1.5.1 Periodic Acid–Thiocarbohydrazide (TCH)–Silver Proteinate

- Fixative—Classical fixatives can be used. Staining is done directly on free sections. Use floating sections. Manipulation is delicate.
- Reagents
  - Thiocarbohydrazide (TCH): See Chapter 12, Section 12.2.28.
  - Periodic acid 1%
  - Silver proteinate 1%
- Protocol
  - Periodic acid, 20 min. Incubation with periodic acid can be prolonged to 25 min.
  - For glycogen, glycosaminoglycans, glycoproteins, 48 to 72 h
  - Distilled water, 3 × 10 min
  - TCH, 30 min. To visualize several kinds of carbohydrates, modify the time of incubation in TCH:
    - For glycogen, 30 to 40 min
    - For glycogen and glycosaminoglycans, 28 to 48 h
  - Acetic acid 10%, 3 × 5 min
  - Distilled water, 3 × 5 min
  - Silver proteinate, 30 min
  - Distilled water, 3 × 5 min
  - Put sections on a grid
- Results—Mucopolysaccharides, glycoproteins, and glycogen are electron dense according to the duration of incubation in TCH (Figure 9.3).
- Controls—It is necessary to perform controls to be sure of the specificity of method. For control, TCH is replaced with acetic acid 20%.

9.1.5.2 Periodic Acid–Thiosemicarbazide (TSC)–Silver Proteinate

- Fixative—Classical fixatives can be used. Staining is done directly on free sections. Use floating sections. Manipulation is delicate.
- Reagents
  - Thiosemicarbazide (TSC): See Chapter 12, Section 12.2.27.
  - Periodic acid 1%
  - Silver proteinate 1%

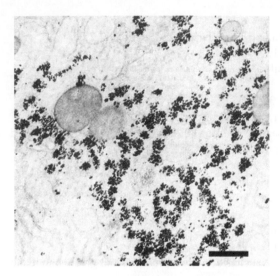

FIGURE 9.3 Visualization of glycogen with thiocarbohydrazide–silver proteinate. Tissue: Rat liver. Preparation: Epon embedding. Fixative: Gutaraldehyde–paraformaldehyde. Observations: Glycogen is observed such as black spots in cytoplasm. (From Pottu-Boumendil, J., 1989, *Microscopie électronique: Principes et méthodes de préparation*, Paris: INSERM. With permission.)

- Protocol
  - Periodic acid, 20 min. Incubation with periodic acid can be prolonged to 25 min.
  - Distilled water, 3 × 10 min
  - TSC, 30 min. To visualize several kinds of carbohydrates, modify the time of incubation in TSC.
    - – For glycogen, 30 to 40 min
    - – For glycogen and glycosaminoglycans, 28 to 48 h
  - For glycogen, glycosaminoglycans, and glycoproteins, 48 to 72 h
  - Acetic acid 10%, 3 × 5 min
  - Distilled water, 3 × 5 min
  - Silver proteinate, 30 min
  - Distilled water, 3 × 5 min
  - Put sections on a grid
- Results—Mucopolysaccharides, glycoproteins, and glycogen are electron dense according to the duration of incubation into TSC.
- Controls—It is necessary to perform controls to be sure of the specificity of method. For control, TSC is replaced with acetic acid 20%.

9.1.5.3 Staining with Lead Citrate
- Fixative—Classical fixatives can be used. Staining is done directly on free sections.
- Reagents
  - Periodic acid 1%. Periodic acid can be used from 0.8% to 3.2% in water. Periodic acid can be replaced with H_2O_2 1.5%.
  - Lead citrate: See Chapter 12, Section 12.2.8.
- Protocol
  - Periodic acid, 30 min; or H_2O_2, 10 to 15 min
  - Distilled water, 3 × 10 min
  - Lead citrate, 10 min. It is possible to add uranyl acetate staining, but this can mask the glycogen.
  - Put sections on a grid
- Results—Glycogen is electron dense.

9.1.6 Visualization of Glycoproteins with Phosphotungstic Acid (PTA)

- Fixative—The technique is done with frozen sections, or with sections fixed with glutaraldehyde, embedded in glycol methacrylate (GMA). It is necessary to use material (grids, fixatives, embedding medium) that do not react with phosphotungstic acid (PTA). Generally, Epon or Araldite embedding do not give good results. If tissues have been postfixed with osmium tetroxide, delete osmium with treatment of sections by hydrogen peroxide 10%.
- Reagents
  - Phosphotungstic acid (PTA): See Chapter 12, Section 12.2.21.
- Protocol
  - PTA, 2 min. Let the grids float on PTA solution. Time can reach 5 min.
  - If floating sections have been stained, put them on a grid.
  - Distilled water, rinse
- Results—Complex carbohydrates and more specifically glycoproteins are intensely contrasted. Several organelles are also visualized. Golgi apparatus, lysosomes, pinocytosis vesicle, and several secretions can be visualized.

9.1.7 Methods with Lead Salts

Staining with lead salts can be used to visualize several kinds of molecules, according to the salt employed. Several methods are more particularly used to visualize polysaccharides. Staining methods with lead are discussed in Section 9.4.

9.1.7.1 Method of Reynolds

Lead citrate gives an intense staining of polysaccharides (Reynolds, 1963). See Section 9.4.3.1 and Chapter 12, Section 12.2.10.

9.1.7.2 Method of Venable and Coggeshall

Lead citrate gives an intense staining of polysaccharides (Venable and Coggeshall, 1965). See Section 9.4.3.2 and Chapter 12, Section 12.2.11.

9.1.7.3 Method of Sato

Lead citrate gives an intense staining of polysaccharides (Sato, 1967). See Section 9.4.3.3 and Chapter 12, Section 12.2.12.

9.1.7.4 Method of Watson

Lead hydroxide gives an intense staining of polysaccharides (Watson, 1958). See Section 9.4.4.1 and Chapter 12, Section 12.2.13.

9.1.7.5 Method of Karnovsky I

Lead citrate gives an intense staining of polysaccharides (Karnovsky, 1961). See Section 9.4.4.2 and Chapter 12, Section 12.2.14.

9.1.7.6 Method of Karnovsky II

Lead citrate gives an intense staining of polysaccharides (Karnovsky, 1961). See Section 9.4.4.3 and Chapter 12, Section 12.2.15.

9.1.7.7 Method of Lever

Lead citrate gives an intense staining of polysaccharides (Lever, 1960). See Section 9.4.4.4 and Chapter 12, Section 12.2.16.

9.1.7.8   Method of Millonig I

Lead citrate gives an intense staining of polysaccharides (Millonig, 1961). See Section 9.4.5.1 and Chapter 12, Section 12.2.17.

9.1.7.9   Method of Millonig II

Lead citrate gives an intense staining of polysaccharides (Millonig, 1961). See Section 9.4.5.2 and Chapter 12, Section 12.2.18.

9.2   VISUALIZATION OF PROTEINS AND COLLAGEN

9.2.1   PHOSPHOTUNGSTIC ACID METHOD

- Fixative—Classical fixatives can be used. Staining is done during dehydration of tissue block.
- Reagents—Phosphotungstic acid: See Chapter 12, Section 12.2.21.
- Protocol
  - Phosphotungstic acid, 30 min
  - Dehydration, impregnation
  - Embedding
  - Sectioning
- Results—Both proteins and collagen are electron dense.

9.2.2   STAINING OF SULFHYDRYLED PROTEINS WITH LEAD

It has been observed that lead components of citrate can be transferred to several molecules such as cysteine. So several methods based upon use of lead citrate can also be used to visualize proteins with thiol functions.

9.2.2.1   Method of Reynolds

See Section 9.4.3.1 and Chapter 12, Section 12.2.10.

9.2.2.2   Method of Venable and Coggeshall

See Section 9.4.3.2 and Chapter 12, Section 12.2.11.

9.2.2.3   Method of Sato

See Section 9.4.3.3 and Chapter 12, Section 12.2.12.

9.3   VISUALIZATION OF NUCLEIC ACIDS

9.3.1   METHOD WITH BISMUTH

- Fixative—Classical fixatives can be used.
- Reagents
  - Metallic bismuth
  - Nitric acid
  - Citric acid
  - Nitric acid–acetate buffer pH 7.0
  - Staining solution, bismuth iodide: See Chapter 12, Section 12.2.1.
- Protocol
  - Nitric acid–sodium acetate buffer
  - Staining solution, 90 min at 4°C
- Results—Nucleic acids are electron dense.

9.3.2 Indium Staining

- Fixative—Classical fixatives can be used. Tissue blocks are stained before embedding.
- Reagents
  - Anhydrous indium trichloride
  - Acetone
  - Lithium borohydride
  - Acetylation mixture
    - Acetic acid saturated with sodium acetate, 40 mL
    - Pyridine, 60 mL
  - Staining solution: See Chapter 12, Section 12.2.6.
- Protocol
  - Pyridine, 3 × 10 min at 4°C
  - Pyridine saturated with lithium borohydride, 2 h at 4°C
  - Pyridine, 3 × 10 min
  - Acetylation mixture, 5 min
  - Pyridine–acetone, 5 min
  - Acetone, 2 × 10 min at 4°C
  - Staining solution, 2 h at 4°C
  - Acetone, 2 × 15 min at 4°C
  - Impregnation
  - Embedding
- Results—Nucleic acids are electron dense.

9.3.3 Methods with Lead Salts

Staining with lead salts can be used to visualize several kinds of molecules, according to the salt employed. Staining methods with lead are discussed in Section 9.4. Several methods are more particularly used to visualize ribonucleoproteins (Dalton and Zeigel, 1960; Bjorkman and Hellström, 1965).

9.3.3.1 Method of Dalton and Zeigel with Lead Citrate
See Section 9.4.2.1.

9.3.3.2 Method of Bjorkman and Hellström with Lead Citrate
See Section 9.4.2.2.

9.4 STAINING WITH LEAD SALTS

9.4.1 General Principles of Staining with Lead

9.4.1.1 Introduction
Staining with lead has a high electron density. The affinity of lead salts with tissue components is wide. So lead staining can be used to visualize proteins belonging to cell membranes, cytoplasms, or nuclei. They can also be used to visualize nucleic acids or carbohydrates as glycogen. Staining methods can be used on sections as well as on blocks before sectioning. Staining with lead can be reinforced with osmium tetroxide or uranyl salts. The exposition of lead salts to air can cause the formation of lead carbonates. Hydrophobic crystals precipitate, giving a bad staining of section with dark deposits.

9.4.1.2   Mechanism of Staining

The affinity of lead salts for tissues and cells depends upon both the reactive groups and pH of staining solution. Lead staining can be independent or dependent of fixative, according to the cell component researched. If tissue is fixed with formaldehyde, glycogen staining with lead appears irregular. If tissue is fixed with osmium tetroxide, glycogen appears as rosettelike particles. Ribosomes and nucleoli appear dense, more particularly with formaldehyde fixation than osmium tetroxide.

At a high pH, staining is more quickly obtained and intense than at a low pH. When pH increases, phosphate, sulfhydryl, tyrosyl, and carbonyl groups are more ionized, and consequently, more lead ions are bound than at a low pH. Staining with lead hydroxide at pH 8.15 stains more intensely tissue components than monobase lead acetate at pH 7.0.

9.4.1.3   General Reaction of Staining

$Pb(OH)_2PbX_2- \rightarrow Pb(OH)_2Pb^{++} + 2X^-$. Element "X" can be Cl. $Pb(OH)2Pb++$ is linked to the cell component.

9.4.1.4   Principle of Membrane Staining

Cell membranes can be visualized with lead after fixation with osmium tetroxide only. Lead cautions are certainly linked to osmium tetroxide combined with membrane phospholipids.

9.4.1.5   Principle of Glycogen Staining

Lead ions are linked to glycogen with hydroxyl groups. Then lead is accumulated to the first lead salts, giving a characteristic rosette aspect to glycogen.

9.4.1.6   Reaction of Lead with Other Tissue Components

Lead salts can combine with –SH chemical groups belonging to cysteine, giving mercaptides. If tissues are not fixed with osmium tetroxide, membranes are not stained with lead salts.

Lead is also used to visualize amino acids, belonging to lateral chains of proteins, zymogene granules, phosphate groups of nucleic acids, and nucleotides. Lead salts can stain proteins with numerous sulfhydryl components (keratin).

Mercaptides are molecules obtained by a combination of a protein with a metallic ion.

9.4.2   Staining with Lead Acetate

9.4.2.1   Method of Dalton and Zeigel

- Fixative—Classical fixatives can be used. Staining is done on grids.
- Reagents
- Lead acetate at saturation in boiled distilled water (pH 5.9). Preparation is done in boiled distilled water. Store in a glass-stopper bottle with crystal of lead citrate.
- Ammonium hydroxide 1%. Ammonium hydroxide can be used from 1% to 5%.
- Protocol
  - Lead acetate, 5 min. Time of staining can be increased to 30 min
  - When lead acetate has been removed from the bottle, fill the bottle to the top with distilled water.
  - Distilled water, 3 min. This bath can be prolonged to 20 min
  - Dry grids by blotting
  - Ammonium hydroxide 1% (vapors), 5 s. Ammonium hydroxide 5% can also be used.
- Results—This method can be used to visualize ribonucleoproteins. At pH 7.0, more carboxyl groups are visualized. At higher pH, proteins are visualized (Dalton and Zeigel, 1960).

9.4.2.2   Method of Bjorkman and Hellström (1965)

- Fixative—Classical fixatives can be used. Staining can be done on grids or blocks before section.
- Reagents
  - Ammonium acetate
  - Saturated lead acetate
  - Lead acetate solution: See Chapter 12, Section 12.2.9.
- Protocol
  - Lead acetate solution, 20 min
  - Distilled water
- Results—This method can be used to visualize ribonucleoproteins (Bjorkman and Hellström, 1965).

9.4.3   STAINING WITH LEAD CITRATE

9.4.3.1   Method of Reynolds

- Fixative—Classical fixatives can be used.
- Reagents
  - Lead nitrate
  - Sodium citrate
  - Sodium hydroxide 1 M
  - Sodium hydroxide 0.02 M
  - Staining solution: See Chapter 12, Section 12.2.10.
- Protocol
  - Prepare several drops of staining solution
  - Float the grid on the drop, 5 min
  - Sodium hydroxide 0.02 M, rinse
  - Distilled water, rinse
- Results—Lead citrate increases staining given by uranyl acetate. It gives an intense staining of polysaccharides (Reynolds, 1963).

9.4.3.2   Method of Venable and Coggeshall

- Fixative—Classical fixatives can be used.
- Reagents
  - Lead citrate
  - Sodium hydroxide 10 M
  - Sodium hydroxide 0.02 M
  - Staining solution: See Chapter 12, Section 12.2.11.
- Protocol
  - Prepare several drops of staining solution
  - Float the grid on the drop, 5 min
  - Sodium hydroxide 0.02 M, rinse
  - Distilled water, rinse
- Results—Lead citrate increases staining given by uranyl acetate. It gives an intense staining of polysaccharides (Venable and Coggeshall, 1965).

9.4.3.3   Method of Sato

- Fixative—Classical fixatives can be used.

- Reagents
  - Lead nitrate
  - Lead citrate
  - Lead acetate
  - Sodium hydroxide 10 M
  - Sodium hydroxide 0.02 M
  - Staining solution: See Chapter 12, Section 12.2.12.
- Protocol
  - Staining solution, 10 min
  - Distilled water, rinse
- Results—Lead citrate increases staining given by uranyl acetate. It gives an intense staining of polysaccharides (Sato, 1967).

9.4.4   STAINING WITH LEAD HYDROXIDE

9.4.4.1   Method of Watson

- Fixative—Classical fixatives can be used.
- Reagents
  - Lead acetate
  - Staining solution: See Chapter 12, Section 12.2.13.
- Protocol
  - Staining solution, 10 min
  - Distilled water, rinse
- Results—Lead hydroxide increases staining given by uranyl acetate. It gives an intense staining of polysaccharides (Watson, 1958).

9.4.4.2   Method of Karnovsky I

- Fixative—Classical fixatives can be used.
- Reagents
  - Lead monoxide
  - Sodium hydroxide
  - Staining solution: See Chapter 12, Section 12.2.14.
- Protocol
  - Prepare several drops of staining solution
  - Float the grid on the drop, 5 min
  - Sodium hydroxide 0.02 M, rinse
  - Distilled water, rinse
- Results—Lead citrate increases staining given by uranyl acetate. It gives an intense staining of polysaccharides (Karnovsky, 1961).

9.4.4.3   Method of Karnovsky II

- Fixative—Classical fixatives can be used.
- Reagents
  - Lead monoxide
  - Sodium cacodylate 10% in water
  - Sodium hydroxide 1 M
  - Staining solution: See Chapter 12, Section 12.2.15.
- Protocol
  - Prepare several drops of staining solution
  - Float the grid on the drop, 5 min

- Sodium hydroxide 0.02 M, rinse
- Distilled water, rinse
- Results—Lead citrate increases staining given by uranyl acetate. It gives an intense staining of polysaccharides (Karnovsky, 1961).

9.4.4.4 Method of Lever

- Fixative—Classical fixatives can be used.
- Reagents
  - Lead hydroxide
  - Potassium hydroxide 1%
  - Staining solution: See Chapter 12, Section 12.2.16.
- Protocol
  - Prepare several drops of staining solution
  - Float the grid on the drop, 5 min
  - Potassium hydroxide 1%, rinse
  - Distilled water, rinse
- Results—Lead citrate increases staining given by uranyl acetate. It gives an intense staining of polysaccharides (Lever, 1960).

9.4.5 STAINING WITH LEAD TARTRATE

9.4.5.1 Method of Millonig I

- Fixative—Classical fixatives can be used.
- Reagents
  - Sodium hydroxide
  - Potassium sodium tartrate
  - Lead hydroxide
  - Staining solution: See Chapter 12, Section 12.2.17.
- Protocol
  - Prepare several drops of staining solution
  - Float the grid on the drop, 5 min
  - Potassium hydroxide 1%, rinse
  - Distilled water, rinse
- Results—Lead citrate increases staining given by uranyl acetate. It gives an intense staining of polysaccharides (Millonig, 1961).

9.4.5.2 Method of Millonig II

- Fixative—Classical fixatives can be used.
- Reagents—Staining solution: See Chapter 12, Section 12.2.18.
- Protocol
  - Staining solution, 5 min
  - Distilled water, rinse
- Results—Lead citrate increases staining given with uranyl acetate. It gives an intense staining of polysaccharides (Millonig, 1961).

9.5 VISUALIZATION OF LIPIDS

9.5.1 GENERAL CONSIDERATIONS

Like it was explained in Chapter 1, lipids are very sensitive to the preparation of tissues. Dehydration with lipid solvents must be avoided. Fixation with osmium tetroxide results in retention of lipids.

Not only the fixative preserves them but, also, it gives a gray staining of these molecules. Several schemes have been proposed for osmium reaction with lipids: Osmium oxides migrate from the hydrophobic region and are deposited at the hydrophilic interface, osmium oxides are moved to the hydrophilic interface by binding of the lipid osmate esters, and unsaturated lipids are cross-linked by means of covalent bonds that do not involve osmium.

9.5.2 METHODS OF VISUALIZATION

9.5.2.1 Staining with Osmium Tetroxide

Lipids are visualized with osmium tetroxide and used like a fixative as well as a staining substance. It is why osmium tetroxide preserves membrane phospholipids so that cells can be observed in electron microscopy (see Chapter 1).

9.5.2.2 Use of Tannic Acid to Visualize Lipids

Tannic acid can be used as a mordant for osmium tetroxide and for lipid visualization (see Chapter 1, Section 1.3.2.4.3).

REFERENCES

Bjorkman, N., and B. Hellström. 1965. Lead-ammonium acetate: A staining medium for electron microscopy free of contamination by carbonate. *Stain Technology* 40:169–171.

Castaing, R. 1951. *Application des sondes électroniques à une méthode d'analyse ponctuelle chimique et cristallographique.* Thèse de doctorat d'état, Université de Paris, 1952.

Dalton, A.J., and R.F. Zeigel. 1960. A simplified method of staining thin sections of biological material with lead hydroxide for electron microscopy. *Journal Biophysics Biochemistry Cytology* 7:409–410.

Hayat, M.A. 1972. *Basic electron microscopy techniques.* Nostrand Rheinhold Company, New York.

Hayat, M.A. 2000. *Principles and techniques of electron microscopy biological applications.* Cambridge University Press, Cambridge.

Karnovsky, M.J. 1961. Simple methods for staining with lead at high pH in electron microscopy. *Journal of Biophysical and Biochemical Cytology* 11:729–732.

Lever, J.D. 1960. A method of staining sectioned tissues with lead for electron microscopy. *Nature* 186:810–811.

Millonig, G. 1961. A modified procedure for lead staining of thin sections. *Journal of Biophysical and Biochemical Cytology* 11:736–739.

Pottu-Boumendil, J. 1989. *Microscopie électronique: Principes et méthodes de préparation.* INSERM, Paris.

Reynolds, E.S. 1963. The use of lead citrate at high pH as an electron-opaque stain in electron microscopy. *Journal of Cell Biology* 17:208–212.

Sato, T. 1967. A modified method for lead staining of thin sections. *Journal of Electron Microscopy* 16:133–141.

Venable, J.H., and R. Coggeshall. 1965. A simplified lead citrate stain for use in electron microscopy. *Journal of Cell Biology* 25:407–408.

Watson, M.L. 1958. Staining of tissue sections for electron microscopy with heavy metals. *Journal of Biophysical and Biochemical Cytology* 4:475–478.

10 Ultrastructural Cytoenzymology

Chantal de Chastellier

CONTENTS

10.1 GENERAL CONSIDERATIONS

The histochemical methods described earlier allow one to determine which cells of a tissue or organ contain specific enzymes under study (see Chapters 8 and 9). Electron microscopy cytoenzymology is the only technique that is able to combine sensitive enzyme detection methods with detailed information on the substructure of intracellular compartments and on the spatial organization of organelles within the cells.

253

10.1.1 Visualization of Enzymes under the Electron Microscope

Enzymes are not electron dense and are, therefore, not visible under the electron microscope (EM) unless enzyme cytochemistry methods are applied. Such methods do not visualize the enzyme itself but the product of the enzymatic reaction. The fact that the latter must be electron dense to be visualized has limited the number of enzymes that can be localized by EM approaches. The localization of phosphatases—such as acid phosphatase, aryl sulfatase, and trimetaphosphatase, which are mostly but not exclusively located in lysosomes, and glucose 6-phosphatase, which is located exclusively in the endoplasmic reticulum—is based on histochemical methods involving precipitation of heavy metal salts, according to the following principle: Phosphate ions liberated during enzymatic hydrolysis of the organic phosphates that serve as substrates are captured by heavy metal ions present in the reaction medium. This leads to the formation of highly insoluble precipitates (see the staining procedures described in Sections 10.2.2 and 10.3.2). The most frequent heavy metal salt used as phosphate capture agent is lead nitrate. Lead phosphate precipitates are easy to visualize as they appear as black deposits under the EM (see the figures cited in Sections 10.2.4 and 10.3.4).

10.1.2 Sample Preparation

Staining for phosphatases can be performed on cell monolayers, cells in suspension or in tissue and organ sections. The methods of preparation of these different materials are the same as those used for enzyme histochemistry and will not be described here.

Preservation of the cell infrastructure is an absolute requirement for analyzing the precise location of an enzyme at the subcellular level. Maintenance of membrane integrity is of the utmost importance for maintaining the reaction product as close as possible to the enzyme under study and in any case in the same organelle. Exposure of live cells or tissue sections to the reaction medium, without prior chemical fixation, most of the time leads to cell or tissue damage. The organelle membranes become porous or display breaks, and they are ultimately lysed. These different events will lead to leakage and diffusion of the reaction product into the surrounding cytoplasm and, ultimately, delivery into the extracellular environment. To maintain membrane integrity, one must therefore fix cells before exposure to the enzyme reaction medium with the knowledge that chemical fixatives inactivate most enzymes. A perpetual dilemma is to find the best fixation conditions to achieve both optimal preservation of ultrastructural details and maintenance of the maximum possible amount of enzyme activity. The quality of the products, the method of preparation and storage of fixatives, and the fixation procedure itself are most important. An appropriate method for fixing cells prior to staining for acid phosphatase or glucose-6-phosphatase is given in this chapter (see Sections 10.2.2 and 10.3.2).

10.1.3 Factors Influencing the Result

To locate phosphatases in tissues or cells, the rate of production of phosphate ions during the enzymatic reaction must be high enough to ensure the formation of insoluble lead phosphate precipitates that are visible under the EM (for more detailed information, see Pearse, 1961). Fixation prior to staining, which cannot be avoided, will necessarily lead to partial (or total) enzyme inactivation depending on the procedure. A lower concentration of fixative, a lower temperature, and a shorter time of incubation in presence of the fixative will generally inactivate enzymes less than the higher ones used for conventional electron microscopy. In what concerns the enzymatic reaction, many substrates are, or have been, used. It is better to choose substrates that are hydrolyzed at a rapid rate. Substrates that are slowly hydrolyzed have one major drawback. A prolonged incubation time becomes necessary and the occurrence of diffusion and other artifacts is thereby significantly raised. Raising or lowering the pH of the enzymatic reaction medium from the optimum pH (for acid phosphatase, pH 5.0) interferes with the result by slowing the rate of the reaction, although within fairly wide limits it does not completely inactivate the enzyme. EM processing of samples following the staining procedure generally has little effect on the lead phosphate precipitates formed during the enzymatic reaction provided samples are processed immediately (see Sections 10.2.3 and 10.3.3).

10.1.4 Interpretation of Results

If, in spite of the many pitfalls previously enumerated, a successful reaction is achieved, one must still interpret the results. It is advisable for a beginner to consult a specialist to avoid misleading or wrong conclusions. The most common problems are listed next.

   False negative reactions—These are defined as failures to produce a positive reaction when the enzyme is known to be present in the tissue or cell under consideration. The two main causes are the presence of the enzyme in subthreshold amounts and factors mentioned earlier that prevent formation of a precipitate.

   False positive reactions—These are defined as reactions similar to the genuine but not due to the enzyme under study contained in the tissue or cell. A frequent case is the staining of phosphatases other than the one under study. The most frequent cause is departure from optimal pH. False-positive staining of the cell nucleus is often observed. Under a variety of conditions, this is due to a diffusion artifact. Electron-dense substances (melanin is one example) present in tissues or cells may be confused with a positive reaction. Free phosphate ions present in certain washing media (phosphate buffered saline [PBS]) or fixation buffers (Sörensen buffer) might precipitate with the heavy metal capture agent in the reaction medium to form insoluble precipitates. The latter are, however, usually much smaller than the genuine precipitates and are distributed at random throughout the cell. Other possible causes are the spontaneous hydrolysis of the substrate and its hydrolysis by bacterial action. The former occurs rarely with the usual substrates and the latter can be avoided by using only freshly prepared reaction medium.

   False localization—This is defined as a positive reaction, due to the specific enzyme under study, appearing at sites other than the primary location. It may be due to (a) diffusion of the enzyme from its original location and its absorption at another or (b) diffusion of the products of enzymatic hydrolysis from their sites of production and their deposition or absorption at other sites. Diffusion will happen if the membrane of the organelle in which the enzyme is contained is not correctly preserved.

10.2 VISUALIZATION OF ACID PHOSPHATASE

The staining procedure can be divided into three major steps as follows:

   1. Fixation of samples prior to staining
   2. Staining for the enzyme
   3. Preparation of samples for observation under the electron microscope (EM)

10.2.1 Fixation Prior to Staining

Several methods have been used depending on whether authors wished to stain tissue or organ preparations, or cells (cell monolayers or cells in suspension). A method that has given satisfactory results for cell monolayers is described next.

10.2.1.1   Reagents

- Glutaraldehyde 25% solution in water, EM grade 1 (from Sigma). Store at –20°C. Once thawed, store at 4°C for at most 1 month. Very toxic.
- Cacodylic acid, sodium salt trihydrate (Sigma or Electron Microscopy Sciences). Store at room temperature (RT). Toxic, contains arsenate.
- Other reagents of common use: $CaCl_2$, $MgCl_2$, sucrose, HCl 1N, ammonium chloride (NH_4Cl).

10.2.1.2   Preparation of Products

- $CaCl_2$ 1M. Store at RT.
  - $CaCl_2$, 14.7 g
  - Distilled water, 100 mL
- $MgCl_2$ 1M. Store at RT.
  - $MgCl_2$, 20.3 g
  - Distilled water, 100 mL
- Ammonium chloride (NH_4Cl) 1 M. Store at RT.
  - NH_4Cl, 5.35 g
  - Distilled water, 100 mL
- Na cacodylate buffer 0.1 M, pH 7.2, containing sucrose 0.1 M, $CaCl_2$ 5 mM, and $MgCl_2$ 5 mM (complete buffer). Store at 4°C, stable for several months, but beware of molds. Toxic.
  - Na cacodylic acid, 5.35 g
  - Distilled water, 200 mL
  - Adjust pH to 7.2 with HCl 1N
  - Complete with distilled water, up to 250 mL
  - $MgCl_2$ 1 M, 1.25 mL
  - $CaCl_2$ 1 M, 1.25 mL
  - Sucrose, 8.5 g
- Complete Na cacodylate buffer containing NH_4Cl 50 mM. Store at 4°C, stable for several months, but beware of molds. Toxic. For 20 mL:
  - NH_4Cl 1 M, 1 mL
  - Complete Na cacodylate buffer, 19 mL
- Glutaraldehyde 1.25% in complete Na cacodylate buffer. Prepare fresh. Can be kept on ice for 3 to 4 h. Toxic: Prepare and use under fume hood. For 10 mL:
  - Glutaraldehyde 25%, 0.5 mL
  - Complete Na cacodylate buffer, 9.5 mL

10.2.1.3   Fixation Procedure

- Wash cells twice with culture medium. Do not wash cells with PBS; free phosphate ions could be captured by Pb^+ to form unspecific precipitates during incubation in the staining medium.
- Remove medium
- Add fixative, that is, glutaraldehyde 1.25% in complete Na cacodylate buffer
- Fix cells 1 h at 4°C. Temperature and fixation time are major factors affecting enzyme activity: low temperature and fixation time ensure better preservation of the enzyme activity.
- Remove fixative
- Wash twice with complete Na cacodylate buffer containing NH_4Cl 50 mM, 2 × 15 min. NH_4Cl 50 mM is added to the first two washes to quench any remaining aldehyde.
- Wash twice with complete Na cacodylate buffer devoid of NH_4Cl 50 mM, 2 × 5 min

10.2.2 Staining Procedure

10.2.2.1 Reagents

- Na acetate, trihydrate reagent ACS (from EMS). Store at RT.
- Acetic acid 10% solution in water. Store at RT. Stable for several months.
- Glycerol 2-phosphate (also called sodium-β glycerophosphate), disodium salt hydrate (from Sigma). Store powder at RT (away from light).
- Ammonium sulfide 20% solution in water (Sigma). Store at RT.
- Sodium fluoride (NaF) (from Sigma). Store powder at RT.
- Water for gradient elution for high-performance liquid chromatography (from Fluka). This quality of water is required for preparation of products.

10.2.2.2 Preparation of Products

All products for the staining medium must be prepared with boiled water (water for gradient elution for high-performance liquid chromatography is even better). Use only disposable plastic, no reusable glassware or glass pipets.

- Na acetate buffer 0.05 M, pH 5.0 containing sucrose 0.1 M. Prepare Na acetate buffer fresh prior to use. For 50 mL:
  - Na acetate, 340 mg
  - Boiled water, 45 mL
  - Adjust pH to 5.0 with acetic acid 10%
  - Complete with boiled water up to 50 mL
  - Sucrose, 1.7 g
- Lead nitrate 6% solution. Prepare fresh prior to use.
  - Lead nitrate, 60 mg
  - Boiled water, 1 mL
- Enzymatic reaction medium (Gomori). Reaction medium must be translucent. If it becomes cloudy, discard and start over. Prewarming the reaction medium at 37°C for 2 to 16 h prior to use improves the staining.
  - Glycerol 2-phosphate, 30 mg
  - Na acetate buffer, 11 mL
  - Mix until properly dissolved
  - Adjust pH to 5.0 with acetic acid 10%, 1 to 2 drops
  - Add lead nitrate 6% solution, 200 μL. Lead nitrate solution must be added dropwise under vigorous agitation.
- Sodium fluoride (NaF) 10 mM
  - NaF, 4.2 mg
  - Boiled water, 10 mL. Prewarming the reaction medium at 37°C for 2 to 16 h prior to use improves the staining.
- Ammonium sulfide 1%. Prepare fresh prior to use.
  - Ammonium sulfide 20%, 100 μL
  - Water, 1.9 mL

10.2.2.3 Staining Procedure

- Remove Na cacodylate buffer (i.e., last wash after fixation step with glutaraldehyde).
- Wash samples twice with Na acetate buffer prewarmed to 37°C, 2 × 5 min
- Remove buffer
- Add prewarmed enzyme reaction medium

- Incubate samples, 30 min at 37°C
- Remove reaction medium
- Wash samples twice with Na acetate buffer, 2 × 2 min
- Remove last wash and process samples for electron microscopy as indicated in Section 10.2.3.3. Before processing for EM, it is advisable to check under the light microscope whether staining was successful. A practical way is to follow the same procedure (with cells stuck onto glass coverslips). After the two washes with Na acetate, wash cells once with distilled water and incubate the coverslips for 1 min over a drop of ammonium sulfide 1% (under fume hood). Wash coverslips in two successive drops of distilled water. Mount on a glass slide. Under the light microscope, the reaction product appears as dark brown deposits. Material treated with ammonium sulfide cannot be processed for EM.

10.2.2.4   Controls

Several controls are necessary for a correct interpretation of results, especially when one is not familiar with the morphological appearance of the electron dense precipitates.

- Addition of NaF 10 mM will inhibit the enzymatic activity. As a result, the cells should not display electron dense deposits. If the cells nonetheless display dense deposits, then the latter could correspond to electron dense substances present in tissues or to free phosphate ions present in one of the buffers or media.
- Removal of substrate or capture agent will also allow one to discriminate between false positives and genuine acid phosphatase activity.
- It is more difficult to discriminate between the genuine activity of acid phosphatase and the activity of other phosphatases, especially if one has no clues about where the enzyme should be located. Varying the pH or the substrate can be helpful.

10.2.3   Processing of Samples for Electron Microscope Observation

A method that has given satisfactory results is described here.

10.2.3.1   Reagents

- Osmium tetroxide 4% solution in water (from EMS or Sigma). Store sealed ampullae at 4°C. Once opened, keep remainder at 4°C in a clean glass vial with a glass stopper. Seal vial with parafilm and store in a plastic container also sealed with parafilm. Stable at 4°C for several months. If solution turns brown or black, discard. Fumes will blacken refrigerator. Avoid storage in same refrigerator as cell culture products as fumes will alter such products. Highly toxic; manipulate with gloves under fume hood.
- Uranyl acetate (from EMS or Merck). Store powder at RT. Radioactive.
- 5,5′-diethylbarbituric acid, sodium salt (from Fluka). Store powder at RT. Highly toxic. Regulations on purchase and delivery are strict in many countries. If unavailable, buy maleic acid (from Sigma).
- Maleic acid (from Sigma). Store powder at RT.
- Agarose low melting point, molecular biology grade (from Sigma). Store powder at RT.
- Molecular sieve dehydrate with indicator for drying solvents (from Fluka). Store at RT. Toxic, carcinogenic by inhalation.
- Ethanol absolute RPE (from Carlo Erba).
- Spurr resin kit (from EMS). Store at RT. Resin is very toxic.
- Other products were described (see Section 10.2.1.1)
- Material—Rubber policeman, microvettes CB 300 Z (from Sartorius, Göttingen, Germany); microfuge with fixed vertical axis or with free angle rotor; glass vials; EM molds for polymerization of resin-embedded samples.

10.2.3.2 Preparation of Products

- Na cacodylate buffer 0.1 M, pH 7.2, containing $CaCl_2$ 5 mM and $MgCl_2$ 5 mM. Preparation was described in Section 10.2.1.2. Do not add sucrose at this step.
- Na veronal buffer, pH 6.0.
    - Solution A
        - Sodium acetate, 1.94 g
        - 5, 5′-diethylbarbituric acid, 2.94 g
        - NaCl, 2.80 g
        - Distilled water, 100 mL
        - Store solution A at 4°C. Stable for several months.
    - Na veronal buffer (complete)
        - Solution A, 20 mL
        - HCl 0.1 N, 28 mL
        - Distilled water, 52 mL
        - Adjust pH to 6.0 with a few drops of 10% acetic acid, if necessary.
        - $CaCl_2$ 1 M, 1 mL
        - Store complete Na veronal buffer at 4°C in dark bottle. Stable for 2 to 3 months. If crystals appear, warm up to 37°C. If the solution becomes opaque, discard.
- NaH–maleate–NaOH buffer 0.05 M, pH 6.0
    - Maleic acid, 0.5 g
    - NaOH, 0.2 g
    - Distilled water, 20 mL
    - Adjust pH to 6.0 with NaOH
    - Complete with distilled water to 100 mL
    - Use this buffer instead of veronal buffer if 5, 5′-diethylbarbituric acid or sodium salt is unavailable. Store at 4°C. Stable for several months.
- Osmium tetroxide (OsO_4) 1% in Na cacodylate buffer devoid of sucrose
    - OsO_4 4% solution, 1 mL
    - Na cacodylate buffer 0.1 M, 3 mL
    - Prepare fresh; warm to RT before use
- Uranyl acetate 1% in Na veronal or maleate buffer
    - Uranyl acetate, 250 mg
    - Na veronal buffer, 25 mL
    - Dissolve uranyl acetate in buffer under mild stirring
    - Prepare in a brown bottle, as product is light sensitive. Store at 4°C. Can usually be kept for 1 month. If crystals form, warm to 37°C before use. Discard if the solution becomes cloudy.
- Agarose 2% in Na cacodylate buffer devoid of sucrose
    - Agarose low melting point, 400 mg
    - Na cacodylate buffer 0.1 M, 20 mL
    - Dissolve in microwave for 1 min, store at 4°C. Prepare small amounts as the solution will concentrate and become viscous over time.
- Graded series of ethanol—Prepare 25%, 50%, 75%, 90%, and 100% solutions in distilled water. The 100% solution must be completely dried as follows: add the commercial solution of ethanol to a glass bottle containing molecular sieve, stir vigorously, and let settle for 24 h before use. All solutions are stable at RT for several months. Handle the 100% solution with care in order to avoid mixing the molecular sieves with the ethanol.
- Spurr resin. Very toxic. Manipulate only under fume hood.
    - NSA, 13.0 g
    - ERL-4221, 5.0 g
    - DER 736 epoxy resin, 3.0 g

- DMAE, 0.2 g
- Add all components to same plastic vial, mix vigorously, let settle for an hour to remove bubbles, use immediately afterward or store in 5 or 10 mL syringes. Add cap to tip to avoid contact with air and store at –20°C. Use only Norm-Ject syringes devoid of black rubber stopper inside the syringe. Once thawed, use content of syringe on the same day; do not refreeze.
- Uranyl acetate 1% in distilled water for staining of thin sections
  - Uranyl acetate, 200 mg
  - Distilled water, 20 mL
  - Dissolve and store in 5 or 10 mL Norm-Ject syringes with a 0.22 μm filter at the tip of the syringe. Store at 4°C. Stable for several months. Light sensitive, cover syringe with aluminum foil. Discard if precipitates appear on thin sections.
- Lead citrate for staining of thin sections
  - Solution A
    - Lead nitrate, 3.3 g
    - Distilled water, 10 mL
  - Solution B
    - Trisodic sodium citrate, 3.6 g
    - Distilled water, 10 mL
  - Solution C
    - NaOH, 1 g
    - Distilled water, 10 mL
- To 16 mL of boiled distilled water, add 3 mL of solution B and mix gently by hand. Add 2 mL of solution A and mix gently by hand. Add, dropwise, 4 mL of solution C while mixing gently by hand. The precipitate must disappear completely. If not, discard and start over. Dispatch in 2 or 5 mL syringes with a 0.22 μm filter at the tip of the syringe. Store at 4°C. The complete solution is stable for several months. Discard if nonspecific fine precipitates appear on thin sections.

10.2.3.3 Processing of Samples

- After the staining procedure described in Section 10.2.2.3, add OsO_4 1% in Na cacodylate buffer 0.1 M devoid of sucrose. Fix cells for 1 h at RT under fume hood.
- Wash cells three times rapidly to remove remaining osmium. Add 1 to 1.5 mL buffer. This step is not necessary if cells are already in suspension or if one is working with thick sections of tissue.
- Scrape cells off the dish with a rubber policeman and put sample in a conical 1.5 mL Eppendorf tube. This step is not necessary if one is working with thick sections of tissue. Go directly to dehydration and embedding in resin.
- Spin down samples in a microcentrifuge (10,000 rpm, 3 min)
- Prepare a water bath at 42°C. Heat 2% agar in microwave. Once liquefied, put in water bath.
- Take samples in turn, remove supernatant, put Eppendorf tube in water bath, add about 0.3 ml agar to Eppendorf, mix cells and agar, then transfer to a microvette. Centrifuge immediately at 10,000 rpm for 2 min. This step must be done rapidly to avoid solidification of agar before the cells have concentrated in the cap at the bottom of the microvette. Put microvette in ice, to harden the agar, for about 5 min.
- Prepare glass vials (2 to 3 ml content) with 1 ml of distilled water for each sample. Water will serve to wash samples briefly before putting them in Na veronal or NaH–maleate–NaOH buffer.
- Remove microvette from ice. With a razor blade, cut along one side of the bottom cap, then remove the cap gently. Gently push through the agar with a match over a glass slide. The sample, easily recognizable by its black color due to osmium fixation, will come out first,

then the agar. Keep 3 to 4 mm of agar to protect the sample during further manipulations, then cut off the remainder. Put the sample in the glass vial. Do not leave samples in water for more than 2 min.

- Remove water. Wash samples for 2 min with 1 ml of Na veronal buffer (or NaH–maleate–NaOH buffer).
- Remove buffer, add 1 ml of 1% uranyl acetate prepared in Na veronal buffer (or NaH–maleate–NaOH buffer). Postfix samples for 1 h at RT. This step serves to improve fixation of membrane phospholipids.
- Dehydrate samples at RT in a graded series of ethanol, that is, 25%, 50%, 75%, 90% for 15 min each followed by three successive 30 min baths in 100% ethanol dried on molecular sieve.
- Embedding in Spurr resin at RT. Prepare 25%, 50% and 75% solutions of Spurr resin (1 ml per sample). The solutions must be prepared with the 100% dried solution of ethanol. Incubate the samples in the successive baths of resin for 30 min each. Then incubate in a first bath of pure resin for 30 min and a second bath for 2 h.
- Put pure resin in molds. Add sample. Do not forget to identify it! (Write with a pencil on a small piece of paper, with number toward manipulator.) Add resin to fill mold.
- Polymerize in a 60°C incubator for 24 h
- Section samples (70 to 75 nm thick sections) with an ultramicrotome.
- Staining of thin sections. Usual case: (1) Deposit grids on a drop of 1% uranyl acetate prepared in distilled water, incubate for 6 min, wash rapidly on five successive drops of distilled water, dry on filter paper; (2) then deposit grids on Reynold's lead citrate for 2.5 minutes, wash rapidly on five successive drops of distilled water. Dry on filter paper. Staining times depend on section thickness. Adapt if material has too little or too much contrast.

10.2.4 Results

Phosphate ions liberated during enzymatic hydrolysis of the organic phosphates that serve as substrates are captured by heavy metal ions present in the reaction medium (usually lead) (de Chastellier and Ryter, 1977; Fréhel et al. 1986; de Chastellier and Berche, 1994). This leads to the formation of insoluble lead phosphate electron-dense precipitates that appear as black deposits on thin sections within organelles containing acid phosphatase. Three examples of a successful reaction are given in Figure 10.1. False-positive results are shown in Figure 10.2. See figure legends for comments. (Also see de Chastellier and Ryter, 1977; Fréhel et al. 1986; de Chastellier and Berche, 1994.)

10.3 VISUALIZATION OF GLUCOSE 6-PHOSPHATASE (G6PASE)

Glucose 6-phosphatase (G6Pase) is a specific marker of the endoplasmic reticulum (ER) compartment. In addition to basic conventional EM, staining for G6Pase can be used to study interactions of this compartment with other cell compartments. It has been used in particular to determine whether vacuoles containing pathogenic bacteria acquire ER characteristics. The procedure as described here was adapted from Griffiths, Quinn, and Warren (1983).

10.3.1 Fixation Prior to Staining

10.3.1.1 Reagents

- Glutaraldehyde 25% solution in water, EM grade 1 (from Sigma)
- Piperazine-N,N′-bis (2-ethanesulfonate) (PIPES) (from Sigma). Store powder at RT.
- Sucrose

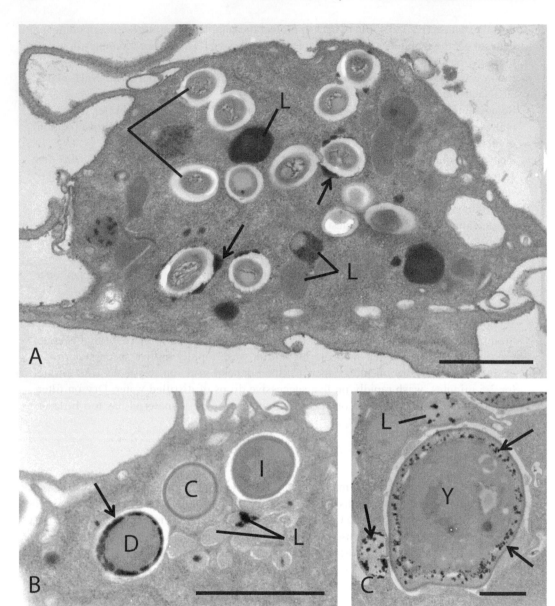

FIGURE 10.1   Staining for acid phosphatase (AcPase): examples of specific staining. (A) Bone-marrow-derived mouse macrophages were infected with *Mycobacterium avium*. Seven days later, cells were fixed, stained for AcPase, and processed for EM. Many lysosomes (L) are labeled as shown by the dark deposits in the entire lumen of the L. In contrast, most of the *M. avium*-containing phagosomes are not stained, thereby indicating that they have not fused with lysosomes. Only a few of them displayed small deposits (arrows). (From de Chastellier, C., 2004, *Cellular microbiology*, 2nd edition, P. Cossart, P. Boquet, S. Normark and R. Rappuoli, eds., Washington, DC: ASM Press. With permission.) (B) Macrophages were infected with *Listeria monocytogenes*. Two hours later, they were fixed, stained for AcPase, and processed for an EM. The phagosome with the damaged (D) bacterium is positively stained for AcPase as shown by the dark deposits within the bacterial wall, whereas the phagosome with the intact (I) one is not. Likewise, the bacterium in the cytoplasm (C) is not stained. (From de Chastellier, C., and P. Berche, 1994, The fate of Listeria monocytogenes in murine macrophages: Evidence for simultaneous killing and survival of intracellular bacteria, *Infection and Immunity* 62:543–553. With permission.) (C) *Dictyostelium discoideum* were given yeast cells (Y), fixed, stained for AcPase, and processed for EM. Both the lysosomes (L) and the yeast contained in the phagosome display dense deposits (arrows). (From Ryter, A., and C. de Chastellier, 1977, Morphometric and cytochemical studies of *Dictyostelium discoideum* in vegetative phase, *Journal of Cell Biology* 75:200–217. With permission.) The thin sections shown in this figure are all devoid of nonspecific deposits. Bar = 1 µm.

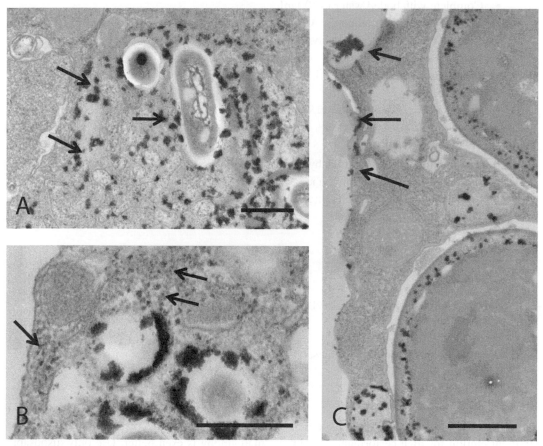

FIGURE 10.2 Staining for acid phosphatase (AcPase): examples of nonspecific staining. (A, B) Bone-marrow-derived mouse macrophages were infected with *M. avium*. Seven days later, cells were fixed, stained for AcPase, and processed for EM. In panel (A), the thin section displays large nonspecific deposits over the cytoplasm and many organelles, including the *M. avium*-containing phagosomes. In panel (B), the deposits are less dense and smaller but equally located all over the cytoplasm. As a result, it is impossible to determine whether the phagosomes display specific deposits, indicative of phagosome maturation and fusion with lysosomes, or whether the deposits in the phagosomes are nonspecific. (C) *Dictyostelium discoideum* were given yeast cells (Y), fixed, stained for AcPase, and processed for EM. This thin section shows dense deposits at the plasma membrane (arrows), presumably resulting from the activity of alkaline phosphatase, in addition to the specific deposits resulting from the activity of acid phosphatase within lysosomes and phagosomes. Bar = 0.5 μm.

10.3.1.2 Preparation of Products

- PIPES 0.1 M, pH 7.0 containing 5% or 10% sucrose
  - PIPES, 3.02 g
  - Boiled water, 75 mL
  - Dissolve. Solution is milky, do not worry.
  - Adjust pH to 7.0 with NaOH 1N, about 15 mL. The solution clears up.
  - Complete with boiled water to 100 mL
  - Split into two 50 mL aliquots. Add either 2.5 g or 5 g of sucrose to obtain PIPES with 5% or 10% sucrose, respectively. Store at 4°C. Use within a week (but better to prepare fresh the day before).
- Glutaraldehyde 1.25% in PIPES 0.1 M, pH 7.0, containing 5% sucrose. For 10 mL:
  - Glutaraldehyde 25%, 0.5 mL
  - PIPES buffer with 5% sucrose, 9.5 mL

10.3.1.3 Fixation Procedure

- Remove medium, add glutaraldehyde 1.25% in PIPES 0.1 M, pH 7.0, containing 5% sucrose and fix cells for 30 min on ice
- Remove fixative and wash cells three times for 3 min each at RT with 0.1 M PIPES, pH 7.0, containing 10% sucrose

10.3.2 STAINING PROCEDURE

10.3.2.1 Reagents

- Water for gradient elution for high-performance liquid chromatography (from Fluka)
- Tris base, purissimo
- Maleic acid
- D-glucose-6-phosphate, disodium salt dehydrate (from Sigma)
- Lead nitrate (from Sigma)

10.3.2.2 Preparation of Products

All buffers as well as the reaction medium must be prepared with boiled distilled water. Water for gradient elution for high-performance liquid chromatography is even better. Use only disposable plastic, no reusable glassware or pipets.

- Tris maleate 0.08 M, pH 6.5
  - Tris base, 485 mg
  - Maleic acid, 465 mg
  - Boiled water, 40 mL
  - Stir until dissolved
  - Adjust pH to 6.5 with NaOH 1N (about 1 to 2 ml)
  - Complete to 50 mL with water. Store Tris maleate at 4°C. Use within a week (better to prepare the day before).
- 6% solution of lead nitrate. Prepare fresh and use immediately in reaction medium. Dissolve:
  - Lead nitrate, 60 mg
  - Boiled water, 1 mL

- 1.25% glutaraldehyde in Na cacodylate buffer 0.1 M, pH 7.2, containing $CaCl_2$ 5 mM and $MgCl_2$ 5 mM. Prepare fresh. Keep on ice for up to 4 h.
  - Glutaraldehyde 25%, 0.5 mL
  - Na cacodylate buffer, 9.5 mL
- Reaction medium
  - Glucose-6-phosphate, 95 mg
  - Tris maleate 0.08 M, pH 6.5, 10 mL
  - Dissolve
  - Adjust pH to 6.5 with a few drops of HCl 1N.
  - 6% solution of lead nitrate: 160 µL. Add, dropwise, under vigorous stirring. Note: The reaction medium is often slightly milky. It is, therefore, preferable to filter it through a 0.22 µm filter. Stable at RT for 2 h.

10.3.2.3 Staining Procedure

- Remove PIPES buffer and wash 1 × 30 sec at RT with 0.08 M Tris maleate buffer, pH 6.5
- Remove buffer, add cytochemical medium, incubate cells for 2 h at 37°C in a CO_2-free incubator
- Remove cytochemical medium, wash cells 3 × 2 min at RT with 0.08 M Tris maleate buffer
- Remove Tris maleate buffer and wash 3 × 2 min at RT with 0.1 M cacodylate buffer, pH 7.2, containing 0.1 M sucrose, 5 mM $CaCl_2$, and 5 mM $MgCl_2$
- Remove last wash, add glutaraldehyde 1.25% in complete Na cacodylate buffer and fix cells for 1 hr at 4°C
- Remove fixative and wash cells 2 × 15 min with complete Na cacodylate buffer

10.3.2.4 Controls

Possible controls: (a) Incubate cells in substrate-free reaction medium or (b) in complete reaction medium containing 10 mM sodium fluoride.

10.3.3 Processing of Samples for Electron Microscope Observation

See Section 10.2.3. Remove last wash and process cells for conventional electron microscopy.

10.3.4 Results

As it is the case for acid phosphatase, phosphate ions liberated during enzymatic hydrolysis of the organic phosphates that serve as substrates are captured by heavy metal ions present in the reaction medium (usually lead). This leads to the formation of insoluble lead phosphate electron-dense precipitates that appear as almond-shaped black deposits on thin sections within organelles containing G6Pase, that is, the lumen of the endoplasmic reticulum and of the nuclear membrane. Hydrolysis of the substrate can also be catalyzed by other phosphatases, including acid phosphatase, which is enriched in lysosomes, or alkaline phosphatase, which is enriched in the plasma membrane. Their contribution should, however, be minor if the cytochemical reaction medium is done properly and at the right pH. Small or large nonspecific deposits can also occur if the reaction medium is not of good quality. It is advisable to consult a specialist if one is not familiar with enzyme cytochemistry. Three examples of a successful reaction are given in Figure 10.3. False-positive results are shown in Figure 10.4. See figure legends for comments. (See also Celli et al., 2003; Touret et al., 2005.)

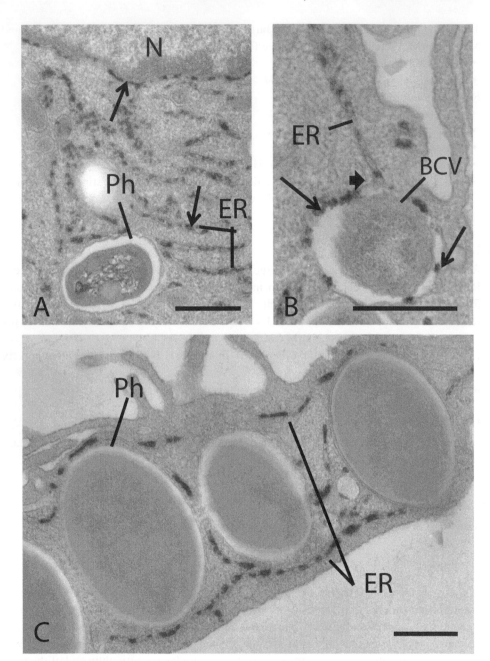

FIGURE 10.3 Staining for glucose 6-phosphatase (G6Pase): examples of specific staining. Bone marrow-derived mouse macrophages were allowed to internalize latex beads or bacteria, fixed at selected time points post-infection, stained for G6Pase, and processed for EM. The endoplasmic reticulum (ER) and the nuclear membrane (N) both display dense deposits (arrows), thereby indicating that they contain G6Pase as expected. Phagosomes containing (A) pathogenic *Mycobacterium avium* or (C) hydrophobic latex beads do not display dark deposits as opposed to (B) vacuoles containing *Brucella abortus* (arrows). A fusion event between the ER and the *Brucella abortus*-containing vacuole can be observed in (B) (arrowhead). These thin sections are devoid of nonspecific deposits. Bar = 0.5 μm. (Panels [A] and [C] from Touret, N., P. Paroutis, M. Terebiznik, R. Harrisson, S. Trombetta, M. Pypaert, A. Chow, et al., 2005, Quantitative and dynamic assessment of the contribution of the endoplasmic reticulum to phagosome formation, *Cell* 123:157–170. With permission.) (Panel [B] from Celli, J., C. de Chastellier, D.-M. Franchini, J. Pizarro-Cerda, E. Moreno, J.P. Gorvel, 2003, *Brucella* evasion of macrophage killing through VirB-dependent sustained interactions with the endoplasmic reticulum, *Journal of Experimental Medicine* 198:545–556. With permission.)

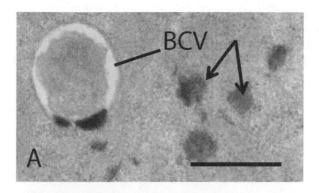

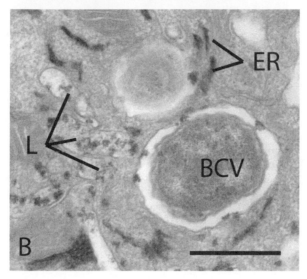

FIGURE 10.4 Staining for glucose 6-phosphatase (G6Pase): examples of nonspecific staining. Bone-marrow-derived mouse macrophages were infected with *Brucella abortus* fixed, stained for G6Pase and processed for an EM. (A) Large nonspecific deposits are observed in the cytoplasm (arrows). (B) Deposits are observed in the ER, as expected, but also in lysosomes (L). The latter result from the activity of AcPase and not G6Pase. Bar = 0.5 μm.

REFERENCES

Celli, J., C. de Chastellier, D.-M. Franchini, J. Pizarro-Cerda, E. Moreno, and J.P. Gorvel. 2003. Brucella evasion of macrophage killing through VirB-dependent sustained interactions with the endoplasmic reticulum. *Journal of Experimental Medicine* 198:545–556.

de Chastellier, C., and P. Berche. 1994. The fate of *Listeria monocytogenes* in murine macrophages: Evidence for simultaneous killing and survival of intracellular bacteria. *Infection and Immunity* 62:543–553.

de Chastellier, C., and A. Ryter. 1977. Changes of the cell surface and of the digestive apparatus of Dictyostelium discoideum during the starvation period triggering aggregation. *Journal of Cell Biology* 75:218–236.

Fréhel, C., C. de Chastellier, T. Lang, and N. Rastogi. 1986. Evidence for inhibition of fusion of the lysosomal and prelysosomal compartments with phagosomes in macrophages infected with pathogenic *Mycobacterium avium*. *Infection and Immunity* 52:252–262.

Griffiths, G., P. Quinn, and G. Warren. 1983. Dissection of the Golgi complex. I Monensin inhibits the transport of viral membrane proteins from medial to trans Golgi cisternae in baby hamster kidney cells infected with Semliki forest virus. *Journal of Cell Biology* 96:835–850.

Pearse, A.G.E. 1961. *Histochemistry, theoretical and applied*. J. & A. Churchill Ltd, London.

Ryter, A., and C. de Chastellier. 1977. Morphometric and cytochemical studies of *Dictyostelium discoideum* in vegetative phase. *Journal of Cell Biology* 75:200–217.

Touret, N., P. Paroutis, M. Terebiznik, R. Harrisson, S. Trombetta, M. Pypaert, A. Chow, et al. 2005. Quantitative and dynamic assessment of the contribution of the endoplasmic reticulum to phagosome formation. *Cell* 123:157–170.

11 The Acetylation Method

Marc Thiry and Nicolas Thelen

CONTENTS

11.1 INTRODUCTION

Like the cytoplasm, the cell nucleus contains various structurally and functionally distinct sub-compartments (Rouquette et al., 2010). The best known is the nucleolus, which is mainly involved in the ribosome biogenesis (Hernandez-Verdun et al., 2010). Although the various structural sub-compartments of the cell nucleus can be recognized by the current classical techniques of electron microscopic cytology, the morphological difference between these nuclear structures is not very striking usually. The acetylation method, originally described by Wassef et al. (1979), is a technique allowing the enhancement of the contrast between the different structures within the cell nucleus; in particular, it offers the possibility to clearly discriminate the different nucleolar components (Thiry et al., 1985; Thiry and Goessens, 1986). It is based on glutaraldehyde fixation and prolonged acetylation in pyridine, usually practiced on the biological material blocks prior to embedding. Staining is carried out as usual with uranyl acetate and lead citrate. The relatively simple procedure can easily be used for the study of nuclear physiology.

11.2 TECHNIQUE

11.2.1 MATERIALS

11.2.1.1 Reagents

- Na_2HPO_4 (Merck, cat. no. 6579)
- NaH_2PO_4 (UCB, cat. no. 1769)

- Glutaraldehyde (70%, 2 ml ampule, Ladd Research, cat. no. 20.100) Glutaraldehyde is toxic (irritant, allergen, carcinogen)
- Acetone (Merck, cat. no. 1.00014.2500)
- Pyridine (Merck, cat. no. 1074622500)
- Acetic anhydride (UCB, cat. no. 94010029)
- Dodecenyl succinic anhydride (DDSA, Ladd research, cat. no. 21340)
- Methyl nadic anhydride (MNA, Ladd research, cat. no. 21350)
- 2,4,6-dimethylaminomethyl phenol (DMP-30, Ladd research, cat. no. 21370)
- LX 112 (Ladd research, cat. no. 21310)
- Uranyl acetate (Fluka, cat. no. 94260). Uranyl is radioactive and should be stored covered by a lead sheathing. Uranyl is toxic; avoid contact with skin and eye.
- Sodium citrate (Merck, cat. no. 6448)
- NaOH pellets (Aldrich, cat. no. 22,146-5)
- Lead nitrate (Merck, cat. no. 7398). Lead nitrate is toxic.

11.2.1.2　Equipment

- Ultramicrotome (Reichert ultracut S, Leica) with diamond knife (Drukker, cat. no. VC 1860200)
- 200 square mesh cupper grids 3.05 mm (Agar scientific, cat. no. G2200PD)
- Anticapillary forceps style 7 and style 5 (Ladd research, cat. nos. 10791 and 10788)
- Filter paper (Whatman, cat. nos. 1440 070 to 1440 240)
- Parafilm (Sigma-Aldrich, cat. no. P 7668-1EA)
- JEOL CX 100 II electron microscope

11.2.1.3　Reagent Setup

- Sörensen's buffer, 0.1 M Na_2HPO_4/NaH_2PO_4, pH 7.4, store at 4°C, stable for up to 1 year
- Glutaraldehyde (70%, 2 ml ampule). Prepare a 1.6% solution in Sörensen's buffer, store at 4°C, stable for several months
- Embedding resin. Mix in a 25 ml Erlenmeyer 5 g of dodecenyl succinic anhydride (DDSA) with 5 g of methyl nadic anhydride (MNA), 0.3 g of 2,4,6-dimethylaminomethyl phenol (DMP-30), and 10 g of LX 112, close the Erlenmeyer, stir until the appearance of bubbles and let the mixture rest until the complete disappearance of bubbles; store at room temperature (~20°C), stable for 2 to 5 days.
- 50% ethanolic uranyl acetate, 0.5 g uranyl acetate, 12.5 ml boiled H_2O, 12.5 ml ethanol; store at 4°C in a brown glass container or otherwise protected from direct light, stable at 4°C for up to one year, filter (0.22 μm in pore size) before use
- Aqueous lead citrate, 4.2% sodium citrate, 2.6% lead nitrate, add concentrated NaOH until clearing up of mixture, stable at 4°C for up to one year, filter (0.22 μm in pore size) before use.

11.2.2　Procedure

11.2.2.1　Fixation

Small organ fragments or cell pellets can be used. Organ or piece dissection before fixation must be done carefully to preserve the integrity of tissue and cells for study. The growing cell cultures are taken when their monolayer density reaches 50% to 60%. Incubate the samples into the fixative as quickly as possible.

- Glutaraldehyde 1.6%, 60 min at 4°C
- Washings
  - Sörensen's buffer, 10 min at 4°C
  - Sörensen's buffer, 10 min at RT
  - Bidistilled water, 10 min at RT

11.2.2.2 Dehydration
- Acetone 25%, 5 min at RT
- Acetone 50%, 5 min at RT
- Acetone 70%, 5 min at RT
- Acetone 90%, 5 min at RT
- Acetone 100%, 3 × 10 min at RT

11.2.2.3 Acetylation
- Acetone 100%/pyridine 1/1, 5 min at RT
- Acetone 100%/pyridine 1/2, 5 min at RT
- Acetone 100%/pyridine 1/3, 5 min at RT
- Acetone 100%/pyridine 1/4, 5 min at RT
- Pyridine/acetic anhydride 6/4, 16 h at 45°C
- Pyridine, 3 × 10 min at RT
- Pyridine/acetone 100% 1/1, 5 min at RT
- Acetone 100%, 3 × 10 min at RT

11.2.2.4 Impregnation
- Acetone/Epon 2/1, 60 min at RT
- Acetone/Epon 1/1, 60 min at RT
- Acetone/Epon 1/2, 60 min at RT
- Pure Epon, overnight at RT

11.2.2.5 Embedding
- Insert samples in embedding molds with fresh resin
- Polymerize resin for 3 to 4 days in an incubator at 42°C

11.2.2.6 Sectioning
- Cut ultrathin sections of Epon-embedded samples on an ultramicrotome with a diamond knife at 60 to 90 nm
- Deposit sections on 200 mesh cupper grids

11.2.2.7 Staining of Sections
Transfer grids into a petri dish with reduced CO_2 concentration (sodium hydroxide pellets in a petri dish). The grids can be stored at room temperature for several months on filter paper placed in a closed petri dish.

- Incubate for 5 min at room temperature in darkness on drops (20 µl) of 50% ethanolic uranyl acetate
- Rinse grids in three 25 ml beaker filled with boiled bidistilled water. The grids can be stored at room temperature for several days or weeks on filter paper placed in a closed petri dish.
- Dry grids on filter paper
- Transfer grids into a Petri dish with reduced CO_2 concentration (sodium hydroxide pellets in a Petri dish) for 5 min at room temperature on drops (20 µl) of aqueous lead citrate
- Rinse grids in three 25 mL beaker filled with boiled bidistilled water
- Dry grids on filter paper

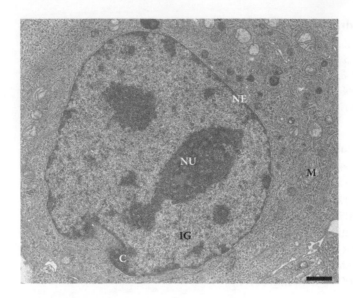

FIGURE 11.1   HEp-2 cell portions as observed in classical electron microscope preparations. Seen are the different structural compartments within the cell nucleus, such as the condensed chromatin (C) blocks, the nucleolus (NU), and the interchromatin granule clusters (IG). M, mitochondria; NE, nuclear envelope. Bar = 1 μm. Compare to Figure 11.2.

11.2.2.8   Electron Microscope Examination

Examine sections on a transmission electron microscope at 60-80 KV.

11.3   RESULTS AND COMMENTS

Compared to the nucleus aspect shown with the usual routine methods for electron microscopy (Figure 11.1), acetylation carried out prior to embedding allows a striking identification of condensed chromatin masses within the nucleoplasm (Figure 11.2). The acetylation can be performed on tissue or cell sections, but better results were observed after en bloc treatment. In particular, in the nucleolus, it allows clear visualization of the clumps of condensed chromatin surrounding the nucleolus and their intranucleolar invaginations (Figure 11.3). The nucleolar components, the fibrillar centers, the dense fibrillar component, and the granular components are also easily distinguished (Figure 11.3). The interchromatin granule clusters and the perichromatin granules are particularly well visible (Figure 11.4). By contrast, the nuclear envelope as well as all the membrane rich cytoplasmic structures (i.e., Golgi apparatus, mitochondria, and endoplasmic reticulum) are unstained or very faintly stained. In the cytoplasm (Figure 11.2), only the ribosomes are clearly obvious.

It is interesting to note that the nucleolar and extranucleolar (interchromatin and perichromatin) granules are still more prominent (Wassef, 1979; Wassef et al., 1979; Thiry, 1995; Charlier et al., 2009) when the sections are stained with the EDTA regressive stain (Bernhard, 1969). This acetylation method has also been combined with other cytochemical techniques such as the Ag-NOR staining method (Ploton et al., 1984; Thiry et al., 1985; Ploton et al., 1987) and immunocytological techniques for RNA or DNA detecting by specific transferases (Thiry, 1992, 1993).

At present, the mechanism of this reaction is not exactly known. Acetylation is an effective procedure to block amino (Lillie, 1958; Watson and Aldridge, 1961, 1964) and hydroxyl (Ganter and Jollès, 1969) groups. The blockage of the –NH$_2$ group could likely suppress the part of the unspecific staining due to uranyl binding to proteins (Lombardi et al., 1971). On the other hand, the

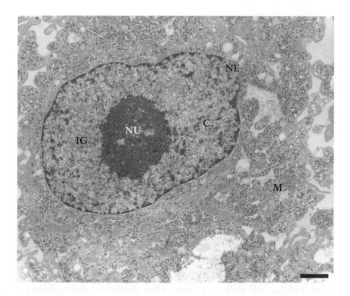

FIGURE 11.2 HEp-2 cell portions as observed after the acetylation method. Seen are the different structural compartments within the cell nucleus such as the condensed chromatin (C) blocks, the nucleolus (NU), and the interchromatin granule clusters (IG). The condensed chromatin clumps are heavily contrasted. M, mitochondria; NE, nuclear envelope. Bars = 1 μm. Compare to Figure 11.1.

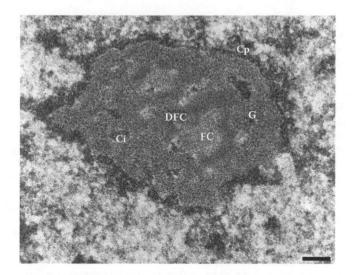

FIGURE 11.3 Nucleolus of HEp-2 cell as revealed by the acetylation method. Under these experimental conditions, the three main nucleolar components are clearly recognized: the fibrillar centers (FC), the dense fibrillar component (DFC), and the granular component (G). The identification of intranucleolar (Ci) and perinucleolar (Cp) condensed chromatin clumps is easy. Bar = 0.2 μm.

increased accessibility of the nucleic acid phosphates freed from their interactions with the amino groups of the basic proteins might contribute to the higher contrast of the nucleic-acid-containing structures. In the same way, the esterification of the –OH group should reduce the nonspecific staining due to lead fixation on proteins by hydrogen bonds while electrostatic binding at the phosphate level is maintained.

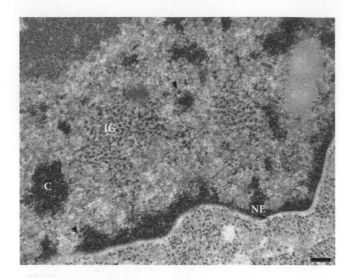

FIGURE 11.4 Portion of a HeLa cell nucleus as seen after the acetylation method. Small clumps of condensed chromatin (C) are present at proximity to an interchromatin granule cluster (IG). Arrows indicate perichromatin granules. NE, nuclear envelope. Bar = 0.2 μm.

REFERENCES

Bernhard, W. 1969. A new staining procedure for electron microscopical cytology. *Journal of Ultrastructural Research* 27:250–265.

Charlier, C., F. Lamaye, N. Thelen, and M. Thiry. 2009. Ultrastructural detection of nucleic acids within heat shock-induced perichromatin granules of HeLa cells by cytochemical and immunocytological methods. *Journal of Structural Biology* 166:329–336.

Ganter, P., and G. Jollès. 1969. *Histochimie normale et pathologique.* Gauthier-Villars, Paris.

Hernandez-Verdun, D., P. Roussel, M. Thiry, V. Sirri, and D.L.J. Lafontaine. 2010. The nucleolus: Structure/ function relationship in RNA metabolism. *WIREs RNA* 1:415–431.

Lillie, R.D. 1958. Acetylation and nitrosation of tissue amines in histochemistry. *Journal of Histochemistry and Cytochemistry* 6:352–362.

Lombardi, L., G. Prenna, L. Okolicsanyi, A. Gautier. 1971. Electron staining with uranyl acetate. Possible role of free amino groups. *Journal of Histochemistry and Cytochemistry* 19:161–168.

Ploton, D., M. Menager, and J.J. Adnet. 1984. Simultaneous high resolution localization of Ag-NOR proteins and nucleoproteins in interphasic and mitotic nuclei. *Histochemical Journal* 16:897–906.

Ploton, D., M. Thiry, M. Menager, A. Lepoint, J.J. Adnet, and G. Goessens. 1987. Behaviour of nucleolus during mitosis. A comparative ultrastructural study of various cancerous cell lines using the Ag-NOR staining procedure. *Chromosoma* 95:95–107.

Rouquette, J., C. Cremer, T. Cremer, and S. Fakan. 2010. Functional nuclear architecture studied by microscopy: Present and future. *International Review of Cellular and Molecular Biology* 282:1–90.

Thiry, M. 1992. Highly sensitive immunodetection of DNA on sections with exogenous terminal deoxynucleotidyl transferase and non-isotopic nucleotide analogues. *Journal of Histochemistry and Cytochemistry* 40:411–419.

Thiry, M., 1993. Immunodetection of RNA on ultra-thin sections incubated with polyadenylate nucleotidyl transferase. *Journal of Histochemistry and Cytochemistry* 41:657–665.

Thiry, M., 1995. Behavior of interchromatin granules during the cell cycle. *European Journal of Cell Biology* 68:14–24.

Thiry, M., and G. Goessens. 1986. Ultrastructural study of the relationships between the various nucleolar components in Ehrlich tumour and HEp-2 cell nucleoli after acetylation. *Experimental Cell Research* 164:232–242.

Thiry, M., A. Lepoint, and G. Goessens. 1985. Re-evaluation of the site of transcription in Ehrlich tumour cell nucleoli. *Biology of the Cell* 54:57–64.

Wassef, M. 1979. A cytochemical study of interchromatin granules. *Journal of Ultrastructural Research* 69:121-133.

Wassef, M., J. Burglen, and W. Bernhard. 1979. A new method for visualization of preribosomal granules in the nucleolus after acetylation. *Biology of the Cell* 34:153–158.

Watson, M.L., and W.G. Aldridge. 1961. Methods for the use of indium as an electron stain for nucleic acids. *Journal of Biophysical and Biochemical Cytology* 11:257–272.

Watson, M.L., and W.G. Aldridge. 1964. Selective electron staining of nucleic acids. *Journal of Histochemistry and Cytochemistry* 12:96–103.

12 Preparation of Products for Electron Microscopy

Jean-Marie Exbrayat

CONTENTS

12.1   FIXATIVES

12.1.1   KARNOWSKY'S FLUID

- Sörensen's buffer 0.2 M, pH = 7.4
  - Solution 1
    - NaH_2PO_4 (monosodic phosphate), 3.12 g
    - Distilled water, 100 mL
  - Solution 2
    - Na_2HPO_4 (disodic phosphate), 7.16 g
    - Distilled water, 100 mL
  - Buffer
    - Solution 1, 19 mL
    - Solution 2, 81 mL
- Glutaraldehyde—From a commercial solution (25% or 50%) prepare glutaraldehyde 4%:
  - Glutaraldehyde at 25%, 6.25 mL
  - Distilled water, 100 mL. For glutaraldehyde 50%, use 200 mL of water.
  - Paraformaldehyde
  - Reserve solution
    - Paraformaldehyde, 4 g
    - Distilled water, 10 mL
  - Working solution
    - Reserve solution, 2 mL
    - Distilled water, 100 mL
- Fixative
  - Sörensen's buffer, 200 mL
  - Glutaraldehyde, 100 mL
  - Paraformaldehyde, 100 mL

12.1.2   PARAFORMALDEHYDE

- Reserve solution
  - Paraformaldehyde, 4 g
  - Distilled water, 10 mL
- Working solution
  - Reserve solution, 2 mL
  - Distilled water, 100 mL

12.1.3   GLUTARALDEHYDE

From a commercial solution (25% or 50%) prepare glutaraldehyde 4%:

- Glutaraldehyde, 6.25 mL
- Distilled water, 100 mL

12.1.4 OSMIUM TETROXIDE

- Osmium tetroxide 2%, 2 g
- Distilled water, 100 mL

12.1.5 BUFFERED OSMIUM TETROXIDE 2%

- Osmium tetroxide 2%, 2 g
- Phosphate-buffered saline (PBS), 100 mL

12.2 STAINING

12.2.1 BISMUTH IODIDE

- Bismuth, 20 mg
- Nitric acid, 0.2 mL
- Distilled water to 80 mL
- Citric acid 0.1 M, 4 mL
- Adjust pH to 7.0 with 1 M sodium hydroxide
- Distilled water to 100 mL
- Sucrose to obtain 0.2 M

12.2.2 COLLOIDAL THORIUM IN ACETIC ACID

- Colloidal thorium, 0.5 g
- Acetic acid 3% (pH 2.5), 50 mL

12.2.3 COLLOIDAL AMMONIUM FERRIC GLYCERATE

- Ferric chloride, 275 g
- Distilled water, 1000 mL
- Glycerin, 400 mL. Add when ferric chloride is totally dissolved.
- Stir
- Ammonium hydroxide, 220 mL. Add gradually to obtain a brown solution.
- Dialysis into cellophane bag. Bag is filled to 40% of its capacity. It is immersed in a large volume of distilled water. Distilled water is changed 10 times during 72 h.
- Acetic acid to adjust to desired pH
- Solution remains stable at RT

12.2.4 DIAMINOBENZIDINE

- Diaminobenzidine, 1 g
- Boric acid 5%, 100 mL

12.2.5 FERRIC CHLORIDE

- Ferric chloride, 0.1 g. Ferric chloride can be replaced with colloidal ammonium chloride glycerate. Quantity of ferric chloride can reach 0.4 g.
- Distilled water, 100 mL. Distilled water can be replaced with buffer.

12.2.6 INDIUM

Used for nucleic acid visualization.

- Indium trichloride, 50 mg
- Acetone, 2 mL

12.2.7 LANTHANUM

- Solution A
  - Glutaraldehyde 25%, 10 mL
  - Lanthanum nitrate, 1 g
  - Cacodylate buffer pH 7.3, 100 mL
- Solution B
  - Lanthanum nitrate, 1 g
  - Cacodylate buffer pH 7.3, 100 mL
- Solution C
  - Osmium tetroxide, 1 g
  - Lanthanum nitrate, 1 g
  - Cacodylate buffer pH 7.3, 100 mL

12.2.8 LEAD CITRATE

This salt is used for standard contrast of ultrathin sections.

- Sodium hydroxide, 2 g
- Distilled water, 50 mL
- Lead citrate, 1 g
- Distilled water, 200 mL
- Mix
- Distilled water complete to 500 mL

12.2.9 METHOD OF BJORKMAN AND HELLSTRÖM

Lead acetate solution:

- Ammonium acetate, 18.5 g
- Saturated lead acetate, 100 mL

12.2.10 METHOD OF REYNOLDS

- Lead nitrate, 1.33 g
- Sodium citrate, 1.76 g
- Distilled water, 30 mL
- Let settle, 20 min
- Sodium hydroxide, 8 mL
- Distilled water, 50 mL

12.2.11 METHOD OF VENABLE AND COGGESHALL

- Lead citrate, 0.4 g
- Sodium hydroxide, 1 mL
- Distilled water, 100 mL

12.2.12 Method of Sato

- Lead nitrate, 1.5 g
- Lead acetate, 1.5 g
- Distilled water, 90 mL
- Heat 1 min at 40°C
- Sodium citrate, 3g
- Stir 1 min
- Sodium hydroxide, 24 mL
- Distilled water, 24 mL

12.2.13 Method of Watson

- Lead acetate, 8.26 g
- Distilled water, 15 mL
- Stir
- Centrifuge
- Sediment in distilled water
- Centrifuge
- Sediment in distilled water
- Centrifuge. Supernatant is used as staining solution.

12.2.14 Method of Karnovsky I

- Stock solution
  - Lead monoxide, excess
  - Sodium hydroxide 1 M, 20 mL
  - Boil, 15 min
- Work solution
  - Stock solution, 1 mL
  - Distilled water, 100 mL

12.2.15 Method of Karnovsky II

- Lead monoxide, excess
- Sodium cacodylate 10%, 20 mL
- Boil, 15 min
- Cool the solution
- Filter
- Filtrate, 2 mL
- Sodium cacodylate 10%, 8 mL
- Sodium hydroxide 1 M, several drops
- Centrifuge

12.2.16 Method of Lever

- Lead hydroxide, 1 g
- Distilled water, 100 mL
- Boil, 15 min
- Cool the solution
- Filter

- Filtrate, 2 mL
- Potassium hydroxide, several drops

12.2.17 METHOD OF MILLONIG I

- Stock solution
  - Sodium hydroxide, 12.5 g
  - Potassium–sodium–tartrate, 5 g
  - Distilled water, 50 mL
- Work solution
  - Stock solution, 0.5 mL
  - Distilled water, 100 mL
  - Heat
  - Lead hydroxide, 1 g
  - Cool solution
  - Filter

12.2.18 METHOD OF MILLONIG II

- Stock solution
  - Sodium hydroxide, 20 g
  - Potassium–sodium–tartrate, 1 g
  - Distilled water to 50 mL
- Work solution
  - Stock solution, 1 mL
  - Lead acetate 20%, 5 mL
  - Distilled water to 50 mL
  - Filter

12.2.19 OSMIUM TETROXIDE

This product is also used for fixation of tissues.

- Osmium tetroxide 2%, 2 g
- Distilled water, 100 mL

12.2.20 PHOSPHOTUNGSTIC ACID I

This technique is used to visualize glycoproteins.

- Phosphotungstic acid (PTA), 2g
- Distilled water, 100 mL
- Centrifuge
- Adjust to pH 7.0

12.2.21 PHOSPHOTUNGSTIC ACID II

- Phosphotungstic acid (PTA), 1 g
- Ethanol 70%, 50 mL

12.2.22 PHOSPHOTUNGSTIC ACID III

This reagent is used for exocytosis visualization.

- Phosphotungstic acid (PTA), 1 g
- Hydrochloric acid M, 100 mL

12.2.23 RUTHENIUM RED

- Solution A
  - Glutaraldehyde 25%, 10 mL
  - Ruthenium red, 50 mg
  - Cacodylate buffer pH 7.3, 100 mL
- Solution B
  - Osmium tetroxide, 1 g
  - Ruthenium red, 1 g
  - Cacodylate buffer pH 7.3, 100 mL

12.2.24 SILVER METHENAMINE SOLUTION

Silver methenamine is used for PAS-like methods used to visualize carbohydrates.

- Hexamethylenetetramine 3%, 18 mL
- Silver nitrate 10%, 2 mL
- Sodium borate 2%, 2 mL

12.2.25 TANNIC ACID (TAGO)

This reagent is used for exocytosis visualization.

- Staining–fixation solution (TAGO)
  - Tannic acid, 0.1 mL
  - Glutaraldehyde, 0.8 mL
  - Cacodylate buffer, 100 mL
- Osmium solution
  - Osmium tetroxide, 1 g
  - Cacodylate buffer, 100 mL

12.2.26 TANNIC ACID (TARI)

This reagent is used for exocytosis visualization.

- Tannic acid, 0.5 g
- Sodium chloride, 30 mM
- Calcium chloride, 4 mM
- Magnesium chloride, 2 mM
- Potassium chloride, 1.5 mM
- Sodium bicarbonate, 18 mM
- Sodium triphosphate, 0.25 mM
- Adjust to obtain pH 6.8 with concentrated sodium hydroxide.

12.2.27   THIOSEMICARBAZIDE (TSC)

In several staining methods, thiosemicarbazide can be replaced with thiocarbohydrazide.

- Thiosemicarbazide, 1g
- Acetic acid 2%, 100 mL

12.2.28   THIOCARBOHYDRAZIDE (TCH)

- Thiocarbohydrazide, 0.2 g
- Acetic acid 10%, 100 mL

12.2.29   URANYL ACETATE

This salt is used for standard contrast of ultrathin sections. It can be also used to visualize Golgi apparatus and glycoproteins.

- Uranyl acetate saturated in ethanol 50%.

12.3   BUFFERS

12.3.1   CACODYLATE–HCL BUFFER (pH 5.0 TO 7.4)

- Solution A
  - Sodium cacodylate, 42.8 g
  - Distilled water, 1000 mL
- Solution B
  - HCl, 10 mL
  - Distilled water, 603 mL
- Buffer
  - Solution A, 50 mL
  - Solution B, 18.3 mL
  - Distilled water to 100 mL

The aforementioned formula is given for pH 6.4. For other pH:

- Solution A, 50 mL
- For pH 6.6
  - Solution B, 13.3 mL
  - Distilled water, 100 mL
- For pH 6.8
  - Solution B, 9.3 mL
  - Distilled water, 100 mL
- For pH 7.0
  - Solution B, 6.3 mL
  - Distilled water, 100 mL
- For pH 7.2
  - Solution B, 4.2 mL
  - Distilled water, 100 mL
- For pH 7.4
  - Solution B, 2.7 mL
  - Distilled water, 100 mL

12.3.2   COLLIDINE BUFFER (pH 7.25 TO 7.59)

- Stock solution
  - Collidine, 5.34 mL
  - Distilled water, 100 mL
  - HCl 1 M, 18 mL
  - Distilled water to 200 mL
- To adjust pH, add HCl
  - pH 7.25, HCl, 22 mL
  - pH 7.33, HCl, 20 mL
  - pH 7.41, HCl, 18 mL
  - pH 7.50, HCl, 16 mL
  - pH 7.59, HCl, 14 mL
  - pH 7.67, HCl, 12 mL
  - pH 7.74, HCl, 10 mL

12.3.3   NITRIC ACID–ACETATE BUFFER

- Nitric acid, 0.2 mL
- Distilled water to 80 mL
- Citric acid 0.1 M, 4 mL
- Adjust pH to 7.0 with 1 M sodium hydroxide
- Distilled water to 100 mL

12.3.4   RINGER'S SOLUTION

- Sodium chloride, 30 mM
- Calcium chloride, 4 mM
- Magnesium chloride, 2 mM
- Potassium chloride, 1.5 mM
- Sodium bicarbonate, 18 mM
- Sodium triphosphate, 0.25 mM
- Adjust to obtain pH 6.8 with concentrated sodium hydroxide

12.3.5   SODIUM CACODYLATE

See Section 12.3.1.

12.3.6   SODIUM PHOSPHATE BUFFER

See Section 12.3.8.

12.3.7   PBS

- NaCl, 0.8 g
- KCl, 0.2 g
- Na_2HPO_4, 1.44 g
- KH_2PO_4, 0.24 g
- Distilled water to 800 mL
- HCl N to pH 7.4

12.3.8 Sörensen's Buffer (2 M; pH = 7.4)

- Solution 1
  - NaH_2PO_4 (monosodic phosphate), 3.12 g
  - Distilled water 100 mL
- Solution 2
  - Na_2HPO_4 (disodic phosphate), 7.16 g
  - Distilled water 100 mL
- Buffer
  - Solution 1, 19 mL
  - Solution 2, 81 mL

Section 3

Image Quantification

13 Image Quantification in Histology and Cytology

Yves Tourneur and Léon Espinosa

CONTENTS

Nowadays more researchers not only take pictures but also apply some of image processing methods to extract information from them. The present chapter aims at explaining the reasons why this practice is growing and will describe the basis of such processes.

13.1   INTRODUCTION

13.1.1   Why Is Quantification Necessary?

The use of images is widespread in experimental sciences and particularly in life sciences. The advent of digital imaging and the large expansion of computer resources availability as well as the need of rigor led to an impressive development of image quantification.

Differences between two samples may be too small to be detected by the eye, and background effects, photobleaching, or autofluorescence of controls may induce errors in fluorescence observation. Quantification allows comparing different groups contained in a big amount of images to monitor the evolution for dynamical phenomena. The multiplication of imaging results implies rigorous analysis providing statistical confirmation of the visual impression.

13.1.2   Methods of Image Analysis

Image analysis techniques arose from the addition of two approaches: signal analysis and mathematical morphology. The birth of signal analysis, during or soon after World War II, comes from the telecommunications field for defense applications on one hand (Norbert Wiener) and from the digital signal transmission field on the other (Claude Shannon and Harry Nyquist). Signal analysis, also called "convolutions," resurrected Fourier analysis (Joseph Fourier in 1822), largely thanks to the so-called fast Fourier transform (FFT) by Cooley and Tukey (1965). Mathematical morphology was invented by Georges Matheron and Jean Serra (1964) for geology purposes. Image analysis initially required computation power. Even a poor quality image such as a TV frame is composed of hundreds of thousands of points, and image analysis software remained until the 1980s limited to military or research applications, and eventually on specialized computers. The power of home computers and the video games market made home computers capable to run any image analysis program.

The basic steps of image analysis include image enhancement, then measurement, preceding the statistics that are out of this focus. The concepts of image analysis can be used in a first step to enhance the image by defining a framework based on a prior hypothesis. The second step of analysis then estimates either relative or absolute value of an identifiable compound (usually a protein) by its intensity, or either counts the number of identified structures. The measurements are processed and are ready for further statistical analysis. The more commonly used workflows are summarized in Figure 13.1.

At the start of the work with images, two basic goals can be sought: (1) either direct quantification of the image intensity as a measure of the amount of colored or fluorescent signal or (2) the identification of particular zones of the images to be analyzed as individual objects with their own specific parameters (position, intensity, color, shape, number, distribution, etc.). Section 13.2 will present the technological overview of monochrome and colored digital imaging, as well as segmentation, which differentiates between objects versus background image zones by applying a binary mask ("all or nothing") of the source image. Sections 13.3, 13.4, and 13.5 will describe in detail the mathematical or computational approaches allowing pertinent segmentations. Finally, two examples of image analysis in cytology and histology are presented.

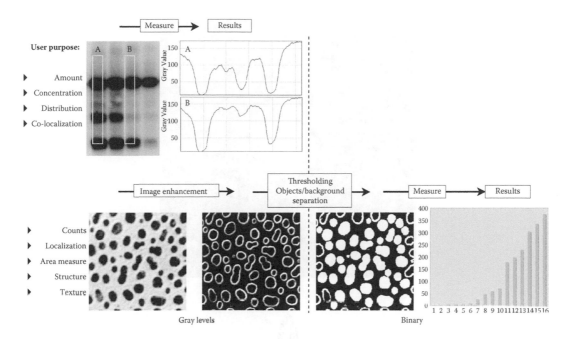

FIGURE 13.1 The two most frequent workflows in image analysis. (Top) Direct quantification from the recorded gray levels. (Bottom) Identification of objects in the image, then computation of properties (level or geometry) of the objects.

13.2 IMAGE RECORDERS AND IMAGE CHARACTERISTICS

Despite their lower resolution, electronic cameras, because of their higher sensitivity, better linear dynamic range, and above all much better convenience, have replaced photographic films. With such devices, an image is directly digitalized into a matrix as an array of numbers. We call "image" a matrix or a collection of matrices stored in the computer memory. The matrix elements are characterized by their positions given in the form of coordinates x and y. Columns are noted x and rows are noted y. We call the elements (x, y) the pixels (for picture element) of the image. The value of each (x, y) position represents the "light" intensity generated by the real object that creates the image. But, according to our definition of the image, intensity may represent something quite different, for example, a sound wave in the case of medical ultrasound sonogram, the velocity of a fluid, or the electrical activity of a tissue. These values are called "gray levels" of the image. (A color image is composed of three gray level layers.) To extract a quantitative information contained in the image we must gather the characteristics of the system that produced it, namely, (1) the spatial scale factor (size of the real object represented by one image pixel), (2) the factors of intensity scale (multiplication of the object intensity by a gain factor and addition or subtraction of a baseline), and (3) linearity of the sensor response; namely, response characteristics of sensors to different wavelengths of light.

13.2.1 OPTICAL RESOLUTION

In optical microscopy, the resolution limit is not set by the magnification factor but depends only on the numerical aperture of the objective. The image of a point is a pattern, described by the English astronomer George Airy (1801–1892), grossly a dome with smaller lobe rings called a point spread function (PSF) or Airy disk (Figure 13.2), described by a Bessel function.

On the image plane, its radius is given by $\sim 0.6 \times G \times \lambda/NA$, where G is the magnification, λ the wavelength, and NA the numerical aperture of the objective. The Rayleigh criterion states that two

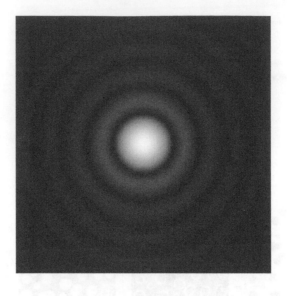

FIGURE 13.2   Shape of the Airy disk, where gray levels are represented in vertical dimension.

points can be distinguished when the first minimum (zero crossing) of one Airy disk is aligned with the maximum of the second Airy disk. In practice, for a 500 nm green-blue wavelength, the dot on the sensor ranges from 14 μm for a 40 × 0.95 plan Apochromatic objective to 33 μm for a 10 × 0.1 Achromatic one.

Since every point of the object produces a fuzzy dot, the PSF in the image, it appears intuitively that the whole image should be fuzzy. Indeed, the intensity of each point in the fuzzy image collects the intensities of its neighbors with a certain factor, described by the shape of the PSF.

13.2.2   SPATIAL SAMPLING OF A MICROSCOPIC IMAGE

The optimal restitution of an optical image requires the size of every element of the sensor (sensor pixel) to separate the smallest observable details, limited by the size of the Airy disk. The exact specification is defined by the Shannon-Nyquist theorem. It states that an image composed of a series of points may contain the same quantity of information as a continuous image. To fulfill this condition, the number of points per surface unit has to be larger than a given value, given by the Shannon criterion.

TABLE 13.1
Example of Recommended Pixel Size for Some Objectives

Objective Plan	G	NA	Cone Angle (Degrees)	Resolution (μm)	Airy Disk on Sensor (μm)	Required Pixel Size (μm)
10X Achromat	10	0.1	5.7°	3.05	30.0	15.0
20X plan Achromat	20	0.4	23.6°	0.76	15.0	7.5
40X plan Apo	40	0.95	71.8°	0.32	12.6	6.3
63X plan Apo	63	1.3	58.8°	0.23	14.5	7.3
60X plan Apo	60	1.4	67.1°	0.22	12.9	6.5
100X plan Apo	100	1.4	67.1°	0.22	21.4	10.7

Notes: Computations are performed for $\lambda = 500$ nm. G, magnification; NA, numerical aperture; cone angle, observation of an object; Resolution, after the Rayleigh criterion; Airy disk radius on the object plane (= $0.61*\lambda/NA$); Airy disk on sensor, preceding column times magnification. Required pixel size: recommended size to fulfill Shannon criterion.

Application of the Shannon criterion to images is more complex than for a signal (Gori, 1993). Given the size of the sensor and the Shannon criterion, the pixel size of the camera should be around one half of the Airy disk radius. A higher number of pixels will not improve the actual resolution. The ratio of the field of image size to the resolution defines the optimal number of pixels, commonly around 1.5 megapixels (see Table 13.1; depending on the sensor size, usually ¼″, ½″, ⅔″, or 1″, an adapting lens should be used).

In transmission or scanning electron microscopy (TEM or SEM), the Airy disk is smaller than the nanometer, and the resolution is limited by spherical aberration. High-resolution cameras with up to 25 megapixels provide in this case useful information. By comparison, the television video standard has a poor resolution of 0.25 megapixels. Table 13.1 lists the optimum camera pixel size in micrometers calculated at 500 nm.

13.2.3 Sensor Characteristics

The rapid progression of imaging technology allows a better response to the requirements of microscopy techniques. The image is acquired either point-by-point in most laser scanning confocal microscopes, or line-by-line, in flat scanners and some fast confocal microscopes, but in most cases by a charge-coupled device (CCD) or complementary metal-oxide semiconductor (CMOS) cameras, described next.

The first solid-state electronic sensor was the charge-coupled device (CCD) invented by Willard Boyle and George E. Smith, who were rewarded with the 2009 Nobel Prize in Physics. It is based on an electronic shift register, which makes a charge transfer from an electrode to the next one. Every pixel is composed of a semiconductor interface between two electrodes. During light exposure, a photon hitting the semiconductor in an electric field creates electron–hole pairs. The electrons (e^-) remain confined in a tiny region below an electrode, called the electron well. Depending on the sensor, an electron well may store thousands up to hundreds of thousands of electrons. After light exposure, during a reading phase, the charge contained in the first well is measured by a charge-to-voltage converter, then all the charges in a line are shifted by one step and the charge under the second pixel is read. This operation is repeated till the end of the line. All the lines are successively read. One drawback of the measurement is the charge-to-voltage uncertainty, called the readout noise, measured in number of electrons, from a few to tens of electrons. The ratio of the well capacity to the readout noise is called dynamic range. For instance, a camera with a full well capacity of 200,000 e^- and a readout noise of 8 e^- has a dynamic range of 200,000/8 = 25,000, which represents how many steps above the read noise floor the CCD chip is capable of recording.

In a modified technology for low-light measurements, electron multiplying CCD (EmCCD), an amplifying series of charge shifts is added before conversion. At every shift, the quantity of charges is slightly enhanced (a few percent at most), but the large number of amplifying elements, for instance, 512, makes the final gain important before the measurement step. The contribution of the readout noise is thus negligible.

An alternative to the CCD sensor is the complementary metal-oxide semiconductor (CMOS). In a CMOS sensor, each pixel possesses its charge-to-voltage converter that can be addressed independently. (As a consequence, a CMOS camera can visualize and store an image from only a region of the sensor.) Initially designed for low-cost or miniature cameras, CMOS sensors are preferred for fast applications, whereas CCD ones, eventually amplified, remain a reference for low-light applications. Presently, both technologies have advantages and disadvantages. A new technology using electron bombardment now reaches sensitivity of one photon, either with CCD or CMOS (Ray et al., 2008).

13.2.4 DIGITAL IMAGE

The voltage representing the measure of light on each pixel is finally converted to a digital number composed of binary units or bits. A number stored on n-bits codes 2^n different values, corresponding to gray levels: 8 bits encode 256 values. Most sensors code using 10 to 16 bits, thus 1,024 to 65,536 values. The number of bits, or depth, has to be related to the dynamic range of the sensor. The images are then stored in bundles of 8 bits, or bytes. A pixel can only occupy an integer number of bytes, so an image coded on 8 bits occupies 1 byte per pixel, whereas an image coded either on 9, 10, or 14 bits occupies the same quantity of memory, 2 bytes per pixel.

13.2.5 STACKS, HYPERSTACKS, AND MULTIDIMENSIONAL ACQUISITION

ImageJ software allows the storage of images in an ordered sequence into a stack. It can represent a temporal (time series) or a spatial sequence (i.e., different focus positions along the vertical axis z), or multiple points in a defined pattern of positions (e.g., microplates). More than one parameter may be sequentially changed (e.g., time and z position); in this case the stack is called a hyperstack by analogy with a hypervolume in a space with more than three spatial coordinates. Acquisitions in which several parameters vary are called multidimensional and the number of dimensions is given by the minimum information needed to locate a particular image in a unique way: position (x, y, z), time, fluorescence channel, and so on. The image stack (or hyperstack) allows the subsequent application of filters, mathematical operation, threshold, and measurement for all the images of the sequence at once. It also allows directly obtaining a graphic curve to follow the evolution of one parameter along one dimension (e.g., intensity with time). Sets of 3D images (x, y, z) may also be used to build a volumetric representation. Depending on what method is used to represent a volume in a 2D image, the reconstruction may be volumic (maximum intensity projection, mean projection, nearest point) or surfacic (isosurface). Many software are specialized in 3D objects reconstruction and measurement.

13.2.6 IMAGE MOSAICING OR STITCHING

Many applications now require images of zones up to several centimeters with the microscope resolution. This is usually performed by moving the object under the field of the objective with a digitally driven microscope stage. Up to thousands of digital images are then used for the reconstruction of a unique large field image (Rankov, Locke, and Edens, 2005).

To be able to do automatic stitching of images, the latter should be partially overlapped. The common regions are identified by the program and fused. This can be followed by a shading correction or an exposure correction thus leading to perfect assembly. Many freeware programs perform these operations, called "mosaicking," "stitching," or "tiling," including in ImageJ or Fiji.

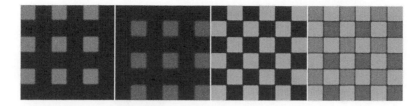

FIGURE 13.3 **(See color insert.)** Representation of the colored filters laid upon the sensor to produce a color image. From left to right, image of the red, blue, and green filters, and of their assembly to create the Bayer matrix.

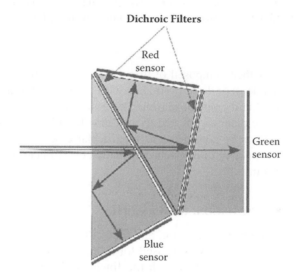

FIGURE 13.4 Representation of a 3CCD camera showing the light path for the three colors. The dichroic mirrors only reflect some colors and are transparent to others.

13.2.7 Color Image

A visible image is produced by an electromagnetic field with a wavelength ranging from 380 to 780 nm. The whole visible spectrum recording was initiated by Lippmann in 1891, Nobel Prize laureate in 1908, and now performed in fluorescence spectral confocal microscopy (Dickinson et al., 2001; Zimmermann, Rietdorf, and Pepperkok, 2003). Conversely, color cameras are based on the principle of only three bandwidth detectors corresponding to the three human cone cells perception. Initially stated in the Young-Helmholtz trichromatic theory, the three components are now specified by the Commission Internationale de l'Eclairage (CIE, 2004). The color image is composed of the superimposition of three color images, called color planes: red, green, and blue (RGB).

The most common color camera technology is based on the Bayer matrix, consisting of a color filter laid upon the sensor (Figure 13.3). The filter is divided into red, green, and blue color filters, every square covering one pixel. It takes advantage of the dominant green vision of the human eye and divides the sensor into 25% red, 25% blue, and 50% green. When reading such an image, the image software fills the missing points in every color plane by interpolation (Alleysson, Süsstrunk, and Hérault, 2005).

Another technology is the 3CCD camera, which uses prisms and filters to apply the image to three sensors for red, blue, and green components (Figure 13.4). The alignment of the sensors has to be tight so that every pixel has its three components.

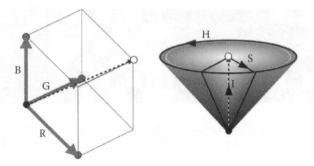

FIGURE 13.5 (Left) RGB (red, green, blue) and (right) HIS (hue, intensity, saturation) space representations. One specific color is localized in these spaces with three coordinates. In each space the gray lines link black to white points.

The digital color image is then composed of three color planes, obtained either by interpolation (from a Bayer matrix camera) or directly (from a 3CCD camera). All image analysis software provides a routine, which splits a color image into three monochrome planes or vice versa. As already stated, a monochrome plane is referred to as a gray-level image, whatever color channel it represents. A color image can be considered as a particular case of spectral (or multispectral) images, composed of a given number of monochrome planes, every plane corresponding to a range of wavelengths in the spectrum.

13.2.8 Color Image Representation: RGB and HIS

The color image representation in red, green, and blue (RGB) is based on human physiology. Other modes of representation are, however, used for image transmission or analysis. Even the physiological feeling of color does not fit with the linear RGB representation. A pure green color is described in the RGB space as (0,255,0) with 3 × 8 bits coding depth. A light green or a grayish green is still considered as the same color with nuances. These colors are composed by addition of different proportions of RGB coordinates. For this purpose, another representation is closer to the physiological feeling. Every color can be described by its coordinates in a 3D space whose axes are its RGB components (Figure 13.5, left panel). In this representation, any color is included inside a cube based on the RGB space. In this cube, the origin (0,0,0) is black, whereas its opposite (255,255,255) is white. The grays are located on a line linking these two points. Any given color (small dot) can be represented in the RGB space by its x,y,z representation. The HIS (hue, intensity, saturation) (Figure 13.5, right panel) representation identifies a color as its:

- Hue as the angle around the line of the grays
- Intensity as the length of the projection along the line of grays
- Saturation as the distance from the line of grays

This representation more accurately identifies objects in a color image, in particular in histology.

13.2.9 Color Deconvolution

Every histological stain such as hematoxylin, for instance, will absorb a given proportion of R, G, and B. Instead of RGB decomposition, Ruifrok and Johnston (2001) described the possibility to decompose a histological image into components corresponding to the different stains. This is illustrated in Figure 13.6. The decomposed images are appropriate for image analysis and quantification.

FIGURE 13.6 Decomposition of a histological image into components corresponding to different stains. Hematoxylin (blue-violet) and peroxidase (brown) linked to an antibody anti-Ki-67 protein, stained the image. (A) Original. (B) Hematoxylin component. (C) Peroxidase component. (D) Eosin component (there was no eosin in the staining).

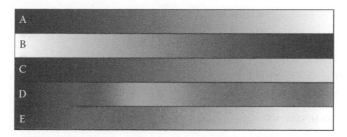

FIGURE 13.7 Representation of colors as a function of the x-axis, from 0 to 255 (8 bits LUT). (A) Normal LUT. (B) Inverted. (C) Sepia LUT. (D) Rainbow LUT. (E) Fire LUT.

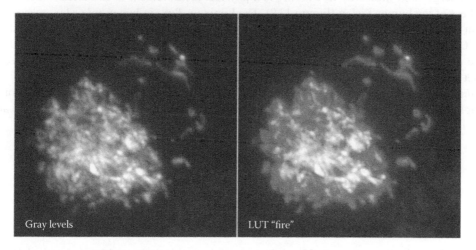

FIGURE 13.8 **(See color insert.)** A LUT applied in a gray-level image transforms low-level intensity differences in differences of colors.

13.2.10 FALSE COLORS REPRESENTATION

Many representations of gray images are colored for altitude, temperature, or hydrology in geography, or intracellular calcium concentration in cell biology, for instance. A false colors scale is used for representation. Such a color representation, called look-up table (LUT), can be stored with the image (Figure 13.7).

LUTs are frequently used in fluorescence to allow the superimposition of different monochrome images, each one corresponding to a fluorophore (Figure 13.8). The use of LUTs is very convenient to visualize subtle differences in gray levels, and a large variety of built-in LUTs are provided with image analysis software.

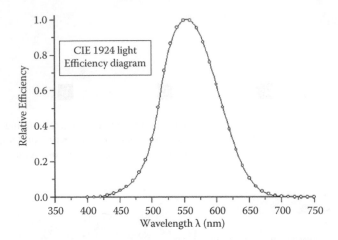

FIGURE 13.9   Spectral sensitivity of human eye (standard eye vision curve). All colors of the visible spectrum do not have the same efficiency to stimulate the human vision.

13.2.11   Light Measurement

For quantitative image analysis, the recorded value in a pixel may be referred to an absolute value of light intensity. Light intensity can be measured in two ways: energetic flux or luminous flux.

The energy flux, measured in watt/m², is a physical measurement of the electromagnetic field regardless of the human vision (Figure 13.9). The luminous flux, in lumen/m² or lux, measures the amount of light actually perceived by a human eye. A standard eye vision curve has been defined by the CIE, from statistical measurements among various human populations (CIE, 1970, 1971).

The quantity of light collected by a sensor is measured by the exposure (H, in lux x sec), the product of the light (in lux), and the exposure time (t, in seconds).

Light transmittance and optical density. The transmittance, τ, is defined as the fraction of incident light that crosses a material and its complement, the absorptance a (Note: it differs from absorbance) is the fraction, which is absorbed:

$$a = 1 - T$$

The optical density (or absorbance) is defined as:

$$OD = -\log_{10}(\tau)$$

The logarithm unit allows the simplification: the superimposition of two objects of optical densities, OD_1 and OD_2, has an optical density $OD = OD_1 + OD_2$.

This led to the Beer-Lambert law to express the optical density (OD) of a solution:

$$OD = \varepsilon \times C \times l$$

where l is the path length in meters, ε is the molar absorptivity of the compound, and C is the concentration.

13.2.12   Relationship between the Recorded Light Level and Compound Concentration in Histology and Cytology

The visualization of a biological compound usually involves colored dyes in histology and fluorescent markers in cytology. Using colored staining, the observer or the detector detects the quantity of

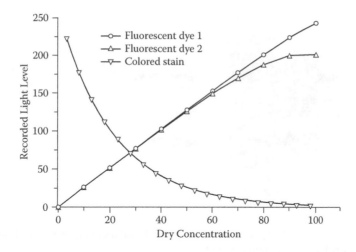

FIGURE 13.10 Schematic representation of the light level function of dye (or stain) concentration. For a perfect fluorescent label (circles), the relation is linear, but fluorescence most likely shows saturation (up triangles) by quenching. For a histological stain (down triangles), light intensity decreases with stain concentration.

light transmitted by the preparation using microscopy. Colored staining follows the Beer-Lambert law: $OD = k \times C$ (the factor k includes here the absorptivity and the thickness of the preparation). Therefore, the recorded light, Φt, is linked to the stain concentration by the equation: $\phi t = \phi i \times 10^{-k \times C}$.

When using fluorescent dyes, the recorded emission light intensity varies with the excitation light in a saturating curve manner, which can even decrease with increasing concentrations due to the autoquenching effect (Figure 13.10). Furthermore, lack of specificity or affinity of the indicator to the compound is responsible for false negative (elements that should have been labeled but that are not) or false positive (elements that appear labeled but that should not).

In conclusion, even assuming that the dye concentration is linear with the protein or target concentration, the image gray level may be nonlinear with the target concentration, which then argues for the use of nonlinear analysis methods, described next.

13.3 SIGNAL ANALYSIS TECHNIQUES

Since a digital color or spectral image can be considered as a series of gray images, we will look into the analysis of a gray image. Different image analysis software are available. The reader may refer to Kherlopian et al. (2008) and Internet resources (http://rsbweb.nih.gov/ij/links.html) for a list of software, including open source programs such as ImageJ. ImageJ, an open source software developed by Wayne Rasband, is becoming a standard, thanks to the numerous add-ins that have been added by the international community, as *plug-in* or as *macros*. A particular distribution containing 3D applications has been developed under the name of Fiji. ImageJ can be downloaded from the National Institutes of Health Web site: http://imagej.nih.gov/ij/.

13.3.1 IMAGE BRIGHTNESS AND CONTRAST

The image quantification aims to overcome the errors induced by visual evaluation and to increase the accuracy of measurement.

The dynamic range of the digitization is the ratio of the brightest to the darkest levels. In an optimal configuration, the dynamic range of digitization should be set to fit the dynamic range of sensor. The human eye is hardly able to identify more than 50 gray levels. In silver photography, the absorptance of a film is not proportional to the exposure: The plot of the optical density to the log of the exposure (Hurter and Driffield, 1890) is an S-shaped curve with a linear fraction whose slope is

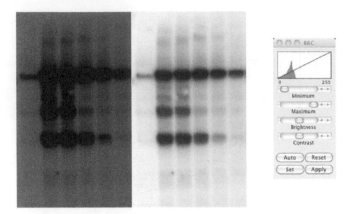

FIGURE 13.11 Enhancement of display contrast by setting the values of the displaying limits (max and min) to the limits of the gray levels histogram of the image.

called gamma (γ), corresponding to an index of contrast. The dynamic range of a film is larger than 1 to 10,000, or 4 OD units.

Electronic scientific cameras are usually linear. In practice, the number of bits of conversion gives the dynamic range: a 12 bit camera stores values between 1 and 4096, corresponding to 3.6 OD units.

The image histogram represents the dynamic range of one image. The brightness and contrast may be adjusted to optimize the display rendering. The pixel values or the display rendering are modified so that the darkest points of the image become black and the brightest become white (Figure 13.11). This is performed automatically by most image analysis software.

13.3.2 DIGITAL FILTERS

Direct observation triggers a cognitive process, which immediately identifies the regions of the image containing the relevant information, such as cells or nuclei. In order to mimic these processes, image analysis software provide different kinds of tools, thus allowing distinguishing among the zones according to their properties.

Three kinds of digital filters are described next: (1) convolution filters are based on linear operations between neighborhood pixels, (2) frequency domain filters operate after decomposing the image as a sum of waves, and (3) morphomathematic operators only use maximum and minimum operators applied to neighborhood pixels.

13.3.3 CONVOLUTION FILTERS

The sharpest picture that can be imagined is a point. It has been previously stated that, through an optical device, the image of a point is a point spread function (PSF). The light intensity at one point will influence its neighbors, so that the whole image is fuzzy. The resulting operation is the convolution, noted with an asterisk (*), whose result is the observed image ImObs, knowing the actual image (ImAct) and the image of a point. It is written:

$$ImObs = ImAct * PSF$$

It results in a suppression of details from the image. The PSF declines to zero but extends to infinity, which means that every point in the observed image depends on all the points in the actual image.

In digital image analysis, digital convolution is defined as being similar to the optical convolution. Convolution is used to suppress details and highlight the major general shape, or by contrast,

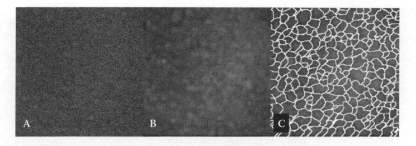

FIGURE 13.12 Illustration of Gaussian filter. (A) Raw confocal image, highly noisy. (B) Effect of 5×5 Gaussian filter revealing structures. (C) Separation of the structures.

TABLE 13.2
Example of Gaussian Kernel

0.0000	0.0005	0.0050	0.0109	0.0050	0.0005	0.0000
0.0001	0.0050	0.0521	0.1139	0.0521	0.0050	0.0001
0.0002	0.0109	0.1139	0.2487	0.1139	0.0109	0.0002
0.0001	0.0050	0.0521	0.1139	0.0521	0.0050	0.0001
0.0000	0.0005	0.0050	0.0109	0.0050	0.0005	0.0000
0.0000	0.0000	0.0001	0.0002	0.0001	0.0000	0.0000

to highlight the details in the image. The major difference with the optical convolution lies in the fact that the digital convolution involves a limited kernel, rather than the PSF extending in the whole plane. The convolution kernel is defined by an odd-sized matrix of points, in most cases of 3×3 to 7×7 coefficients, and admits a rotational or a translational symmetry.

Figure 13.12 shows how the internal structure of a noisy image can be revealed by a Gaussian filter, showing a granular structure, which can later be used to separate these objects (by a watershed algorithm method described later). An illustration of the 7×7 Gaussian filter (0.8 pixel standard deviation) is provided in Table 13.2 (7×7 Gaussian filter kernel). As is usually the case, the coefficients are normalized so that their sum is one.

13.3.4 USE OF CONVOLUTION TO HIGHLIGHT CONTOURS

Other convolution filters can underline details, in particular the contours of a structure. The Sobel is a typical filter used to underline contours. From an initial image, it creates two images relative to the horizontal and vertical contours, and then a result by quadratic sum of the two contours. In Figure 13.13, we see the application of the Sobel filter to the image of a square and to an actual biology image (It is used in ImageJ for "Process/Find Edges.") Other filters (Roberts and Prewitt) provide similar though noisier images. A more sophisticated version of the contour filter is the Canny operator consisting of a Gaussian filter preceding a contour filter such as the Sobel.

13.3.5 FAST FOURIER TRANSFORM (FFT) AND WAVELETS

The fast Fourier transform (FFT) is now used as a synonym of the so-called Fourier analysis developed by Joseph Fourier between 1807 and 1822 for heat conduction. Its principle can be illustrated simply for a signal in one dimension: the sound of music played by a loudspeaker is transformed into an electric signal decomposed into three signals for the woofer, midrange speakers, and tweeter, or in five into the subwoofer, woofer, midrange speakers, tweeter, and super-tweeters. Every speaker,

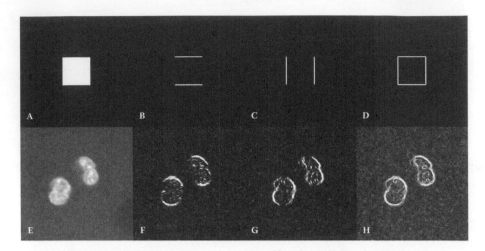

FIGURE 13.13 Example of Sobel filter applied to a model (A to D) or an actual image (E and F). (A and E) Initial image. (B and F) Horizontal contour h. (C and G) Vertical v. (D and H) Result of Sobel filter $\sqrt{(h^2 + v^2)}$.

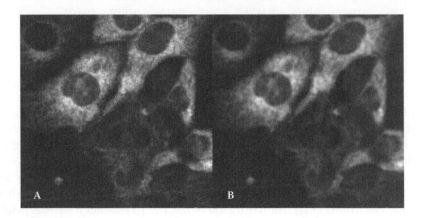

FIGURE 13.14 Use of FFT low pass filter to smooth an image.

or acoustic transducer, transmits a range of frequencies of the sound. Assuming that there are an infinite number of loudspeakers, every device only transmits one frequency, or a sinus function. For the listener, no difference can be made, which means that the sound can be decomposed into sinus waves that are perfectly reconstructed into the original sound by the ear. In a similar way, the Fourier theorem can be applied to decompose an image, which is a two-dimensions signal. In this case, the sinus has to be replaced by waves of intensity in a surface. A sum of waves can reproduce any two-dimensional gray image. As sound can then appear lower or shriller by suppression of the treble or the bass. Similarly, suppression of high or low frequencies will either smooth an image or underline its details.

The Fourier decomposition aims at obtaining these waves by computation. The numerical decomposition was slow before the implementation of a variant of a Gauss method, called FFT. FFT is now implemented in all analysis software. In principle, it only analyzes a square image whose dimension is a power of 2 (128×128, 2048×2048, etc.), but most programs overcome this requirement. FFT can efficiently selectively remove either the slow or the fast waves, or favor or suppress the waves in a given direction.

In Figure 13.14, fluorescent cells acquired with a confocal microscope show details that are actually due to the noise of the photomultiplier and do not carry any information. Removal of "short waves" (low pass FFT filter) smoothes the image and its effect is similar to a convolution filter.

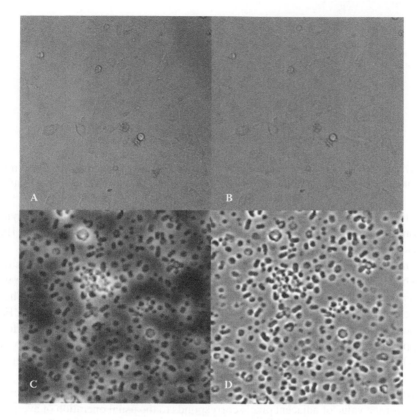

FIGURE 13.15 Use of FFT high pass filter to remove background nonuniformity.

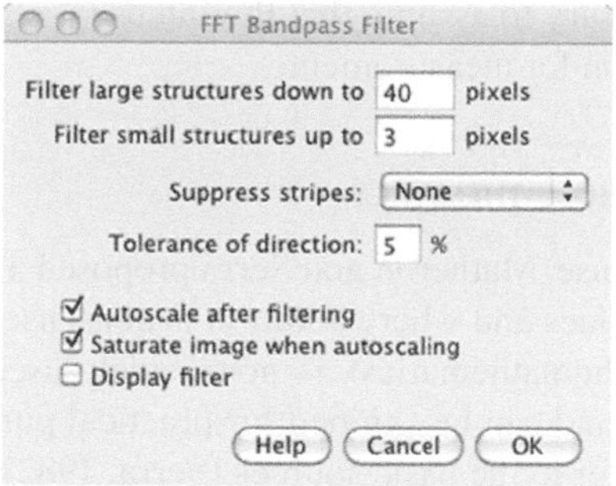

FIGURE 13.16 Using ImageJ menu for FFT either high pass or low pass filter.

On the other hand, the removal of "long waves" (high pass FFT filter) highlights high frequencies (Figure 13.15). This kind of procedure is particularly useful to analyze images obtained with transmission microscopy when the background is not uniform, for example, when the light was not correctly tuned.

In ImageJ, the size of the waves to keep in the image is directly given, between the size of large structures and the size of small structures in pixels, or in other units (e.g., micrometers), if the image has been calibrated (Figure 13.16).

Note: In theory, convolution filters and Fourier domain filters are equivalent. This is not the case in a limited domain such as an image. In practice, a convolution smoothing (low pass filter) provides satisfactory results, but it may be easier to remove nonuniform background (high pass filter) with a FFT filter.

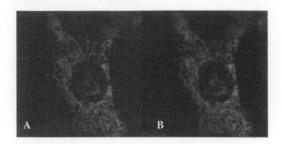

FIGURE 13.17 Example of mitochondria observed in confocal microscopy in a living cell using TMRM dye. (A) Raw image. (B) Filtered image using *à trous* wavelet filter (plug-in developed by J. Mutterer).

Usually, in the image analysis software, Fourier filtering is applied without displaying the frequency decomposition of the image, and the filtered result is directly displayed. The JPEG image format is based on FFT decomposition of the image by square elements.

Wavelets and wavelet filtering. The major drawback of Fourier analysis lies in its decomposition of the whole image into waves, whereas only some regions may appear wavy. The wavelets, in contrast, offer the possibility of the decomposition of the image into waves by region. The interested reader may refer to the mathematical concepts (Addison, 2002) out of the present field. For this, the 2010 Friedrich Gauss prize was awarded to Yves Meyer. Image analysis software offers the possibility to apply wavelet transform for smoothing (in the given example, redundant wavelets, called *à trous*) or efficient compression (Figure 13.17). The JPEG 2000 image format is based on wavelet compression.

The convolution kernel implies the sum of different neighborhood pixels, with weighing coefficients. The average value of adjacent pixels corresponds to the usual operation of average of different measurements. However, we have seen that the gray level of a pixel may reflect a measurement of a nonlinear process. A linear averaging between neighborhood points is therefore strictly not legitimate. It is, however, necessary to assume that the intensity is strictly monotonous (increasing or decreasing) with concentration for measurement.

13.3.6 PRINCIPLES OF MORPHOMATHEMATICS

Based on the nonlinearity response, Matheron and Serra proposed a mathematical formalism based solely on comparison among values and where addition is nonsense. Their technique, called mathematical morphology (or morphomathematics), is now widely used, although their formalism of "lattice" is rather cumbersome and can be skipped for practical purpose. The reader familiar with mathematical formalism can refer to the basic sources (Serra, 1982).

13.3.6.1 Basic Operations in Mathematical Morphology

Mathematical morphology applies either to a gray (eventually color) image or to the binary image. The morphological operations are based on a small image element called structuring element, and a point that is called the center (Figure 13.18).

The operation of erosion transforms one image into an eroded image by placing the center of the structuring element over every point of the source image, and attributing to the eroded image the minimum value of all the points covered by the structuring element. Gray image erosion darkens the image and binary image erosion produces a squeezing of the objects (Figure 13.19). On the binary image, the erosion suppresses the small objects. The major effect of erosion is then to suppress the details.

The dilation of an image is the opposite operation. It transforms one image into a dilated image by passing the center of the structuring element over every point of the image, and attributing to the dilated image the maximum value of all the points covered by the structuring element. The effect of a binary dilation is shown in Figure 13.20: it thickens the structures and links objects.

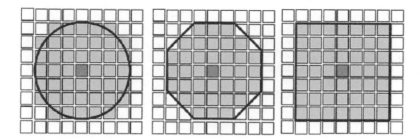

FIGURE 13.18 The structuring element is a collection of pixel positions relative to a particular pixel called "center." These positions usually draw a regular polygon geometrically centered on the "center" of the structuring element.

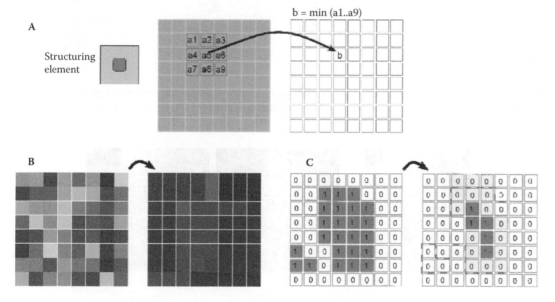

FIGURE 13.19 Principle of erosion. (A) Structuring element and minimum operation applied to 3×3 matrix. The result value is set to the center position in the eroded image. (B) Gray image erosion. (C) Binary image erosion.

Erosion and dilation allow defining new morphological operators, opening and closing. Opening is the result of the dilation applied after an erosion of the image (Figure 13.21). Similarly, the closing is defined as erosion applied after dilation (Figure 13.22). The major effect of closing is to link small objects. Similarly, a gray image closing may be a useful tool to clarify the general structure of an image for contour determination.

13.3.6.2 Geodesic Dilation

Geodesic dilation, sometimes called build operation, uses two binary images, one called mask and the other seed. The objects of the seed must be part of the ones in the mask. The seed image is usually obtained by modifications of the mask image. These modifications intend to remove undesired elements, but with the other objects reconstructed as the original (Figure 13.23).

Often the seed image was obtained by modifications of the mask image. These modifications intend to remove undesired elements, but with the other objects reconstructed as the original. Figure 13.24 illustrates this method, adapted from Kammerer, Zolda, and Sablatnig (2003) and implemented as a macro ImageJ code by Gabriel Landini in its excellent collection "Morphology." The task is to remove the thin lines without changing the shape of thick structures. The first operation

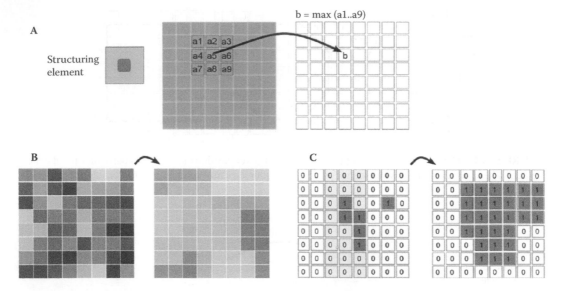

FIGURE 13.20 Principle of dilation. (A) Structuring element and maximum operation applied to the 3×3 matrix. The result value is set to the center position in the opened image. (B) Gray image dilation. (C) Binary image dilation.

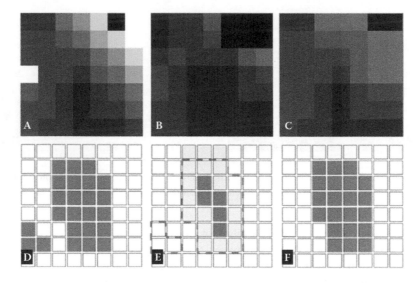

FIGURE 13.21 Principle of opening. (A–C) Gray image, eroded image, and then opening, that is, dilation applied to the eroded image. (D–F) Binary image, eroded image, and then opening, that is, dilation applied to the eroded image. In this case, the operation restored the large object and suppressed the small one.

is a 10 pixels erosion, only thick structures remain but with a modified area. Then iterative dilations are applied to this image with the conditional test that the entire dilated object must be included in the mask (original image).

13.4 MEASUREMENTS AND QUANTIFICATION

Manual measurements are generally separated from automated measurements. All software offer tools to select an area of the image and measure its characteristics: mean intensity, area, length,

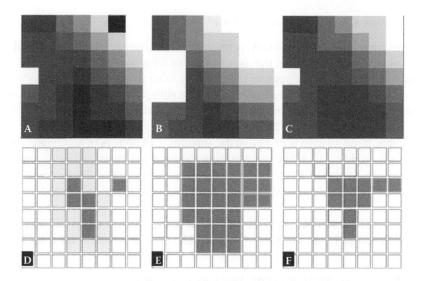

FIGURE 13.22 Principle of closing. (A–C) Gray image, dilated image, and then closing, which is erosion applied to the dilated image. (D–F) Binary image, dilated image, and then closing, which is erosion applied to the dilated image. The operation had an effect to link separated elements.

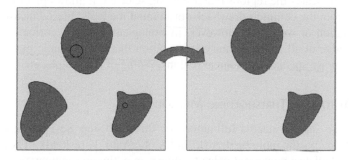

FIGURE 13.23 Geodesic dilation of black regions in the left panel ("seed" image) constrained by a "mask" image (gray regions in left panel). The result in the right panel shows that the seed image was dilated to obtain only the gray objects having a seed.

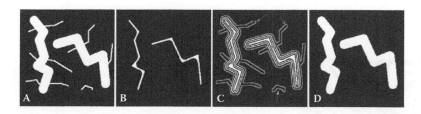

FIGURE 13.24 Example of geodesic reconstruction to eliminate thin structure. (A) Original image. (B) Eroded image. (C) Reconstruction. (D) Final image without the small details.

angle, number, and so forth. In some cases the selection can be done semiautomatically, such as the functions of "wand," which detects and selects all pixels adjacent to the point where the user has clicked with a given intensity range. However, if the number of measurements to carry out becomes very large or if the user may induce a bias in the measurements, it is often best to use one of the automatic detection methods of objects.

13.4.1 NOTION OF OBJECTS OR BINARY IMAGE

The aim of most operations in image analysis requires identifying elements such as cells and nuclei from what is initially a digital image, that is a matrix, or for color images, a series of matrix. The notion of object implies that every point of the image is either inside or outside, in other words that the initial image has been transformed into an "all or nothing" image, a binary image. Usually, binary images are 8 bit images with only two values: a black object, which has the level zero, and a white object, which has level one (or 255, in ImageJ).

We will explore three ways to accomplish the separation of objects in an image. The first is the threshold intensity, which is to define two intensity values, a low threshold value and high threshold value, and assign a value of 255 to pixels of the image whose value lies between the lower and high thresholds. All other pixels are set to zero. A second set of methods is based on the modification of gray values in the image in order to dramatically increase the contrast between objects and background. We will distinguish methods that increase the contrast edges of objects from those that increase the contrast of the surfaces of objects. The third approach is for color images.

13.4.2 AREA-ORIENTED SEGMENTATION: THRESHOLD METHODS

The manual threshold is based on the visual control by the user of regions in the image that are selected as object versus background. The software proposes a setup window with the histogram, a slider to set the low value, and a slider to set the high value. When the values of the sliders change, an overlay directly indicates the regions that are being selected. It allows precise and fast segmentation but suffers from the common drawbacks of manual methods: biased measures and prevents a complete automation of workflow. However, in biological samples, correct segmentation may require the knowledge of all supplementary information that no automatic method could retrieve (i.e., staining conditions, special histological regions, fading of the sample, etc.).

13.4.3 SOME AUTOMATIC THRESHOLDING METHODS

Many cases in image analysis need a full automatic threshold step. Several algorithms have been developed (see http://pacific.mpi-cbg.de/wiki/index.php/Auto_Threshold). They use histogram characteristics to calculate a numerical value to obtain an optimum segmentation.

Here are two methods: IsoData and Mean (Figure 13.25). Mean sets threshold as the mean value for all gray levels of the image. This method can be used for images with high contrast and very low background. The more precise IsoData method calculates the numeric value that satisfies the equation

$$G = \frac{\text{mean}\left(\text{foreground}_G\right) + \text{mean}\left(\text{background}_G\right)}{2}$$

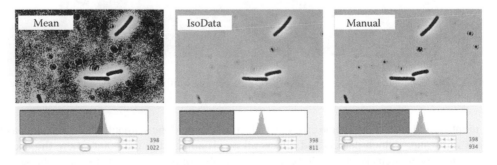

FIGURE 13.25 Three values for threshold of the same image obtained with Mean, IsoData, and Manual methods (ImageJ software interface).

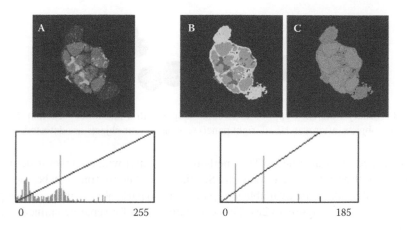

FIGURE 13.26 Segmentation by k-means clustering. (A) Source image (fly brain) with a continuous histogram showing four peaks. (B) Four regions clustering transform (fire LUT in panel C) with discrete histogram. All the values of pixels are set to the closest center of cluster.

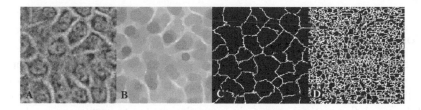

FIGURE 13.27 Segmentation by watershed transform. (A) The source image needs to be simplified (closing filter, panel B) before applying the (C) watershed transform to overcome (D) the usual oversampling of the watershed.

This method is robust and yields good enough results in many cases.

k-Means clustering. The k-means clustering consists of distinguishing the different zones by their level (Jain and Dubes, 1988). An ImageJ plug-in was developed by Jarek Sacha (http://ij-plugins.sourceforge.net/). The principle of this algorithm is to divide the image into clusters, pixels are grouped by their proximity to a cluster's centroids, and their value is set to the cluster number (usually less than five) (Figure 13.26). The algorithm minimizes the variance between the attributed value and the original values in the whole image. Sacha's plug-in implements a heuristic method.

13.4.4 THRESHOLD BASED ON THE GRAY IMAGE LEVELS: EDGES-ORIENTED SEGMENTATION

In order to obtain the objects, a direct transformation can be the spatial derivation or gradient (see the convolution filters: underlining contour). However, in most cases, biological samples do not have enough straight edges or the background is not smooth enough to yield satisfactory results.

The classical approach is the watershed transform (Figure 13.27). Let us consider the image as a representation of a landscape, with the brightest points being the hills and the darkest, the valleys. The watershed operation will transform an image in a series of regions identified by flooding. Let us imagine that from every bottom of a valley will spring water that will progressively flood the whole region. Every time water flows from a valley to another, a dam is built to prevent flooding.

The final all-or-nothing image of the dams is the watershed-transformed image. Watershed is one of the most used algorithms to detect cell contours in an image. The watershed separates the different elements of an image by a single line of unit width. In other situations, such a single line is required either to represent an object or to separate objects (see Daniel Sage plug-in, http://

FIGURE 13.28 Binary watershed applied to circular shapes with a middle constriction. In the left panel there are only two objects, after watershed (right panel) separation four objects are detected.

bigwww.epfl.ch/sage/soft/watershed/). This method works better with compact structures (typically images of tissues with a "cobblestone" pattern). Similar watershed routines can be applied on binary images. Different algorithms are available to separate in two parts objects with a shape presenting a region with a constriction (Figure 13.28). This is usually useful for separate stained nucleus if they are contiguous.

The edge-oriented segmentation identifies dissimilarities among regions (frontiers), whereas the area-oriented segmentation identifies similarities (regions). The watershed transform produced all-or-nothing lines isolating different zones of the image and is part of the edges-oriented segmentation methods.

13.4.5 Color Image Thresholding

We reported examples of histology images that can be decomposed into different components corresponding to different stains. The images obtained are actually monochrome with a LUT corresponding to the different dyes. Segmentation can then use the previously described methods. The color-based deconvolution method cannot always apply; one has to perform manual segmentation in three planes, either in RGB or HIS representation. In most situations, the best identification of objects is obtained first in HIS in the hue plane (see Task 2 in Section 13.5.2).

13.4.6 Quantification and Measure

13.4.6.1 Intensity Measures

The binary image can be used as a mask to select the pixels of any image with identical size and measure the values "under" the area occupied by the objects. It is possible to measure a set of intensity parameters as: mean, max, min, standard deviation, center of mass (same as centroid but brightness weighted), integrated density, modal value, median, and so on. A special case is the object represented by a line that allows obtaining a profile of intensity and is very useful in image analysis.

13.4.6.2 Geometric Measures of Objects

Software dedicated to image analysis offer the tools to measure several parameters of binary masks. The most used parameters are number of objects, area, centroid (average of the x and y coordinates of all pixels in the object), shape descriptors (circularity, aspect ratio, etc.), area fraction (ratio between background area and objects area), perimeter, angle of major axis, fit ellipse, and Feret.

Some common formulae are (coming from ImageJ implementation):

Circularity (dimensionless parameter between 0 and 1): $4\pi \times ([Area])/([Perimeter]^2)$. A value of 1.0 indicates a perfect disk. As the value approaches 0.0 it indicates a very elongated or very irregular shape. (Some programs define circularity as the inverse of this value.)

Aspect ratio: ([*Major Axis*])/([*Minor Axis*]) taken from a fitted ellipse.

Feret (and min Feret): The longest (shortest) distance between two points of the boundary, also known as maximum caliper.

13.5 DETAILED WORKFLOW OF TWO CLASSICAL TASKS

The following is an illustration of two image analysis tasks.

13.5.1 TASK 1

The aim of this image analysis task is to quantify the cytoplasm to nucleus translocation of the transcription factor NFκB in MCF7 (human breast adenocarcinoma cell line) and A549 (human alveolar basal epithelial) cells in response to TNFα. For this, the SBS Vitra CNT image set, provided by Ilya Ravkin and available from the Broad Bioimage Benchmark Collection (www.broad.mit.edu/bbbc), was used.

Acquisition Two channels were acquired (green = NFκB; blue = DAPI). Objective = 10× (Automatized system) in order to record the maximum of translocations events. Condition 1 = MCF7 cells stimulated by TNFα. Superimposed image shows cyan nucleus indicating a spectral mixing between the blue and green fluorescence. There could be artifacts in the image.	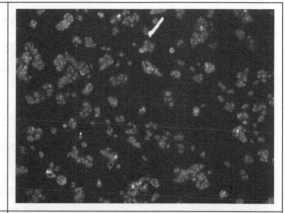
Acquisition Condition 2 = A543 cells stimulated by TNFα. Task Quantify the mean ratio NFκB nucleus over NFκB in the cytoplasm.	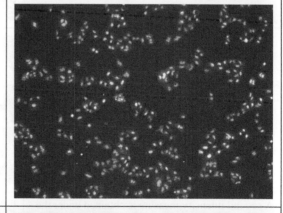
Segmentation—The image of the green channel is thresholded (by hand or automatic method). Morphomathematical filters can be applied to clean the binary image obtained after threshold. Some manual editing may be necessary to remove artifacts.	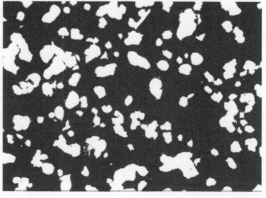

Segmentation—The image of the blue channel is also thresholded.	
Binary operations—An XOR operator is applied to "green" mask and blue "mask" in order to obtain a specific mask of the cytoplasmic regions.	
The binary mask for cytoplasm allows measuring specific fluorescence intensity. The results obtained are stored in a table.	
A second intensity measure is obtained under the binary mask for nucleus. The results obtained are also stored in a table.	

Computations—All results may be imported in a worksheet and processed. The ratio between the fluorescence of cytoplasm and nucleus indicates the nuclear translocation of NFκB in response to TNFα and compares the two kinds of cells.

13.5.2 Task 2

The aim of this task is to quantify steatosis (abnormal retention of lipids within a cell) in pig liver from a frozen preparation. Lipids are stained with red oil; cell nuclei are blue. The diagnosis depends on the relative size of cell nuclei and lipids droplets. (Thank you to Sylviane Guerret, NOVOTEC, Lyon, France.)

Image acquisition—The nuclei will be detected by color threshold in the HIS space. The nuclei have a blue hue, a mild saturation, and high intensity. The threshold operation is, as usual, followed by a binary opening to remove unwanted points. The black-and-white image permits to evaluate the mean size of nuclei.

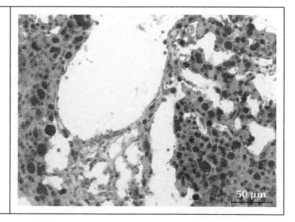

Segmentation and computation

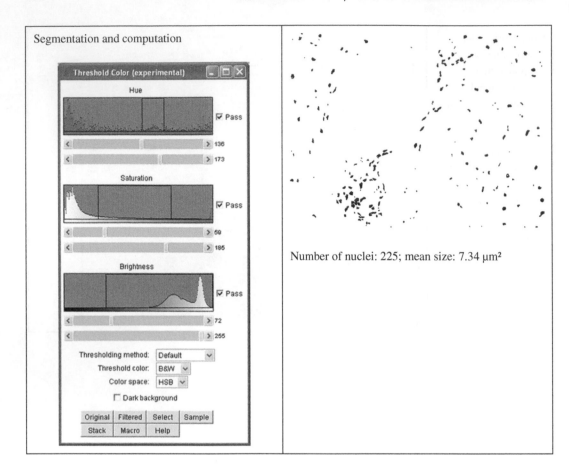

Number of nuclei: 225; mean size: 7.34 μm²

Segmentation—Similarly, the lipids dots are detected in the red range of hue. Here, the threshold is not taken between two cursors, but out of range* (pass checkbox unselected). This operation is followed by an opening.

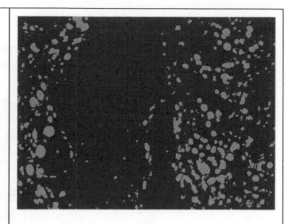

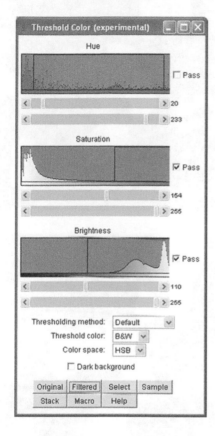

---

* Since hue is an angle, this function is periodic so that we must imagine both ends meeting. Software then allows selection of the region between two cursors, but also conversely the region excluded by the cursors.

A binary watershed splits fused objects of droplets (detail of the image, enlarged twice)	Raw after watershed
	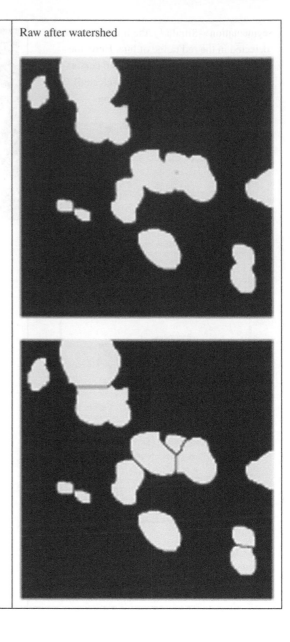

Computation—Counting droplets larger than average nuclei size (7.34 µm²) 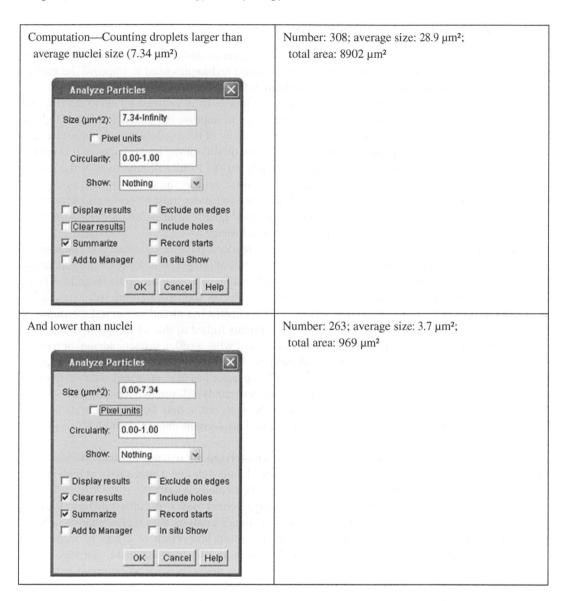	Number: 308; average size: 28.9 µm²; total area: 8902 µm²
And lower than nuclei	Number: 263; average size: 3.7 µm²; total area: 969 µm²

13.6 CONCLUSION AND PERSPECTIVES

Microscopy was originally a qualitative technique of observation and tends to become a quantitative science. This chapter describes some of the most common techniques used to improve the quality of an image for a given focus as well as the quantitation tools.

It is common to focus on either large objects in the image, or in contrast on small details, and image analysis provides automated tools for that. It is easy to improve the nonuniformity of light or to remove dust, for instance. Despite the power and simplicity of the available tools, it must be kept in mind that the first step in image analysis is the care for quality of the image caption. It is important to start with a well-centered light; to have correctly set the Köhler illumination of the microscope; and to have carefully adjusted the DIC (Nomarski) contrast, the phase anneals of a phase contrast, or the masks of a Hoffmann contrast. The lenses and slides must be perfectly cleaned.

A good image analysis can only be performed from good quality images. No digital modification can improve the quality and it cannot restore an incorrect image.

Powerful image analysis tools, in particular ImageJ, render it possible to isolate and count spots, mitochondria, cells, or tissue in an image, or in a large series of images. It is, however, disappointing for many people to discover that among the objects selected by the program, some should not be or conversely some of them are missing. This means that however sophisticated the software can be, it cannot reproduce the knowledge of the operator. The scientist has already observed thousands of samples. He or she has a culture of events and experiments linked to that observation, which cannot be translated to a computer. A computer has no idea of what a cell, a mitochondrion, or a tissue is, and can only operate by approximation and analogy. Assigning a task to a program represents a huge gain of time, but at the expense of accuracy. The operator has to control and verify every step of the operations. If hundreds of measurements have been analyzed, random sampling of some of the results can do this. If only a few images have to be analyzed, it may be more accurate to draw by hand some contours that the computer did not take into account than to spend a lot of time to enhance the analysis program.

The advances in computer-aided microscopy, on a motorized microscope or other specific apparatus, allow the complete scan of a preparation into one digital slide and a huge image at high resolution. The amount of information in such data can only be treated digitally.

The high throughput screening methods can analyze either slides or culture dishes at a high rate and can operate continuously night and day. The image analysis uses a combination of tools described in this chapter. This analysis generates a certain proportion of errors in identifying objects, but the large amount of treated data makes them (or aim to make them) statistically not significant. The introduction of new "intelligent" analysis methods, such as neuronal networks, reduces the errors in identification of objects, but there is still ground for improvement.

It remains clear that although automated image analysis allows a powerful set of tools to provide quantitative data, the observation and interpretation of a biological image by an expert scientist remains crucial and a source of wonder.

REFERENCES

Addison, P.S. 2002. *The illustrated wavelet transform handbook: Introductory theory and applications in science, engineering, medicine and finance*. Institute of Physics Publishing, London.

Alleysson, D., S. Süsstrunk, and J. Hérault. 2005. Linear demosaicing inspired by the human visual system. *IEEE Transactions on Image Processing: A Publication of the IEEE Signal Processing Society* 14:439–449.

Cooley, J.W., and J.W. Tukey. 1965. An algorithm for the machine calculation of complex Fourier series. *Mathematics of Computation* 19:297—301.

Dickinson, M., G. Bearman, S. Tille, and R. Lansford. 2001. Multi-spectral imaging and linear unmixing add a whole new dimension to laser scanning fluorescence microscopy. *Biotechniques* 31:1272, 1274–1276, 1278.

Gori, F. 1993. Sampling in optics. In *Advanced topics in Shannon sampling and interpolation theory*, ed. R.J. Marks. Springer, New York.

Hurter, F., and V.C. Driffield. 1890. Photochemical investigations and a new method of determination of the sensitiveness of photographic plates. *Journal of Chemical Society* 31:455–469.

Jacobs, G.H. 2009. Evolution of colour vision in mammals. *Philosophical Transactions of the Royal Society of London B Biological Sciences* 364:2957–2967.

Jain, A.K., and R.C. Dubes. 1988. *Algorithms for clustering data*. Prentice Hall, Englewood Cliffs, NJ.

Kammerer, P., E. Zolda, and R. Sablatnig. 2003. Computer aided analysis of underdrawings in infrared reflectograms. 4th International Symposium on Virtual Reality, Archaeology and Intelligent Cultural Heritage, 1–9.

Kherlopian, A.R., T. Song, Q. Duan, M.A. Neimark, M.J. Po, J.K. Gohagan, and A.F. Laine. 2008. A review of imaging techniques for systems biology. *BMC Systems Biology* 2:74.

Rasband, W.S. 1997–2011. ImageJ, http://rsb.info.nih.gov/ij/, U.S. National Institutes of Health, Bethesda, MD.

Ray, V., R. Subramanian, P. Bhadrachalam, L.-C. Ma, C.-U. Kim, and S.J. Koh. 2008. CMOS-compatible fabrication of room-temperature single-electron devices. *Nature Nanotechnology* 3:603–608.

Ridler, T.W., and S. Calvard. 1978. Picture thresholding using an iterative selection method. *IEEE Transactions on Systems, Man and Cybernetics* 8:630–632.

Ruifrok, A.C., and D.A. Johnston. 2001. Quantification of histochemical staining by color deconvolution. *Analytical and Quantitative Cytolology and Histology* 23:291–299.

Serra, J. 1982. *Image analysis and mathematical morphology*. Academic Press, London.

Zimmermann, T., J. Rietdorf, and R. Pepperkok. 2003. Spectral imaging and its applications in live cell microscopy. *FEBS Letters* 546:87–92.

Index

Printed and bound by CPI Group (UK) Ltd, Croydon, CR0 4YY

21/10/2024

01777040-0010